TRANSPORTATION ISSUES, POLICIES AND R&D

MOTOR VEHICLE CRASHES

ECONOMIC AND SOCIETAL IMPACT, IN DEPTH

TRANSPORTATION ISSUES, POLICIES AND R&D

Additional books in this series can be found on Nova's website under the Series tab.

Additional e-books in this series can be found on Nova's website under the e-book tab.

TRANSPORTATION ISSUES, POLICIES AND R&D

MOTOR VEHICLE CRASHES

ECONOMIC AND SOCIETAL IMPACT, IN DEPTH

ANGIE KIPLING
EDITOR

nova publishers
New York

For permission to use material from this book please contact us:
Telephone 631-231-7269; Fax 631-231-8175
Web Site: http://www.novapublishers.com

Library of Congress Cataloging-in-Publication Data

ISBN: 978-1-61209-225-6

Published by Nova Science Publishers, Inc. † New York

CONTENTS

Preface **vii**

Chapter 1 The Economic and Societal Impact of Motor Vehicle Crashes, 2010 **1**
Lawrence Blincoe, Ted R. Miller, Eduard Zaloshnja and Bruce A. Lawrence

Chapter 2 The Impact of Safety Standards and Behavioral Trends on Motor Vehicle Fatality Rates **287**
Lawrence Blincoe and Umesh Shankar

Index **335**

PREFACE

In 2010 the total economic cost of motor vehicle crashes in the United States was $277 billion. This represents the present value of lifetime economic costs for 32,999 fatalities, 3.9 million non-fatal injuries, and 24 million damaged vehicles. These figures include both police-reported and unreported crashes. The National Highway Traffic Safety Administration (NHTSA) is charged with improving the safety of our Nation's roadways. The agency accomplishes this by regulating vehicle safety performance, enforcing existing safety requirements, and promoting programs that influence driver behavior. This book describes the process, assumptions, and methods used by the agency to estimate the impact of its safety regulations and behavioral programs on fatality rates, and measures the impact of these programs on those rates.

Chapter 1 – In 2010, there were 32,999 people killed, 3.9 million were injured, and 24 million vehicles were damaged in motor vehicle crashes in the United States. The economic costs of these crashes totaled $277 billion. Included in these losses are lost productivity, medical costs, legal and court costs, emergency service costs (EMS), insurance administration costs, congestion costs, property damage, and workplace losses. The $277 billion cost of motor vehicle crashes represents the equivalent of nearly $897 for each of the 308.7 million people living in the United States, and 1.9 percent of the $14.96 trillion real U.S. Gross Domestic Product for 2010. These figures include both police-reported and unreported crashes. When quality of life valuations are considered, the total value of societal harm from motor vehicle crashes in 2010 was $871 billion. Lost market and household productivity accounted for $93 billion of the total $277 billion economic costs, while property damage accounted for $76 billion. Medical expenses totaled $35 billion. Congestion caused by crashes, including travel delay, excess fuel consumption, greenhouse gases and criteria pollutants accounted for $28 billion. Each fatality resulted in an average discounted lifetime cost of $1.4 million. Public revenues paid for roughly 9 percent of all motor vehicle crash costs, costing tax payers $24 billion in 2010, the equivalent of over $200 in added taxes for every household in the United States. Alcohol involved crashes accounted for $59 billion or 21 percent of all economic costs, and 84 percent of these costs occurred in crashes where a driver or non-occupant had a blood alcohol concentration (BAC) of .08 grams per deciliter or greater. Alcohol was the cause of the crash in roughly 82 percent of these cases, causing $49 billion in costs. Crashes in which alcohol levels are BAC of .08 or higher are responsible for over 90 percent of the economic costs and societal harm that occurs in crashes attributable to alcohol use. Crashes in which police indicate that at least one driver was exceeding the legal speed limit or driving too fast for conditions cost $59 billion in 2010. Seat belt use prevented

12,500 fatalities, 308,000 serious injuries, and $69 billion in injury related costs in 2010, but the failure of a substantial portion of the driving population to buckle up caused 3,350 unnecessary fatalities, 54,300 serious injuries, and cost society $14 billion in easily preventable injury related costs. Crashes in which at least one driver was identified as being distracted cost $46 billion in 2010. The report also includes data on the costs associated with motorcycle crashes, failure to wear motorcycle helmets, pedestrian crash, bicyclist crashes, and numerous different roadway designation crashes.

Chapter 2 – The National Highway Traffic Safety Administration (NHTSA) is charged with improving the safety of our Nation's roadways. The agency accomplishes this by regulating vehicle safety performance, enforcing existing safety requirements, and promoting programs that influence driver behavior. As a routine part of the agency's planning process, the agency sets goals that reflect its current and future programs. These goals give the agency a focus for setting priorities in its regulatory and behavioral efforts, as well as a basis for measuring the success of these efforts. Typically, the metrics the agency uses to set goals are fatality rates based on exposure to risk. This paper describes the process, assumptions, and methods used by the agency to estimate the impact of its safety regulations and behavioral programs on fatality rates, and measures the impact of these programs on those rates through the year 2020. The study projects an increase in seat belt use rates from 81 percent in 2006 to 94 percent by 2020. This would prevent a growing number of fatalities over this period starting with over 400 additional lives saved in 2007 and increasing to over 3,500 additional lives saved in 2020. Similar savings could come from a continuation of trends in alcohol-involved crashes. Together, these behavioral programs could save over 5,800 additional lives annually by 2020, and reduce the fatality rate by 0.16. Federal requirements for electronic stability control, improved side impact protection, ejection mitigation, and heavy-truck stopping distance will prevent over 3,500 additional fatalities annually by 2020 and reduce the fatality rate by 0.10. If behavioral trends continue, NHTSA's *combined program activity*, including both motor vehicle safety regulations and behavioral programs, will prevent approximately 1,500 additional fatalities by 2008, and nearly 10,000 additional fatalities by 2020, and reduce the fatality rate by 0.27.

In: Motor Vehicle Crashes
Editor: Angie Kipling

ISBN: 978-1-61209-225-6

Chapter 1

THE ECONOMIC AND SOCIETAL IMPACT OF MOTOR VEHICLE CRASHES, 2010*

Lawrence Blincoe, Ted R. Miller, Eduard Zaloshnja and Bruce A. Lawrence

ABSTRACT

In 2010, there were 32,999 people killed, 3.9 million were injured, and 24 million vehicles were damaged in motor vehicle crashes in the United States. The economic costs of these crashes totaled $277 billion. Included in these losses are lost productivity, medical costs, legal and court costs, emergency service costs (EMS), insurance administration costs, congestion costs, property damage, and workplace losses. The $277 billion cost of motor vehicle crashes represents the equivalent of nearly $897 for each of the 308.7 million people living in the United States, and 1.9 percent of the $14.96 trillion real U.S. Gross Domestic Product for 2010. These figures include both police-reported and unreported crashes. When quality of life valuations are considered, the total value of societal harm from motor vehicle crashes in 2010 was $871 billion. Lost market and household productivity accounted for $93 billion of the total $277 billion economic costs, while property damage accounted for $76 billion. Medical expenses totaled $35 billion. Congestion caused by crashes, including travel delay, excess fuel consumption, greenhouse gases and criteria pollutants accounted for $28 billion. Each fatality resulted in an average discounted lifetime cost of $1.4 million. Public revenues paid for roughly 9 percent of all motor vehicle crash costs, costing tax payers $24 billion in 2010, the equivalent of over $200 in added taxes for every household in the United States. Alcohol involved crashes accounted for $59 billion or 21 percent of all economic costs, and 84 percent of these costs occurred in crashes where a driver or non-occupant had a blood alcohol concentration (BAC) of .08 grams per deciliter or greater. Alcohol was the cause of the crash in roughly 82 percent of these cases, causing $49 billion in costs. Crashes in which alcohol levels are BAC of .08 or higher are responsible for over 90 percent of the economic costs and societal harm that occurs in crashes attributable to alcohol use.

---

* This is an edited, reformatted and augmented version of a National Highway Traffic Safety Administration sponsored document, DOT HS 812 013, issued May 2014.

Crashes in which police indicate that at least one driver was exceeding the legal speed limit or driving too fast for conditions cost $59 billion in 2010. Seat belt use prevented 12,500 fatalities, 308,000 serious injuries, and $69 billion in injury related costs in 2010, but the failure of a substantial portion of the driving population to buckle up caused 3,350 unnecessary fatalities, 54,300 serious injuries, and cost society $14 billion in easily preventable injury related costs. Crashes in which at least one driver was identified as being distracted cost $46 billion in 2010. The report also includes data on the costs associated with motorcycle crashes, failure to wear motorcycle helmets, pedestrian crash, bicyclist crashes, and numerous different roadway designation crashes.

EXECUTIVE SUMMARY

In 2010 the total economic cost of motor vehicle crashes in the United States was $277 billion. This represents the present value of lifetime economic costs for 32,999 fatalities, 3.9 million non-fatal injuries, and 24 million damaged vehicles. These figures include both police-reported and unreported crashes. When quality-of-life valuations are considered, the total value of societal harm from motor vehicle crashes in 2010 was $871 billion.

All costs in this report are expressed in year 2010 economics using a 3-percent discount rate. Nonfatal injury costs are stratified by severity level based on the Abbreviated Injury Scale,[1] but unit costs based on the KABCO scale commonly used in police reports are also supplied in an appendix. The cost components include productivity losses, property damage, medical costs, rehabilitation costs, congestion costs, legal and court costs, emergency services such as medical, police, and fire services, insurance administration costs, and the costs to employers. Values for more intangible consequences such as physical pain or lost quality-of-life are also examined in estimates of comprehensive costs, which include both economic cost components and quality-of-life valuations.

Economic Impact of Crashes

- The economic cost of motor vehicle crashes that occurred in 2010 totaled $277.0 billion. This is equivalent to approximately $897 for every person living in the United States and 1.9 percent of the U.S. Gross Domestic Product.
- The lifetime economic cost to society for each fatality is $1.4 million. Over 90 percent of this amount is attributable to lost workplace and household productivity and legal costs.
- Each critically injured survivor (using the MAIS 5 scale) cost an average of $1.1 million. Medical costs and lost productivity accounted for 82 percent of the cost for this most serious level of non-fatal injury.
- Lost workplace productivity costs totaled $70.2 billion, which equaled 25 percent of the total costs. Lost household productivity totaled $22.9 billion, representing 8 percent of the total economic costs.

- Total property damage costs for all crash types (fatal, injury, and property damage only [PDO]) totaled $76.1 billion and accounted for 28 percent of all economic costs.
- Property-damage-only crashes (in which vehicles were damaged but nobody was injured) cost $71.5 billion and accounted for 26 percent of total economic motor vehicle crash costs.
- Present and future medical costs due to injuries occurring in 2000 were $34.9 billion, representing 13 percent of the total costs. Medical costs accounted for 24 percent of costs from non-fatal injuries.
- Congestion costs, including travel delay, added fuel usage, and adverse environmental impacts cost $28 billion, or 10 percent of total economic crash costs.
- Police-reported crashes account for 83 percent of the economic costs and 89 percent of total societal harm that occurs from traffic crashes. Crashes that are not reported to the police account for 17 percent of economic costs and 11 percent of total societal harm.
- Approximately 9 percent of all motor vehicle crash costs are paid from public revenues. Federal revenues accounted for 5 percent and States and localities paid for approximately 3 percent. An additional 1 percent is from programs that are heavily subsidized by public revenues, but for which the exact source could not be determined. Private insurers pay approximately 52 percent of all costs. Individual crash victims pay approximately 25 percent while third parties such as uninvolved motorists delayed in traffic, charities, and health care providers pay about 14 percent. Overall, those not directly involved in crashes pay for nearly three-quarters of all crash costs, primarily through insurance premiums, taxes and congestion related costs such as travel delay, excess fuel consumption, and increased environmental impacts. In 2010 these costs, borne by society rather than by crash victims, totaled over $200 billion.

Incidence of Crashes

- Some 3.9 million people were injured in 13.6 million motor vehicle crashes in 2010, including 32,999 fatalities. Twenty-four percent of these injuries occurred in crashes that were not reported to police.
- About 23.9 million vehicles were damaged in motor vehicle crashes in 2010; 18.5 million or 77 percent of these vehicles were damaged in incidents that incurred property damage only. The remaining 23 percent involved injuries to occupants of the vehicle, or to nonoccupants such as pedestrians or bicyclists.
- Approximately 60 percent of property-damage-only crashes and 24 percent of all injury crashes are not reported to the police. Unreported injury crashes tend to involve only minor or moderate injuries.

Alcohol Involvement in Crashes

- Alcohol-involved crashes resulted in 13,323 fatalities, 430,000 nonfatal injuries, and $59.4 billion in economic costs in 2010, accounting for 21 percent of all crash costs.
- Crashes involving drivers or nonoccupants with a blood alcohol concentration of .08 grams per deciliter or higher (the legal definition of impairment in all States) accounted for 84 percent of the total economic cost of all alcohol-involved crashes.
- The impact of alcohol involvement increases with injury severity. Alcohol-involved crashes accounted for 14 percent of property-damage-only crash costs, 17 percent of nonfatal injury crash costs; and 48 percent of fatal injury crash costs.
- Although drinking drivers may experience impaired judgment, perceptions, and reaction times, not all crashes in which alcohol was present were caused by alcohol. Crashes in which alcohol was the cause resulted in 11,226 fatalities, 326,000 nonfatal injuries, and $49 billion in economic costs. This is approximately 84 percent of the alcohol-related fatalities and 82 percent of alcohol-related economic costs. It represents 34 percent of all fatalities and 18 percent of all costs from motor vehicle crashes.

Impact of Speed-Related Crashes

- Crashes in which at least one driver was exceeding the legal speed limit or driving too fast for conditions cost $59.1 billion in 2010.
- Speed-related crashes are associated with 10,536 fatalities, 800,000 nonfatal injuries and damage to 3.0 million vehicles in property-damage-only crashes. This represents 32 percent of all fatalities; 20 percent of all nonfatal injuries, and 16 percent of all property-damage-only crashes.
- Speed-related crashes cost an average of $191 for every person in the United States.

Seat Belt Use

- In 2010, seat belts prevented 12,500 fatalities and 308,000 serious injuries, saving $69 billion in medical care, lost productivity, and other injury-related costs.
- Seat belt non-use represents an enormous lost opportunity for injury prevention. In 2010 alone, over 3,350 people were killed and 54,300 were seriously injured unnecessarily because they failed to wear their seat belts, costing society $13.8 billion.
- Over the past 36 years, since FARS started collecting data in 1975, seat belts have prevented over 280,000 fatalities and 7.2 million serious injuries. This saved society $1.6 trillion in medical care, lost productivity, and other injury-related economic costs. During the same time period, nearly 367,000 additional fatalities and 5.8 million additional serious injuries could have been prevented by seat belts if all

occupants had used them. This represents an economic loss of roughly $1.5 trillion in unnecessary expenses and lost productivity.

Distracted Driving Crashes

- Crashes in which at least one driver was identified as being distracted resulted in 3,267 fatalities, 735,000 nonfatal injuries and damaged 3.3 million vehicles in property-damage-only crashes in 2010. This represents about 10 percent of all motor vehicle fatalities and 18 percent of all nonfatal crashes. These crashes cost $45.8 billion in 2010, roughly 17 percent of all economic costs from motor vehicle crashes.

Societal Impacts of Crashes (Comprehensive Costs)

- The value of societal harm from motor vehicle crashes, which includes both economic impacts and valuation for lost quality-of-life, was $870.8 billion in 2010. Sixty-eight percent of this value represents lost quality-of-life, while 32 percent is economic impacts.
- The lifetime comprehensive cost to society for each fatality is $9.1 million. Eighty-five percent of this amount is attributable to lost quality-of-life.
- Each critically injured survivor (MAIS 5) has comprehensive costs that average of $5.7 million. Lost quality-of-life accounted for 81 percent of the total harm for this most serious level of non-fatal injury.
- Alcohol-involved crashes resulted in $242.6 billion in comprehensive costs in 2010, accounting for 28 percent of all societal harm from motor vehicle crashes. Eighty-five percent of these costs occurred in crashes where one driver had a BAC of .08 g/dL or greater.
- Although drinking drivers may experience impaired judgment, perceptions, and reaction times, not all crashes in which alcohol was present were caused by alcohol. Crashes in which alcohol was the cause resulted in $196 billion in societal harm in 2010. This represents 22 percent of all societal harm from motor vehicle crashes. Ninety-eight percent of societal harm from crashes caused by alcohol occurs in crashes where drivers had BACs of .08 or greater.
- Crashes in which at least one driver was exceeding the legal speed limit or driving too fast for conditions caused $210.3 billion in comprehensive costs in 2010. This represents 24 percent of all societal harm from motor vehicle crashes.
- Crashes in which at least one driver was identified as being distracted caused $129.5 billion in comprehensive costs in 2010, causing roughly 15 percent of all societal harm from motor vehicle crashes.
- In 2010, seat belts prevented $349.0 billion in comprehensive costs to society. Over the last 36 years, seat belts have prevented over $8 trillion in societal harm, resulting in lower economic costs to society and improved quality-of-life for millions of motor vehicle occupants.

1. Introduction

In 2010, there were 32,999 people killed, 3.9 million were injured, and 24 million vehicles were damaged in motor vehicle crashes in the United States. The economic costs of these crashes totaled $277.0 billion. Included in these losses are lost productivity, medical costs, legal and court costs, emergency service costs (EMS), insurance administration costs, congestion costs, property damage, and workplace losses. The $277 billion cost of motor vehicle crashes represents the equivalent of nearly $897 for each of the 308.7 million people living in the United States, and 1.9 percent of the $14.96 trillion real Gross Domestic Product for 2010.

All levels of society -- the individual crash victims and their families, their employers, and society at large -- are affected by motor vehicle crashes in many ways. The cost of medical care is borne by the individual in the form of payments for insurance, deductibles, uncovered costs, and uninsured expenses. It is borne by society through higher insurance premiums and through the diversion of medical resources away from other medical needs, such as medical research, disease prevention and control, and basic public health needs. There are also significant costs associated with the lost productivity experienced by an individual and others when the victim dies prematurely or experiences a short or long-term disability. The victim's dependents suffer immediate economic hardship in the loss of the victim's income and other contributions, while society is burdened by the necessity to support the victim or their dependents and through foregone contributions to the Nation's productivity. Aside from these economic consequences, victims suffer from physical pain, disability, and emotional impacts that can greatly reduce the quality of their lives.

This report examines these and other costs resulting from motor vehicle crashes. The purpose of presenting these costs is to place in perspective the economic losses and societal harm that result from these crashes, and to provide information to government and private sector officials for use in structuring programs to reduce or prevent these losses.

Economic Impacts

Total economic costs are summarized in Table 1-1. The total economic cost of motor vehicle crashes in 2010 is estimated to have been $277.0 billion. Of this total, medical costs were responsible for $34.9 billion, property damage losses for $76.1 billion, lost productivity (both market and household) for $93.1 billion, and congestion impacts for $28 billion. All other crash related costs totaled $45 billion.

The most significant costs were property damage and lost market productivity, which accounted for 28 and 25 percent, respectively, of the total economic costs in 2010. For lost productivity, these high costs are a function of the level of disability that has been documented for crashes involving injury and death. For property damage, costs are primarily a function of the very high incidence of minor crashes in which injury does not occur or is negligible. Medical care costs and emergency services (which include police and fire services) are responsible for about 13 percent of the total. Travel delay, added fuel consumption, and pollution impacts caused by congestion at the crash site accounts for 10 percent.

The value of lost household productivity accounts for 8 percent of total costs. Legal and court costs account for 5 percent and insurance administration costs for about 9 percent of the total. These costs are summarized in Tables 1-1, 1-2, and Figure 1-A. The incidence of injuries and crashes that produced these costs is summarized in Table 3.

Approximately 9 percent of all motor vehicle crash costs are paid from public revenues. Federal revenues accounted for 5 percent and States and localities paid for approximately 3 percent. An additional 1 percent is from programs that are heavily subsidized by public revenues, but for which the exact source could not be determined. Private insurers pay approximately 52 percent of all costs. Individual crash victims pay approximately 25 percent while third parties such as uninvolved motorists delayed in traffic, charities, and health care providers pay about 14 percent. Overall, those not directly involved in crashes pay for nearly three-quarters of all crash costs, primarily through insurance premiums, taxes and congestion related costs such as travel delay, excess fuel consumption, and increased environmental impacts. In 2010 these costs, borne by society rather than by crash victims, totaled over $200 billion. Figure 1-A summarizes these illustrates these cost distributions.

From Table 1-3, over half of all PDO crashes and about a quarter of all non-fatal injury crashes are not reported to police. However, analyses of safety countermeasures frequently rely only on police-reported incidence data. Crashes that get reported to police are likely to be more severe than unreported crashes because vehicles are more likely to require towing and occupants are more likely to require hospitalization or emergency services. These crashes are typically also likely to require more time to investigate and clear from roadways than unreported crashes. Analysis based solely on police-reported crashes should thus be based on unit costs that are specific to police-reported crashes. For injury related costs, this is more or less automatically accounted for by the shift in the injury severity profile. Unreported crashes have a lower average severity profile than do reported crashes. However, for non- injury related cost components – property damage and congestion costs – there is no profile to shift. In addition, police-reported crashes have higher response rates for emergency services.

For this report, costs specific to police-reported and unreported crashes have been developed. The results of this analysis are presented in Tables 1-4, 1-5, 1-6, and 1-7. The differences seem negligible at the more severe injury levels due to the overwhelming costs of factors such as lost productivity and medical care which do not vary by reporting status, except through the shift in injury profiles. However, at lower severity levels the unit costs are significant. For PDO vehicles and MAIS0 cases, police-reported crashes have costs that are three times those of unreported crashes. For minor (MAIS1) injuries, reported crashes cost 40 percent more than unreported crashes. These ratios decline as injury severity increases. Note that for MAIS4s, MAIS5s, and fatalities, property damage costs are identical under both reported and unreported cases. Virtually all injuries at these levels are believed to be reported to police, and the original property damage cost estimate is thus assumed to represent police-reported cases. These same costs are thus listed under both scenarios.

Figure 1-B shows the proportion of each cost category that is accounted for by police-reported crashes. For most categories, the portions vary due to the differing proportions of incidence across the various injury levels. For congestion, property damage, and emergency services, differing unit costs are involved as well. Overall, police-reported crashes are estimated to account for 83 percent of the economic costs that are incurred from traffic crashes.

NHTSA last examined the cost of motor vehicle crashes in 2002. At that time the report was based on 2000 data. The current report indicates a total cost from traffic crashes in 2010 of $277.0 billion, approximately 20-percent higher than our previous estimate of $230.6 billion in 2000. The higher estimate is attributable to a number of factors. Inflation accounts for an overall rise in the cost of goods and services of approximately 27 percent, but incidence of fatalities and injuries has declined over the past decade due to a combination of technological, behavioral, infrastructure, and economic factors. In 2010 there were 32,999 fatalities in motor vehicle crashes, a decline of 21 percent from 2000. At the same time, the number of police-reported injuries reported in the General Estimates System (GES) dropped by 30 percent. These declines reflect significant safety improvements in the on-road vehicle fleet. Since 2000 a number of significant technological safety improvements have been phased in to the vehicle fleet. These include advanced air bags, better side impact protection, tire pressure monitoring systems, interior padding, improved seat belts, improved vehicle conspicuity, antilock brake systems, and electronic stability control. Seat belt use has also increased over this decade, rising from 73 percent in 2000 to 85 percent in 2010, due in part to enforcement of primary belt use laws and to public education programs that educate drivers to the importance of belt use. These and other factors such as improved roadways have helped drive the fatality rate/hundred million VMT down from 1.53 deaths per hundred million vehicle miles travelled (VMT) to 1.11 in 2010. The economic recession likely had some impact as well, although VMT did not decline significantly. However, fatalities usually decline during periods of economic uncertainty, possibly due to a more sober attitude on the part of drivers. Cost shifts also occurred because most cost factors were re-examined based on more recent data sources and this caused shifts in unit costs that impacted the overall estimate in a variety of ways. The specifics of these changes are described in the body of this report. Note also that lifetime cost impacts such as lost productivity and medical care for serious injury are measured using a 3-percent discount rate, whereas the previous report used a 4-percent rate. The shift to the 3-percent rate reflects lower real investment returns over the past decade and has been established as the appropriate value to represent the social rate of time preference by the Office of Management and Budget. This accounts for a small portion of the difference in costs as well.

Alcohol consumption remains a major cause of motor vehicle crashes; 2010 data shows that alcohol involved crashes declined slightly in incidence. Historically, approximately half of all motor vehicle fatalities have occurred in crashes where the drivers or nonoccupants had been drinking, but this number has gradually declined in recent years to about 40 percent. Alcohol is involved in crashes that account for 21 percent of all economic costs, with 84 percent of these costs involving crashes where a driver or nonoccupant was legally intoxicated (illegal per se), defined as a blood alcohol concentration (BAC) of .08 grams per deciliter (g/dL) or higher.

The report indicates that while alcohol-involved crashes are more costly than in 2000, they account for a smaller portion of the overall crash cost. This reflects the impact of efforts at Federal, State, and local levels to reduce the incidence of drunk driving. The report also estimates the portion of alcohol-involved crash costs that were actually caused by impaired driving. Although drinking drivers may experience impaired judgment, perceptions and reaction times, not all crashes in which alcohol was present were caused by alcohol. For example, a driver with a BAC of .04 g/dL could be stopped at a light and run into by a texting driver. Crashes caused by alcohol accounted for 82 percent of all economic costs from crashes where at least one driver or nonoccupant had been drinking. The portion attributable to alcohol rises dramatically as BAC increases, with only 6 percent attribution at low BAC levels (BAC=.01 to .04), but 94 percent attribution at legally impaired (illegal per se) levels (BAC>=.08). Crashes caused by legally impaired drivers with BACs in excess of .08 g/dL account for over 90 percent of the economic and societal harm that results from all alcohol caused crashes.

The report also analyzes the impact of seat belt use as well as the cost the Nation incurs from failure to wear seat belts. Over the last 35 years, seat belts have prevented over 280,000 fatalities and 7.2 million serious nonfatal injuries, which saved $1.6 trillion in economic costs (in 2010 dollars). During this same period, the failure of a substantial portion of the driving population to wear belts caused 367,000 unnecessary deaths and 5.8 million nonfatal injuries, costing the Nation $1.5 trillion in preventable medical costs, lost productivity, and other injury related expenditures.

The Abbreviated Injury Scale (AIS) used in this report provides the basis for stratifying societal costs by injury severity. Significant sources of economic loss, such as medical costs and lost productivity, are highly dependent on injury outcome. AIS codes are primarily oriented toward the immediate threat to life resulting from the injury, and are estimated soon after a crash occurs. Although the more serious injuries tend to have more serious outcomes, AIS codes are not always accurate predictors of long-term injury outcomes. Some injuries with low AIS codes, such as lower extremity injuries, can actually result in serious and expensive long-term outcomes. There is currently no incidence database organized by injury outcome. The development and use of such a database could improve the accuracy of economic cost estimates, and might result in a significant shift in the relative number of injuries regarded as serious.

This report focuses on "average" costs for injuries of different severity. While this approach is valid for computing combined costs at a nationwide level, the costs of individual cases at different injury levels can vary quite dramatically. The average costs outlined in this report are significant; however, in individual cases they can be exceeded by a factor of three or more. There is considerable evidence to indicate that the most serious injuries are not adequately covered by insurance. Depending on the financial ability and insurance coverage of the individual crash victims, the medical and rehabilitation costs, as well as the loss in wages resulting from serious injury, can be catastrophic to the victim's economic wellbeing in addition to their physical and emotional condition.

When using this report for the analysis of crash impact and injury countermeasures, it is important to include only those cost elements that are applicable to the specific programs addressed. For example, programs that encourage seat belt use may reduce costs associated with injuries, but would not have an effect on property-damage or congestion costs. Therefore, careful consideration should be given to the nature of the benefits from any proposal before incorporating the results of this report into analyses or recommendations. Economic costs represent only one aspect of the consequences of motor vehicle crashes. People injured in these crashes often suffer physical pain and emotional anguish that is beyond any economic recompense. The permanent disability of spinal cord damage, loss of mobility, loss of eyesight, and serious brain injury can profoundly limit a person's life, and can result in dependence on others for routine physical care. More common, but less serious injuries, can cause physical pain and limit a victim's physical activities for years after the crash. Serious burns or lacerations can lead to long- term discomfort and the emotional trauma associated with permanent disfigurement. For an individual, these non-monetary outcomes can be the most devastating aspect of a motor vehicle crash.

The family and friends of the victim feel the psychological repercussions of the victim's injury acutely as well. Caring for an injured family member can be very demanding for others in the family, resulting in economic loss and emotional burdens for all parties concerned. It can change the very nature of their family life; the emotional difficulties of the victim can affect other family members and the cohesiveness of the family unit. When a crash leads to death, the emotional damage is even more intense, affecting family and friends for years afterward and sometimes leading to the breakup of previously stable family units.

Action taken by society to alleviate the individual suffering of its members can be justified in and of itself; in order to increase the overall quality-of-life for individual citizens. In this context, economic benefits from such actions are useful to determine the net cost to society of programs that are primarily based on humane considerations. If the focus of policy decisions was purely on the economic consequences of motor vehicle crashes, the most tragic, and, in both individual and societal terms, possibly the most costly aspect of such crashes would be overlooked.

Societal Impacts

Previous versions of this report have focused on the economic impact of motor vehicle crashes – the societal losses that can be directly measured in economic terms. However, these costs do not represent the more intangible consequences of these events and should not, therefore, be used alone to produce cost-benefit ratios. Measurement of the dollar value of intangible consequences such as pain and suffering has been undertaken in numerous studies. These studies have estimated values based on wages for high-risk occupations and prices paid in the market place for safety products, among other measurement techniques. These "willingness to pay" based estimates of how society values risk reduction capture valuations not associated with direct monetary consequences. Most researchers agree that the value of fatal risk reduction falls in the range of $5 to $15 million per life saved. In this study, comprehensive costs, which include both the economic impacts of crashes and valuation of lost quality-of-life, are also examined. Comprehensive costs represent the value of the total societal harm that results from traffic crashes. The basis for these estimates is the most recent guidance issued by the U.S. Department of Transportation for valuing risk reduction. This guidance, which was issued in February 2013, establishes a new value of a statistical life (VSL) at $9.1 million in 2012 economics ($8.86 million in 2010 economics). In addition, it establishes new relative disutility factors stratified by injury severity level to estimate the lost quality-of-life for nonfatal injuries. These factors were derived in a research contract designed specifically for this current cost study. More detailed discussion of comprehensive costs is included in Chapter 4 of this report. The total societal harm from motor vehicle crashes as measured by comprehensive costs, is shown in Tables 1-4 and 1-5, and Figure 1-C.

From Table 1-4, the total societal harm from motor vehicle crashes in 2010 is estimated to have been $870.8 billion, roughly three times the value measured by economic impacts alone. Of this total, 68 percent represents lost quality-of-life, dwarfing the contribution of all other cost categories. This highlights the importance of accounting for all societal impacts when measuring costs and benefits from motor vehicle safety countermeasures. However, the literature on VSL estimates indicates a wide range of measured estimates of VSLs – some as low as a few million dollars, some as high as over $30 million. The U. S. DOT guidance memorandum (U.S. Department of Transportation (2013), Guidance on Treatment of the Economic Value of a Statistical Life in U.S. Department of Transportation Analyses, Memorandum from the Office of the Secretary of Transportation, U.S. Department of Transportation. Available at: http://www.dot.gov/sites/dot.gov/files/docs/VSL%20 Guidance_2013.) discusses a feasible range of VSLs for sensitivity analysis from $5.2 million to $12.9 million. There is thus far more uncertainty regarding the accuracy of estimates of lost quality-of-life than there is regarding economic costs. In Appendix A comprehensive costs are estimated based on this range. The results indicate a feasible range of societal harm from motor vehicle crashes of from $583 billion to $1.16 trillion in 2010, with lost quality-of-life accounting for between half and three-quarters of all societal harm respectively.

Tables 10 to 13 examine the comprehensive costs of police-reported and unreported crashes. Roughly 89 percent of aggregate societal harm from motor vehicle crashes occurs in police-reported crashes. This is somewhat higher than the 83 percent for economic costs. The difference is due to the impact of quality-of-life valuations on fatalities and the most serious injuries (MAIS4+), which are all police- reported.

Overview

Table 1-14 summarizes both the economic and comprehensive costs of selected crash categories examined in this study. Nonfatal injuries were the most costly severity outcome, accounting for over half of both economic costs and societal harm. Damage to vehicles in which no injury occurred was the second highest economic cost outcome due to the high frequency of these low impact crashes. However, in terms of societal harm, fatalities were the second most costly outcome due to the inclusion of lost quality-of-life for the life years that fatal crash victims lose.

This report examined five different types of adverse driver behavior - alcohol use, speeding, distracted driving, failure to wear seat belts, and riding a motorcycle without a helmet. The most costly of these involved alcohol use. Alcohol-involved crashes, in which drivers or pedestrians had some level of alcohol in their bloodstreams, accounted for 21 percent of economic costs and 28 percent of societal harm. However, crashes in which alcohol was a likely cause of the crashes accounted for 18 percent of economic costs and 23 percent of societal harm. Over 80 percent of this toll occurred in crashes where the drivers were legally intoxicated.

Crashes in which one or more drivers were exceeding the legal speed limit or driving too fast for conditions caused 21 percent of economic costs and 24 percent of societal harm. The extent to which speed actually caused these crashes is uncertain, but higher speeds leave less time for drivers to react to emergency situations.

Distracted driving, which includes talking on cell phones, texting, eating, and other non-driving activities, was a factor in crashes that caused 17 percent of economic costs and 15 percent of societal harm. However, distracted driving is difficult to detect and it is likely that distraction plays an even larger role in causing crashes and their resulting impacts on society.

The failure of some vehicle occupants to use their seat belts accounts for roughly 5 percent of economic costs and 8 percent of societal harm. While these portions seem relatively small, they represent economic costs of $14 billion and societal harm of $72 billion annually. Likewise, failure to wear motorcycle helmets causes a small portion of the overall total, but has serious economic and quality-of- life consequences for the injured riders and their families.

Injuries to nonoccupants also have significant economic and societal impacts. Motorcyclist injuries cause 5 percent of the economic costs and 8 percent of societal harm from traffic crashes. Injuries to pedestrians and bicyclists cause 7 percent of the economic costs and 10 percent of the societal harm.

The report also examines crash costs for various roadway types and crash configurations. Among its findings, crashes on interstate highways account for roughly 10 percent of both economic costs and societal harm, while the more frequent but generally less serious crashes at intersections account for 50 percent of economic costs and 45 percent of societal harm. Crashes on urban roadways account for roughly 62 percent of all economic and 56 percent of all societal harm, while crashes on rural roadways account for roughly 38 percent of economic impacts and 44 percent of societal harm.

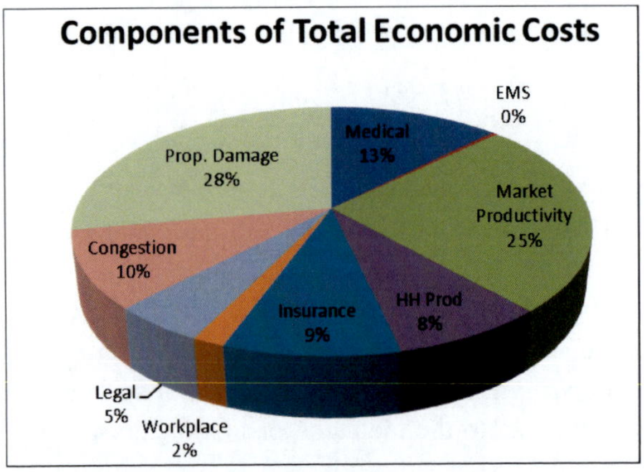

Figure 1-A. Components of Total Economic Costs.

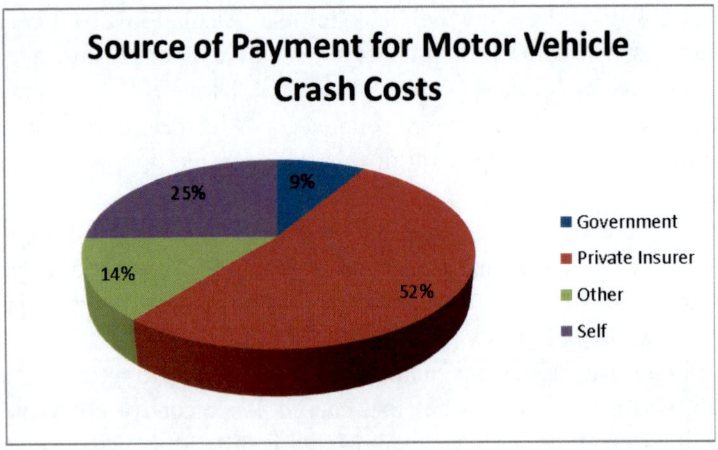

Figure 1-B. Source of Payment for Motor Vehicle Crash Costs.

Table 1-1. Summary of Total Economic Costs, (Millions of 2010 Dollars)

	PDO Vehicle	MAIS0	MAIS1	MAIS2	MAIS3	MAIS4	MAIS5	Fatal	Total	% Total
Medical	$0	$0	$13,148	$8,292	$7,143	$3,364	$2,539	$373	$34,860	12.6%
EMS	$518	$96	$308	$66	$42	$14	$5	$30	$1,079	0.4%
Market Prd.	$0	$0	$9,886	$13,426	$11,090	$2,985	$2,025	$30,797	$70,210	25.3%
Household	$1,111	$206	$3,255	$4,038	$3,375	$762	$616	$9,567	$22,929	8.3%
Ins. Admin.	$3,535	$655	$13,612	$3,174	$2,453	$639	$460	$935	$25,462	9.2%
Workplace	$1,148	$211	$1,180	$896	$582	$109	$64	$389	$4,577	1.7%
Legal	$0	$0	$4,877	$2,283	$1,979	$603	$524	$3,514	$13,781	5.0%
Subtotal	$6,311	$1,169	$46,267	$32,175	$26,664	$8,477	$6,232	$44,991	$172,898	62.4%
Congestion	$19,934	$3,483	$3,836	$405	$144	$26	$9	$189	$28,027	10.1%
Prop. Dmg.	$45,235	$8,378	$18,694	$1,957	$1,096	$279	$87	$370	$76,096	27.5%
Subtotal	$65,169	$11,861	$22,530	$2,363	$1,241	$305	$96	$559	$104,123	37.6%
Total	$71,480	$13,030	$68,797	$34,537	$27,905	$8,781	$6,327	$46,163	$277,020	100.0%
% Total	25.8%	4.7%	24.8%	12.5%	10.1%	3.2%	2.3%	16.7%	100.0%	0.0%

Table 1-2. Summary of Unit Costs and Police-Reported and Unreported Crashes, 2010 Dollars

	PDO Vehicle	MAIS0	MAIS1	MAIS2	MAIS3	MAIS4	MAIS5	Fatal
Medical Care	$0	$0	$3,801	$24,481	$70,902	$196,911	$441,618	$11,317
EMS	$28	$21	$89	$194	$416	$838	$855	$902
Market Prod.	$0	$0	$2,858	$39,637	$110,088	$174,728	$352,178	$933,262
Household Prod.	$60	$45	$941	$11,921	$33,505	$44,593	$107,070	$289,910
Insurance Adm.	$191	$143	$3,935	$9,370	$24,348	$37,372	$79,967	$28,322
Workplace Costs	$62	$46	$341	$2,644	$5,776	$6,361	$11,091	$11,783
Legal Costs	$0	$0	$1,410	$6,739	$19,645	$35,307	$91,197	$106,488

Table 1-2. (Continued)

	PDO Vehicle	MAIS0	MAIS1	MAIS2	MAIS3	MAIS4	MAIS5	Fatal
Subtotal Injury	$341	$255	$13,375	$94,986	$264,680	$496,110	$1,083,976	$1,381,984
Congestion	$1,077	$760	$1,109	$1,197	$1,434	$1,511	$1,529	$5,720
Prop. Damage	$2,444	$1,828	$5,404	$5,778	$10,882	$16,328	$15,092	$11,212
Subtotal Non-Inj.	$3,521	$2,588	$6,513	$6,975	$12,316	$17,839	$16,621	$16,932
Total	$3,862	$2,843	$19,888	$101,961	$276,996	$513,949	$1,100,597	$1,398,916

* Note: Unit costs are expressed on a per-person basis for all injury levels. PDO costs are expressed on a per-damaged-vehicle basis.

Table 1-3. Incidence Summary – 2010 Total Police-Reported and Unreported Injuries

Severity	Police- reported	Not Police- reported	Total	Percent Unreported
Vehicles				
Injury Vehicles	3,225,839	2,121,769	5,347,608	39.7%
PDO Vehicles	7,454,761	11,053,871	18,508,632	59.7%
Total Vehicles	10,680,601	13,175,640	23,856,241	55.2%
People in Injury Crashes				
MAIS0	2,147,857	2,435,409	4,583,265	53.1%
MAIS1	2,578,993	880,207	3,459,200	25.4%
MAIS2	271,160	67,570	338,730	19.9%
MAIS3	96,397	4,343	100,740	4.3%
MAIS4	17,086	0	17,086	0.0%
MAIS5	5,749	0	5,749	0.0%
Fatal	32,999	0	32,999	0.0%
Total	5,150,241	3,387,528	8,537,770	39.7%
Total Injuries	3,002,385	952,120	3,954,504	24.1%
Crashes				
PDO	4,255,495	6,310,019	10,565,514	59.7%
Injury	1,791,572	1,178,391	2,969,963	39.7%
Fatal	30,296	0	30,296	0.0%
Total Crashes	6,077,362	7,488,411	13,565,773	55.2%

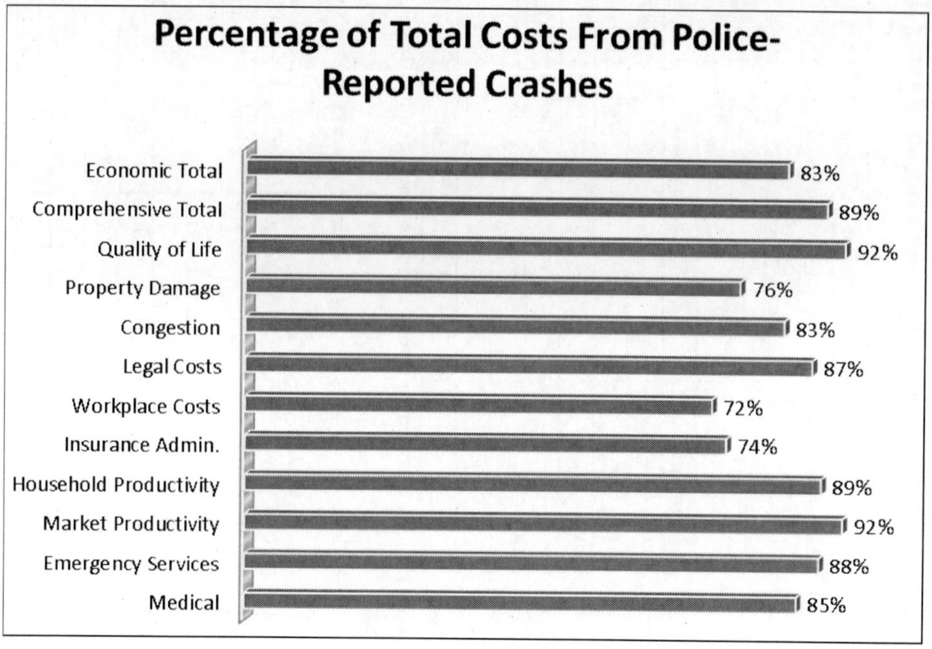

Figure 1-C. Percentage of Total Costs From Police-Reported Crashes.

Table 1-4. Summary of Unit Costs, Police-Reported Crashes, 2010 Dollars

	PDO Vehicle	MAIS0	MAIS1	MAIS2	MAIS3	MAIS4	MAIS5	Fatal
Medical	$0	$0	$3,801	$24,481	$70,902	$196,911	$441,618	$11,317
EMS	$59	$38	$109	$221	$416	$838	$855	$902
Market	$0	$0	$2,858	$39,637	$110,088	$174,728	$352,178	$933,262
Household	$60	$45	$941	$11,921	$33,505	$44,593	$107,070	$289,910
Insurance	$191	$143	$3,935	$9,370	$24,348	$37,372	$79,967	$28,322
Workplace	$62	$46	$341	$2,644	$5,776	$6,361	$11,091	$11,783
Legal Costs	$0	$0	$1,410	$6,739	$19,645	$35,307	$91,197	$106,488
Subtotal	$372	$272	$13,395	$95,013	$264,680	$496,110	$1,083,976	$1,381,984
Congestion	$2,104	$1,416	$1,426	$1,450	$1,490	$1,511	$1,529	$5,720
Prop. Damage	$3,599	$2,692	$7,959	$8,510	$16,027	$16,328	$15,092	$11,212
Subtotal	$5,704	$4,108	$9,385	$9,960	$17,517	$17,839	$16,621	$16,932
Total	$6,076	$4,380	$22,779	$104,974	$282,197	$513,949	$1,100,597	$1,398,916

*Note: Unit costs are expressed on a per-person basis for all injury levels. PDO costs are expressed on a per-damaged-vehicle basis.

Table 1-5. Summary of Unit Costs, Unreported Crashes, 2010 Dollars

	PDO Vehicle*	MAIS0	MAIS1	MAIS2	MAIS3	MAIS4*	MAIS5*	Fatal*
Medical	$0	$0	$3,801	$24,481	$70,902	$196,911	$441,618	$11,317
EMS	$7	$6	$32	$84	$416	$838	$855	$902
Market	$0	$0	$2,858	$39,637	$110,088	$174,728	$352,178	$933,262
Household	$60	$45	$941	$11,921	$33,505	$44,593	$107,070	$289,910
Insurance	$191	$143	$3,935	$9,370	$24,348	$37,372	$79,967	$28,322
Workplace	$62	$46	$341	$2,644	$5,776	$6,361	$11,091	$11,783
Legal Costs	$0	$0	$1,410	$6,739	$19,645	$35,307	$91,197	$106,488

	PDO Vehicle*	MAIS0	MAIS1	MAIS2	MAIS3	MAIS4*	MAIS5*	Fatal*
Subtotal	$320	$240	$13,318	$94,876	$264,680	$496,110	$1,083,976	$1,381,984
Congestion	$384	$180	$180	$180	$180	$180	$180	$458
Prop. Damage	$1,224	$916	$2,707	$2,894	$5,451	$16,328	$15,092	$11,212
Subtotal	$1,609	$1,096	$2,888	$3,075	$5,632	$16,508	$15,272	$11,670
Total	$1,928	$1,337	$16,205	$97,951	$270,312	$512,618	$1,099,248	$1,393,654

*Note: Unit costs are expressed on a per-person basis for all injury levels. PDO costs are expressed on a per-damaged-vehicle basis. Generally, all MAIS 4, 5, and fatal injuries are believed to be police-reported. Values are still included here for reference to cover any exceptional case where unreported crashes might be found for these injury severity categories.

Table 1-6. Summary of Total Economic Costs in Police-Reported Crashes, Millions of 2010 Dollars

	PDO Vehicle	MAIS0	MAIS1	MAIS2	MAIS3	MAIS4	MAIS5	Fatal	Total	% Total
Medical	$0	$0	$9,803	$6,638	$6,835	$3,364	$2,539	$373	$29,552	12.8%
EMS	$443	$81	$280	$60	$40	$14	$5	$30	$952	0.4%
Market	$0	$0	$7,371	$10,748	$10,612	$2,985	$2,025	$30,797	$64,538	28.0%
Household	$447	$97	$2,427	$3,232	$3,230	$762	$616	$9,567	$20,377	8.8%
Insurance	$1,424	$307	$10,148	$2,541	$2,347	$639	$460	$935	$18,800	8.2%
Workplace	$462	$99	$879	$717	$557	$109	$64	$389	$3,275	1.4%
Legal Costs	$0	$0	$3,636	$1,827	$1,894	$603	$524	$3,514	$11,999	5.2%
Subtotal	$2,776	$583	$34,544	$25,764	$25,514	$8,477	$6,232	$45,604	$149,494	64.9%
Congestion	$15,687	$3,042	$3,677	$393	$144	$26	$9	$189	$23,167	10.1%
Prop. Damage	$26,833	$5,783	$20,526	$2,308	$1,545	$279	$87	$370	$57,730	25.1%
Subtotal	$42,521	$8,825	$24,203	$2,701	$1,689	$305	$96	$559	$80,898	35.1%
Total	$45,297	$9,408	$58,748	$28,465	$27,203	$8,781	$6,327	$46,163	$230,392	100.0%
% Total	19.7%	4.1%	25.5%	12.4%	11.8%	3.8%	2.7%	20.0%	100.0%	0.0%

Table 1-7. Summary of Total Economic Costs in Unreported Crashes, Millions of 2010 Dollars

	PDO Vehicle	MAIS0	MAIS1	MAIS2	MAIS3	MAIS4	MAIS5	Fatal	Total	% Total
Medical	$0	$0	$3,346	$1,654	$308	$0	$0	$0	$5,308	11.4%
EMS	$76	$16	$28	$6	$2	$0	$0	$0	$127	0.3%
Market	$0	$0	$2,516	$2,678	$478	$0	$0	$0	$5,672	12.2%
Household	$663	$110	$828	$806	$146	$0	$0	$0	$2,552	5.5%
Insurance	$2,111	$348	$3,464	$633	$106	$0	$0	$0	$6,662	14.3%
Workplace	$685	$112	$300	$179	$25	$0	$0	$0	$1,301	2.8%
Legal Costs	$0	$0	$1,241	$455	$85	$0	$0	$0	$1,782	3.8%
Subtotal	$3,536	$586	$11,722	$6,411	$1,150	$0	$0	$0	$23,404	50.2%
Congestion	$4,248	$439	$159	$12	$1	$0	$0	$0	$4,859	10.4%
Prop. Damage	$13,534	$2,230	$2,383	$196	$24	$0	$0	$0	$18,366	39.4%
Subtotal	$17,782	$2,670	$2,542	$208	$24	$0	$0	$0	$23,225	49.8%
Total	$21,317	$3,255	$14,264	$6,619	$1,174	$0	$0	$0	$46,629	100.0%
% Total	45.7%	7.0%	30.6%	14.2%	2.5%	0.0%	0.0%	0.0%	100.0%	0.0%

Table 1-8. Summary of Total Comprehensive Costs, Reported and Unreported Crashes, Millions of 2010 Dollars

	PDO Vehicle	MAIS0	MAIS1	MAIS2	MAIS3	MAIS4	MAIS5	Fatal	Total	% Total
Medical	$0	$0	$13,148	$8,292	$7,143	$3,364	$2,539	$373	$34,860	4.0%
EMS	$518	$96	$308	$66	$42	$14	$5	$30	$1,079	0.1%
Market Prod.	$0	$0	$9,886	$13,426	$11,090	$2,985	$2,025	$30,797	$70,210	8.1%
Household	$1,111	$206	$3,255	$4,038	$3,375	$762	$616	$9,567	$22,929	2.6%
Insurance	$3,535	$655	$13,612	$3,174	$2,453	$639	$460	$935	$25,462	2.9%
Workplace	$1,148	$211	$1,180	$896	$582	$109	$64	$389	$4,577	0.5%
Legal Costs	$0	$0	$4,877	$2,283	$1,979	$603	$524	$3,514	$13,781	1.6%
Subtotal	$6,311	$1,169	$46,267	$32,175	$26,664	$8,477	$6,232	$45,604	$172,898	19.9%

	PDO Vehicle	MAIS0	MAIS1	MAIS2	MAIS3	MAIS4	MAIS5	Fatal	Total	% Total
Congestion	$19,934	$3,483	$3,836	$405	$144	$26	$9	$189	$28,027	3.2%
Prop. Damage	$45,235	$8,378	$18,694	$1,957	$1,096	$279	$87	$370	$76,096	8.7%
Subtotal	$65,169	$11,861	$22,530	$2,363	$1,241	$305	$96	$559	$104,123	12.0%
Total	$71,480	$13,030	$68,797	$34,537	$27,905	$8,781	$6,327	$46,163	$277,020	31.8%
QALYs	$0	$0	$80,395	$115,464	$81,166	$34,812	$26,322	$255,646	$593,805	68.2%
Comp. Total	$71,480	$13,030	$149,192	$150,001	$109,070	$43,594	$32,649	$301,809	$870,826	100.0%
% Total	8.2%	1.5%	17.1%	17.2%	12.5%	5.0%	3.7%	34.7%	100.0%	0.0%

Table 1-9. Summary of Comprehensive Unit Costs, Reported and Unreported Crashes, 2010 Dollars

	PDO Vehicle	MAIS0	MAIS1	MAIS2	MAIS3	MAIS4	MAIS5	Fatal
Medical	$0	$0	$3,801	$24,481	$70,902	$196,911	$441,618	$11,317
EMS	$28	$21	$89	$194	$416	$838	$855	$902
Market	$0	$0	$2,858	$39,637	$110,088	$174,728	$352,178	$933,262
Household	$60	$45	$941	$11,921	$33,505	$44,593	$107,070	$289,910
Insurance	$191	$143	$3,935	$9,370	$24,348	$37,372	$79,967	$28,322
Workplace	$62	$46	$341	$2,644	$5,776	$6,361	$11,091	$11,783
Legal Costs	$0	$0	$1,410	$6,739	$19,645	$35,307	$91,197	$106,488
Subtotal	$341	$255	$13,375	$94,986	$264,680	$496,110	$1,083,976	$1,381,984
Congestion	$1,077	$760	$1,109	$1,197	$1,434	$1,511	$1,529	$5,720
Prop. Damage	$2,444	$1,828	$5,404	$5,778	$10,882	$16,328	$15,092	$11,212
Subtotal	$3,521	$2,588	$6,513	$6,975	$12,316	$17,839	$16,621	$16,932
Total Econ.	$3,862	$2,843	$19,888	$101,961	$276,996	$513,949	$1,100,597	$1,398,916
QALYs	$0	$0	$23,241	$340,872	$805,697	$2,037,483	$4,578,525	$7,747,082
Comp. Total	$3,862	$2,843	$43,129	$442,833	$1,082,693	$2,551,432	$5,679,122	$9,145,998

* Note: Unit costs are expressed on a per-person basis for all injury levels. PDO costs are expressed on a per-damaged-vehicle basis.

Table 1-10. Summary of Comprehensive Unit Costs, Police-Reported Crashes, 2010 Dollars

	PDO Vehicle	MAIS0	MAIS1	MAIS2	MAIS3	MAIS4	MAIS5	Fatal
Medical	$0	$0	$3,801	$24,481	$70,902	$196,911	$441,618	$11,317
EMS	$59	$38	$109	$221	$416	$838	$855	$902
Market	$0	$0	$2,858	$39,637	$110,088	$174,728	$352,178	$933,262
Household	$60	$45	$941	$11,921	$33,505	$44,593	$107,070	$289,910
Insurance	$191	$143	$3,935	$9,370	$24,348	$37,372	$79,967	$28,322
Workplace	$62	$46	$341	$2,644	$5,776	$6,361	$11,091	$11,783
Legal Costs	$0	$0	$1,410	$6,739	$19,645	$35,307	$91,197	$106,488
Subtotal	$372	$272	$13,395	$95,013	$264,680	$496,110	$1,083,976	$1,381,984
Congestion	$2,104	$1,416	$1,426	$1,450	$1,490	$1,511	$1,529	$5,720
Prop. Damage	$3,599	$2,692	$7,959	$8,510	$16,027	$16,328	$15,092	$11,212
Subtotal	$5,704	$4,108	$9,385	$9,960	$17,517	$17,839	$16,621	$16,932
Total Economic	$6,076	$4,380	$22,779	$104,974	$282,197	$513,949	$1,100,597	$1,398,916
QALYs	$0	$0	$23,241	$340,872	$805,697	$2,037,483	$4,578,525	$7,747,082
Total Compr.	$6,076	$4,380	$46,020	$445,846	$1,087,894	$2,551,432	$5,679,122	$9,145,998

* Note: Unit costs are expressed on a per-person basis for all injury levels. PDO costs areexpressed on a per-damaged-vehicle basis.

Table 1-11. Summary of Comprehensive Unit Costs, Unreported Crashes, 2010 Dollars

	PDO Vehicle	MAIS0	MAIS1	MAIS2	MAIS3	MAIS4	MAIS5	Fatal
Medical	$0	$0	$3,801	$24,481	$70,902	$196,911	$441,618	$11,317
EMS	$7	$6	$32	$84	$416	$838	$855	$902
Market	$0	$0	$2,858	$39,637	$110,088	$174,728	$352,178	$933,262
Household	$60	$45	$941	$11,921	$33,505	$44,593	$107,070	$289,910
Insurance	$191	$143	$3,935	$9,370	$24,348	$37,372	$79,967	$28,322
Workplace	$62	$46	$341	$2,644	$5,776	$6,361	$11,091	$11,783
Legal Costs	$0	$0	$1,410	$6,739	$19,645	$35,307	$91,197	$106,488

	PDO Vehicle	MAIS0	MAIS1	MAIS2	MAIS3	MAIS4	MAIS5	Fatal
Subtotal	$320	$240	$13,318	$94,876	$264,680	$496,110	$1,083,976	$1,381,984
Congestion	$384	$180	$180	$180	$180	$180	$180	$458
Prop. Damage	$1,224	$916	$2,707	$2,894	$5,451	$16,328	$15,092	$11,212
Subtotal	$1,609	$1,096	$2,888	$3,075	$5,632	$16,508	$15,272	$11,670
Total Economic	$1,928	$1,337	$16,205	$97,951	$270,312	$512,618	$1,099,248	$1,393,654
QALYs	$0	$0	$23,241	$340,872	$805,697	$2,037,483	$4,578,525	$7,747,082
Comp. Total	$1,928	$1,337	$39,446	$438,823	$1,076,009	$2,550,101	$5,677,773	$9,140,736

* Note: Unit costs are expressed on a per-person basis for all injury levels. PDO costs are expressed on a per-damaged-vehicle basis. Generally, all MAIS 4, 5, and fatal injuries are believed to be police-reported. Values are still included here for reference to cover any exceptional case where unreported crashes might be found for these injury severity categories.

Table 1-12. Summary of Total Comprehensive Costs, Police-Reported Crashes, Millions of 2010 Dollars

	PDO	MAIS0	MAIS1	MAIS2	MAIS3	MAIS4	MAIS5	Fatal	Total	% Total
Medical	$0	$0	$9,803	$6,638	$6,835	$3,364	$2,539	$373	$29,552	3.8%
EMS	$443	$81	$280	$60	$40	$14	$5	$30	$952	0.1%
Market	$0	$0	$7,371	$10,748	$10,612	$2,985	$2,025	$30,797	$64,538	8.3%
Household	$447	$97	$2,427	$3,232	$3,230	$762	$616	$9,567	$20,377	2.6%
Insurance	$1,424	$307	$10,148	$2,541	$2,347	$639	$460	$935	$18,800	2.4%
Workplace	$462	$99	$879	$717	$557	$109	$64	$389	$3,275	0.4%
Legal Costs	$0	$0	$3,636	$1,827	$1,894	$603	$524	$3,514	$11,999	1.5%
Subtotal	$2,776	$583	$34,544	$25,764	$25,514	$8,477	$6,232	$45,604	$149,494	19.2%
Congestion	$15,687	$3,042	$3,677	$393	$144	$26	$9	$189	$23,167	3.0%
Prop. Damage	$26,833	$5,783	$20,526	$2,308	$1,545	$279	$87	$370	$57,730	7.4%
Subtotal	$42,521	$8,825	$24,203	$2,701	$1,689	$305	$96	$559	$80,898	10.4%
Total Economic	$45,297	$9,408	$58,748	$28,465	$27,203	$8,781	$6,327	$46,163	$230,392	29.6%
QALYs	$0	$0	$59,938	$92,431	$77,667	$34,812	$26,322	$255,646	$546,816	70.4%
Comp. Total	$45,297	$9,408	$118,686	$120,896	$104,870	$43,594	$32,649	$301,809	$777,208	100.0%
% Total	5.8%	1.2%	15.3%	15.6%	13.5%	5.6%	4.2%	38.8%	100.0%	0.0%

Table 1-13. Summary of Total Comprehensive Costs, Unreported Crashes (Millions of 2010 Dollars)

	PDO	MAIS0	MAIS1	MAIS2	MAIS3	MAIS4	MAIS5	Fatal	Total	% Total
Medical	$0	$0	$3,346	$1,654	$308	$0	$0	$0	$5,308	5.7%
EMS	$76	$16	$28	$6	$2	$0	$0	$0	$127	0.1%
Market	$0	$0	$2,516	$2,678	$478	$0	$0	$0	$5,672	6.1%
Household	$663	$110	$828	$806	$146	$0	$0	$0	$2,552	2.7%
Insurance	$2,111	$348	$3,464	$633	$106	$0	$0	$0	$6,662	7.1%
Workplace	$685	$112	$300	$179	$25	$0	$0	$0	$1,301	1.4%
Legal Costs	$0	$0	$1,241	$455	$85	$0	$0	$0	$1,782	1.9%
Subtotal	$3,536	$586	$11,722	$6,411	$1,150	$0	$0	$0	$23,404	25.0%
Congestion	$4,248	$439	$159	$12	$1	$0	$0	$0	$4,859	5.2%
Prop. Damage	$13,534	$2,230	$2,383	$196	$24	$0	$0	$0	$18,366	19.6%
Subtotal	$17,782	$2,670	$2,542	$208	$24	$0	$0	$0	$23,225	24.8%
Total	$21,317	$3,255	$14,264	$6,619	$1,174	$0	$0	$0	$46,629	49.8%
QALYs	$0	$0	$20,457	$23,033	$3,499	$0	$0	$0	$46,989	50.2%
Comp. Total	$21,317	$3,255	$34,721	$29,651	$4,673	$0	$0	$0	$93,617	100.0%
% Total	22.8%	3.5%	37.1%	31.7%	5.0%	0.0%	0.0%	0.0%	100.0%	0.0%

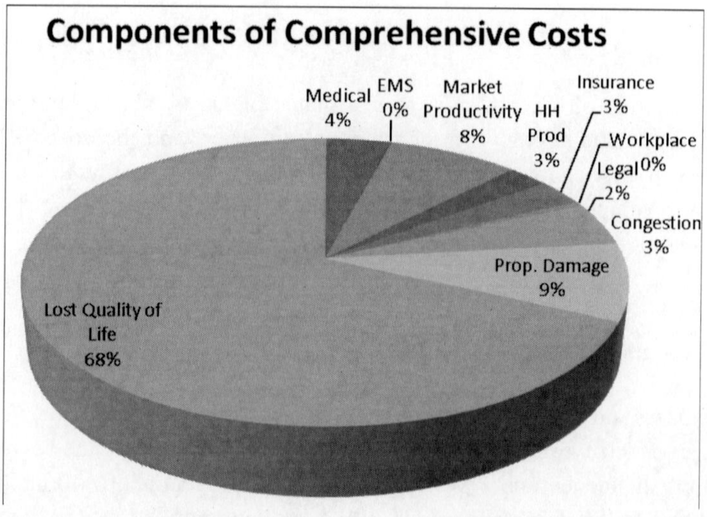

Figure 1-D. Components of Comprehensive Costs.

Table 1-14. Economic and Societal Costs for Selected Crash Types

	Economic Cost (Millions of 2010 Dollars)	% Total	Comprehensive Cost (Millions of 2010 Dollars)	% Total
Outcome Severity:				
Fatalities	$46,163	16.7%	$301,809	34.7%
Nonfatal Injuries	$146,347	52.8%	$484,506	55.6%
PDO Vehicles	$71,480	25.8%	$71,480	8.2%
Uninjured (MAIS0)	$13,030	4.7%	$13,030	1.5%
Total	$277,020	100.0%	$870,826	100.0%
Adverse Driver Behavior:				
Seat Belt Non-use	$13,819	5.0%	$71,984	8.3%
Helmet non-use	$1,251	0.5%	$7,606	0.9%
Distraction	$45,763	16.5%	$129,452	14.9%
Alcohol Involvement	$59,362	21.4%	$242,602	27.9%
Alcohol Causation	$48,858	17.6%	$199,346	22.9%
Speed	$59,078	21.3%	$210,342	24.2%
Nonoccupants:				
Motorcycles	$13,537	4.9%	$66,380	7.6%
Pedestrian/Cyclist	$19,443	7.0%	$90,025	10.3%
Crash Types:				
Roadway Departure Crashes	$73,313	26.5%	$307,022	35.3%
Single-Vehicle Crashes	$88,605	32.0%	$357,052	41.0%
Crash Location:				
Interstate Highway Crashes	$27,983	10.1%	$88,203	10.1%
Intersection Crashes	$139,535	50.4%	$390,512	44.8%
Urban Roadways	$170,553	61.6%	$491,064	56.4%
Rural Roadways	$106,467	38.4%	$379,762	43.6%

2. Human Capital Costs

Estimating the cost of a crash requires estimates of the number of people and vehicles involved in the crash, the severity of each person's injuries, and the costs of those injuries. The first section of this chapter describes the methods used to estimate the incidence and severity of motor vehicle crashes. The succeeding sections explain how the unit costs of injuries were estimated and present those estimates.

I. Crash Data and Severity Estimation

Crash databases do not accurately describe the severity of motor vehicle crashes. Accordingly, we made several adjustments to more accurately reflect the severity of crashes. To estimate injury incidence and severity, we followed procedures developed by Miller and Blincoe (1994) and Miller, Galbraith, et al. (1995) and later applied in Blincoe (1996); Miller, Levy, et al. (1998); Miller, Lestina, and Spicer (1998); Miller, Spicer, et al. (1999); Blincoe et al. (2002); and Zaloshnja et al. (2004). Below we summarize the procedures and describe the adjustments.

NHTSA's General Estimates System (GES) provides a sample of U.S. crashes by police-reported severity for all crash types. GES records injury severity by crash victim on the KABCO scale (National Safety Council, 1990) from police crash reports. Police reports in almost every State use KABCO to classify crash victims as K–killed, A–incapacitating injury, B–non-incapacitating injury, C–possible injury, or O–no apparent injury.

KABCO ratings are coarse and inconsistently coded between States and over time. The codes are selected by police officers without medical training, typically without benefit of a hands-on examination. Some victims are transported from the scene before the police officer who completes the crash report even arrives. Miller, Viner, et al. (1991) and Blincoe and Faigin (1992) documented the great diversity in KABCO coding across cases. O'Day (1993) more carefully quantified the wide variability in use of the A- injury code between States. Viner and Conley (1994) explained the contribution to this variability of differing State definitions of A-injury. Miller, Whiting, et al. (1987) found that police-reported injury counts by KABCO severity systematically varied between States because of differing State crash reporting thresholds (the rules governing which crashes should be reported to the police). Miller and Blincoe (1994) found that State reporting thresholds often changed over time.

Thus police reporting does not accurately describe injuries medically. To minimize the effects of variability in severity definitions by State, reporting threshold, and police perception of injury severity, we turned to NHTSA data sets that included both police-reported KABCO and medical descriptions of injury in the Occupant Injury Coding system (OIC; AAAM, 1990, 1985). OIC codes include AIS severity score and body region, plus more detailed injury descriptors. We used both 2008–2010 Crashworthiness Data System (CDS) and 1984–1986 National Accident Sampling System (NASS; NHTSA, 1987) data. CDS describes injuries to passenger vehicle occupants involved in tow-away crashes. The 1984–1986 NASS data provide the most recent medical description available of injuries to medium/heavy truck and bus occupants, nonoccupants, and other non-CDS crash victims. The

NASS data was coded with the 1980 version of AIS, which differs slightly from the 1985 version; but NHTSA made most AIS-85 changes well before their formal adoption. CDS data was coded in AIS-90/98 with coding shifting to AIS-2005, Update 2008, in 2011. We differentiated our analysis of the two versions of AIS because AIS-90/98 scores and OIC codes differ greatly from codes and scores in AIS-85, especially for brain and severe lower limb injury. Garthe, Ferguson, and Early (1996) find that AIS scores shifted for roughly 25 percent of all OICs between AIS-85 and AIS-90/98.

We used weighted, annualized 2008–2010 GES counts to reweight the CDS and NASS data so that they represent the estimated GES injury victim counts in motor vehicle crashes during 2008–2010. In applying the GES counts to adjust old NASS weights at the person level, we controlled for police-reported injury severity, restraint use, alcohol involvement, and occupant type (CDS occupant, non-CDS occupant, and nonoccupant). All cells had at least 10 cases. Weighting the NASS data to GES restraint use and alcohol involvement levels updates the NASS injury profile to reflect contemporary belt use and alcohol- involvement levels, although it is imperfect in terms of its representation of airbag use in non-tow-away crashes. At the completion of the weighting process, we had a hybrid CDS/NASS casualty-level file—that is, we had an appropriately reweighted NASS record for each injury victim in each non-CDS crash. Similarly, we reweighted the 2008–2010 CDS file to match GES counts in order to get appropriately weighted unit records for the CDS sample strata.

Unit Cost Estimates

The second step required to estimate average crash costs was to generate costs per crash victim by maximum AIS (MAIS), body part, and whether the victim suffered a fracture or dislocation. A 41-level body part descriptor was created based on information provided by the NASS/CDS variables describing the body region, system/organ, lesion, and aspect of each injury. Burns were classified as a separate category due to the lack of location information for burn injuries.

The sections that follow describe unit medical costs, work loss costs, and selected ancillary costs. Appendix A describes the costing methods. Medical and work loss costs cover three mutually exclusive categories that reflect injury severity: (1) injuries resulting in death, including post-injury deaths in a healthcare setting; (2) injuries resulting in hospitalization with survival to discharge; and (3) injuries requiring an emergency department visit not resulting in hospitalization (ED-treated injuries). For injuries treated only in doctor's offices or outpatient departments, we used prior estimates of unit costs (Finkelstein et al., 2006), properly inflated. To estimate mean costs across all surviving crash victims, we needed to add costs for cases treated only in physicians' offices or outpatient departments to the cost for cases treated in hospital emergency departments or admitted to hospitals. To do so, we multiplied unit costs for ED-treated injuries by body part and nature of injury (as per the Barell injury-diagnosis matrix) times ratios of ED-treated injuries versus injuries treated only in doctor's offices or outpatient departments found in Finkelstein et al. (2006). We then took averages across treatment settings. We computed costs from a societal perspective, which means we included all costs regardless of who paid for them.

We estimated mean costs per surviving victim by maximum AIS (MAIS), body part, and fracture/dislocation involvement from combined Healthcare Cost and Utilization Project (HCUP) Nationwide Inpatient Sample (NIS) and Nationwide Emergency Department Sample (NEDS) files. (For descriptions of these files, see Appendix E.) We used ICDMAP-90 software (Johns Hopkins University and Tri-Analytics Inc., 1997) to assign MAIS-90 scores to cases.[2] We assigned AIS-85 scores with mappings developed by Miller et al. (1991). After assigning AIS scores to each injury, we determined the MAIS for each person. We estimated standard errors of means with the SURVEYMEANS command in SAS 9.2, which accounts for sample stratification. Appendix B presents unit costs and standard errors at different discount rates.

Merging HCUP-Based Costs onto the Reweighted NASS/CDS Injury File

Typically, motor vehicle crash patients suffer multiple injuries. In the HCUP-based data, when a victim had two injuries of maximum AIS, we assigned the body part of the more costly injury. In merging costs onto the re-weighted NASS/CDS injury level file (NASS/CDS lists up to six injuries per injury victim) we merged medical and work loss costs separately. In each case, we assigned the cost for the injury with the highest cost for that cost component. Thus if a victim's ruptured spleen had the highest medical cost and her broken leg had the highest work loss cost, this hybrid set of costs was assigned to the case. This will result in conservative cost estimates since it assumes that secondary injury conditions do not result in additional costs.

To estimate the standard error of the mean cost per victim in the reweighted NASS/CDS file we used the following procedure. Based on the standard errors estimated from the HCUP files we estimated the upper and lower levels of the confidence interval for the unit costs at significance level $\alpha=0.1$. We then merged these two levels onto the reweighted NASS/CDS casualty-level file, following the same procedures as above. For each level we estimated the upper and lower levels of the confidence interval for the unit costs at significance level $\alpha=0.1$, separately for CDS and non-CDS strata. Again, to estimate these intervals, we used the SURVEYMEANS command in SAS 9.2, which takes into account the sample stratification. At the end of the process, we had a combined confidence interval at significance level $\alpha=0.01$ ($0.1\times0.1=0.01$); or, to put it differently, the 99-percent confidence interval of the mean unit costs. Assuming a normal distribution of the combined sampling errors, we estimated implied standard errors based on the 99-percent confidence interval of the mean unit costs, by dividing the difference between the 99-percent upper limit and the mean by 2.7045 (the multiplier of the standard error for the 99% confidence interval, assuming a normal distribution for the sampling errors).

Unit Costs Estimated from the Reweighted NASS/CDS File

Table 2-1 presents NASS/CDS crash costs per surviving victim at a 3-percent discount rate by MAIS separately for CDS and non-CDS strata. A paucity of MAIS-6 cases dictated collapsing MAIS-5 and MAIS-6 into a single category. Unit costs generally are higher for crash survivors in CDS than non-CDS strata. The difference results in part from differences between the 1985 and 1990 versions of the AIS coding system (Zaloshnja et al., 2001). However, comparing HCUP-based unit cost estimates by MAIS (Table 2-2), with each crash survivor scored both in AIS-85 and AIS-90 (i.e., keeping the injury mix constant), indicates that the non-CDS strata injury mix drives the cost difference. Table 2-3 presents the NASS/CDS crash costs per surviving victim and fatality at 3 percent and 4-percent discount rates by MAIS, regardless of the AIS version. At a 3-percent discount rate, the average crash fatality involves an estimated $11,317 in medical spending (with a standard error of $100 based on 33,932 crash deaths in 2010), $933,262 in wage and fringe benefit losses (standard error $3,282 based on 32,885 crash deaths in 2010), and $289,910 in household work losses (standard error $631, also based on 32,885 crash deaths in 2010).

Tables 2-4 to 2-6 present NASS/CDS crash costs per surviving victim at 3-percent discount rate by body region, fracture/dislocation involvement, and MAIS. Appendix B provides detailed unit costs by body part, fracture/dislocation involvement, and MAIS, at different discount rates.

A major limitation of the costs presented is that some cost components are unavoidably quite old. In particular, no recent source exists for the percentage of lifetime medical costs that is incurred more than 18 months post-injury, probabilities of permanent disability by detailed diagnosis and whether hospital admitted, or the ratio of household work days lost to wage work days lost.

Table 2-1. 2008–2010 NASS/CDS-based crash costs per surviving victim at 3-percent discount rate by MAIS (2010 Dollars)

MAIS	Non-CDS strata (in AIS85 scale)				CDS stratum (in AIS-90 scale)			
	Mean	Implied std. error	99% Conf. interval		Mean	Implied std. error	99% Conf. interval	
Medical cost per victim								
1	2,706	63	2,535	2,877	4,546	985	1,882	7,211
2	11,341	999	8,639	14,043	29,143	4,764	16,259	42,027
3	54,764	4,406	42,848	66,680	79,317	4,330	67,607	91,026
4	97,788	9,662	71,658	123,918	213,984	19,340	161,681	266,288
5&6	494,443	67,350	312,296	676,590	424,332	18,104	375,370	473,293
Wage loss per victim								
1	2,555	133	2,195	2,915	3,073	448	1,861	4,285
2	21,817	1,834	16,856	26,777	46,156	5,884	30,242	62,069
3	76,427	6,228	59,584	93,270	128,380	8,343	105,816	150,944
4	122,659	8,778	98,918	146,399	183,871	16,852	138,294	229,448
5&6	246,845	31,990	160,330	333,359	385,958	16,498	341,339	430,576
Household productivity loss per victim								
1	1,029	66	851	1,207	878	127	534	1,222
2	8,248	639	6,519	9,977	13,264	1,357	9,596	16,933
3	29,402	1,773	24,608	34,196	35,735	1,325	32,151	39,320
4	49,381	3,533	39,827	58,935	43,753	3,081	35,420	52,086
5&6	89,749	5,540	74,766	104,731	112,625	3,034	104,421	120,830

Table 2-2. HCUP-based crash costs per surviving victim at 3-percent discount rate by MAIS; AIS-85 versus AIS-90 (2010 dollars)

Cost category	Scored in AIS-85	Scored in AIS-90
MAIS 1		
Medical cost per victim	2,303	2,481
Earnings loss per victim	2,362	2,436
Household production loss per victim	1,142	1,154
MAIS 2		
Medical cost per victim	8,215	11,672
Earnings loss per victim	11,176	16,839
Household production loss per victim	4,226	6,324
MAIS 3		
Medical cost per victim	48,269	44,949
Earnings loss per victim	63,794	56,525
Household production loss per victim	22,376	21,218
MAIS 4		
Medical cost per victim	136,897	131,595
Earnings loss per victim	111,664	141,669
Household production loss per victim	39,186	45,765
MAIS 5		
Medical cost per victim	460,042	323,511
Earnings loss per victim	348,419	268,461
Household production loss per victim	100,144	78,035
MAIS 6		
Medical cost per victim	N/A	482,964
Earnings loss per victim	N/A	387,240
Household production loss per victim	N/A	112,880

Table 2-3. 2008–2010 NASS/CDS-based crash costs per victim at 3-percent and 4-percent discount rates by MAI-

MAIS	Medical costs		Earnings loss		Household production loss	
	Discounted @ 3%	Discounted @ 4%	Discounted @ 3%	Discounted @ 4%	Discounted @ 3%	Discounted @ 4%
1	3,784	3,784	2,858	2,484	941	829
2	24,375	24,375	39,637	34,273	11,921	10,324
3	70,672	70,672	110,088	96,463	33,505	29,084
4	196,629	195,646	174,728	147,950	44,593	37,812
5&6	441,356	437,089	352,178	332,027	107,070	102,143
Fatality	11,317	11317	933,262	799,270	289,910	246,559

Table 2-4. 2008–2010 NASS/CDS-based medical costs per surviving victim at a 3-percent discount rate, by body region and MAIS (2010 dollars)[3]

Body region	Fracture/ dislocation	MAIS	Non-CDS strata (in AIS-85 scale)				CDS stratum (in AIS-90 scale)			
			Mean	Implied std. error	99% confidence interval		Mean	Implied std. error	99% confidence interval	
Spinal Cord Injury	No	3	254,058	75,152	50,812	457,304	218,326	53,394	73,923	362,728
		4	347,380	89,912	104,214	590,546	498,562	20,344	443,542	553,583
		5&6	974,741	1,816	969,829	979,654	1,090,124	10,229	1,062,458	1,117,789
Traumatic Brain Injury	No	1	3,967	72	3,772	4,162	3,769	18	3,719	3,818
		2	5,423	26	5,354	5,492	10,361	1,922	5,163	15,559
		3	69,736	2,890	61,920	77,551	41,167	1,701	36,567	45,767
		4	122,964	3,636	113,131	132,797	126,226	2,624	119,130	133,322
		5&6	401,223	15,216	360,072	442,374	349,353	14,102	311,214	387,491
Lower extremity	No	1	1,872	22	1,813	1,930	2,602	81	2,382	2,821
		2	15,583	1,008	12,856	18,310	8,075	1,165	4,924	11,226
		3	37,906	2,971	29,871	45,940	70,801	8,096	48,906	92,696
		4	127,358	17,414	80,262	174,454	138,409	18,945	87,174	189,644
		5	N/A	N/A	N/A	N/A	N/A	N/A	N/A	N/A
	Yes	1	2,239	126	1,898	2,579	2,359	94	2,105	2,613
		2	14,986	710	13,066	16,906	25,061	4,536	12,793	37,329
		3	61,367	1,625	56,973	65,761	68,497	2,928	60,579	76,414
		4	46,803	12,036	14,253	79,353	116,872	3,609	107,112	126,632
		5	N/A	N/A	N/A	N/A	183,024	14,008	145,139	220,909
Upper extremity	No	1	1,560	19	1,509	1,611	1,789	21	1,733	1,846
		2	4,021	210	3,452	4,590	10,089	3,530	543	19,636
		3	37,030	5,732	21,528	52,532	41,431	9,908	14,635	68,227
	Yes	1	2,030	87	1,795	2,265	2,391	75	2,187	2,595
		2	6,175	228	5,560	6,791	7,345	1,999	1,939	12,751

Table 2-4. (Continued)

Body region	Fracture/ dislocation	MAIS	Non-CDS strata (in AIS-85 scale)				CDS stratum (in AIS-90 scale)			
			Mean	Implied std. error	99% confidence interval		Mean	Implied std. error	99% confidence interval	
		3	43,603	1,354	39,941	47,265	41,751	1,545	37,574	45,929
Trunk/ abdomen	No	1	3,072	97	2,811	3,333	12,089	2,819	4,465	19,714
		2	16,346	4,835	3,269	29,422	17,399	5,529	2,447	32,352
		3	57,386	13,920	19,739	95,033	64,810	8,797	41,019	88,601
		4	70,901	22,488	10,082	131,720	85,454	10,962	55,808	115,101
		5	161,528	30,047	80,267	242,788	190,945	32,189	103,889	278,000
	Yes	1	4,444	61	4,280	4,608	3,415	259	2,716	4,115
		2	9,876	455	8,646	11,106	16,096	2,709	8,770	23,421
		3	26,730	1,171	23,564	29,896	78,704	4,104	67,604	89,804
		4	81,361	4,720	68,596	94,126	66,758	4,299	55,132	78,383
		5	N/A	N/A	N/A	N/A	205,873	60,151	43,196	368,550
Other head	No	1	2,293	12	2,261	2,324	3,857	61	3,694	4,021
		2	5,725	741	3,721	7,728	18,421	5,573	3,350	33,493
		3	29,023	4,659	16,423	41,623	39,794	3,804	29,506	50,082
	Yes	1	4,444	61	4,280	4,608	7,187	110	6,891	7,484
		2	14,362	636	12,640	16,083	25,799	2,481	19,090	32,508
		3	91,164	5,461	76,395	105,933	103,322	9,280	78,224	128,420
		4	177,917	21,276	120,377	235,457	185,255	3,420	176,005	194,505
Burns	No	1	1,889	116	1,577	2,201	2,443	410	1,334	3,552
		2	17,878	3,041	9,653	26,102	30,112	4,834	17,037	43,186
		5	251,671	32,167	164,676	338,666	N/A	N/A	N/A	N/A
Minor external	No	1	3,313	29	3,235	3,391	3,896	35	3,803	3,989

Table 2-5. 2008–2010 NASS/CDS-based earnings loss per surviving victim at a 3-percent discount rate, by body region and MAIS (2010 dollars)[4]

Body region	Fracture/ dislocation	MAIS	Non-CDS strata (in AIS-85 scale)				CDS stratum (in AIS-90 scale)			
			Mean	Implied std. error	99% confidence interval		Mean	Implied std. error	99% confidence interval	
Spinal Cord Injury	No	3	94,869	22,174	34,899	154,838	237,487	51,676	97,732	377,242
		4	189,924	21,655	131,358	248,489	212,687	9,947	185,785	239,588
		5&6	89,564	11,208	59,254	119,875	123,976	30,740	40,840	207,113
Traumatic Brain Injury	No	1	3,714	164	3,271	4,157	3,349	49	3,216	3,481
		2	7,798	78	7,586	8,010	28,886	3,849	18,476	39,297
		3	132,791	1,802	127,918	137,664	94,391	1,898	89,257	99,525
		4	145,272	3,197	136,626	153,918	177,952	2,598	170,927	184,977
		5&6	365,430	8,026	343,725	387,135	402,422	8,998	378,087	426,756
Lower extremity	No	1	1,319	19	1,268	1,371	2,186	147	1,787	2,584
		2	25,407	1,830	20,457	30,356	18,840	2,474	12,150	25,530
		3	24,864	3,826	14,516	35,211	100,316	11,218	69,978	130,655
		4	209,495	9,905	182,708	236,282	218,075	11,891	185,916	250,234
		5	N/A	N/A	N/A	N/A	N/A	N/A	N/A	N/A
	Yes	1	3,275	116	2,961	3,589	3,576	55	3,428	3,723
		2	29,854	978	27,208	32,499	63,520	9,925	36,679	90,362
		3	77,272	1,241	73,915	80,628	86,202	4,524	73,966	98,438
		4	87,050	20,028	32,884	141,216	88,675	2,273	82,529	94,821
		5	N/A	N/A	N/A	N/A	86,311	17,200	39,795	132,827
Upper extremity	No	1	1,233	23	1,171	1,295	1,462	19	1,410	1,515
		2	5,789	525	4,369	7,210	19,966	5,462	5,193	34,738
		3	73,408	12,832	38,704	108,113	73,692	21,076	16,693	130,691
	Yes	1	5,947	92	5,697	6,197	6,566	202	6,019	7,113
		2	16,526	677	14,694	18,358	21,121	4,143	9,917	32,325

Table 2-5. (Continued)

Body region	Fracture/ dislocation	MAIS	Non-CDS strata (in AIS-85 scale)				CDS stratum (in AIS-90 scale)			
			Mean	Implied std. error	99% confidence interval		Mean	Implied std. error	99% confidence interval	
		3	59,195	1,817	54,281	64,110	67,381	4,538	55,108	79,655
Trunk/ abdomen	No	1	4,739	163	4,298	5,181	42,732	5,074	29,010	56,454
		2	34,033	8,870	10,045	58,020	41,737	9,573	15,847	67,627
		3	81,107	27,246	7,419	154,794	116,052	21,942	56,711	175,393
		4	81,197	20,530	25,676	136,719	96,674	14,629	57,111	136,236
		5	175,858	25,791	106,108	245,608	198,500	21,649	139,950	257,050
	Yes	1	4,099	48	3,970	4,228	4,427	327	3,542	5,313
		2	25,072	761	23,013	27,130	41,515	5,572	26,445	56,584
		3	45,234	1,672	40,712	49,755	115,912	10,104	88,585	143,238
		4	57,074	7,547	36,664	77,485	69,420	7,376	49,472	89,368
		5	N/A	N/A	N/A	N/A	290,421	96,587	29,203	551,639
Other head	No	1	2,223	19	2,172	2,274	4,339	206	3,780	4,897
		2	16,546	5,058	2,866	30,227	37,363	3,103	28,970	45,755
		3	34,953	8,963	10,712	59,194	73,394	6,347	56,229	90,558
	Yes	1	4,099	48	3,970	4,228	6,901	98	6,637	7,165
		2	16,426	424	15,279	17,574	40,340	4,809	27,334	53,346
		3	115,266	6,073	98,841	131,690	122,690	5,912	106,700	138,680
		4	177,451	15,663	135,090	219,812	209,400	2,612	202,335	216,465
Burns	No	1	3,525	147	3,127	3,924	3,731	129	3,383	4,080
		2	10,343	622	8,661	12,025	19,238	1,637	14,812	23,664
		5	105,044	32,695	16,621	193,467	N/A	N/A	N/A	N/A
Minor external	No	1	1,618	17	1,571	1,666	1,221	17	1,175	1,267

Table 2-6. 2008–2010 NASS/CDS-based household production per surviving victim at 3-percent discount rate, by body region and MAIS (2010 dollars)[5]

Body region	Fracture/ dislocation	MAIS	Non-CDS strata (in AIS-85 scale)				CDS stratum (in AIS-90 scale)			
			Mean	Implied std.error	99% confidence interval		Mean	Implied std.error	99% confidence interval	
Spinal Cord Injury	No	3	29,717	6,938	10,952	48,481	78,510	16,616	33,571	123,449
		4	58,306	8,741	34,666	81,945	72,541	2,316	66,277	78,805
		5&6	80,793	417	79,666	81,921	91,380	2,161	85,536	97,224
Traumatic Brain Injury	No	1	1,159	42	1,046	1,273	1,226	17	1,180	1,272
		2	2,829	26	2,760	2,898	7,826	946	5,268	10,384
		3	44,940	468	43,675	46,205	35,225	539	33,766	36,684
		4	54,669	772	52,580	56,758	60,579	621	58,899	62,259
		5&6	109,361	2,166	103,503	115,219	119,789	2,279	113,626	125,952
Lower extremity	No	1	540	6	523	556	836	53	692	979
		2	8,142	550	6,655	9,630	5,189	618	3,518	6,859
		3	15,212	967	12,598	17,826	34,678	2,873	26,909	42,447
		4	60,451	1,745	55,732	65,170	61,817	1,723	57,156	66,478
		5	N/A	N/A	N/A	N/A	N/A	N/A	N/A	N/A
	Yes	1	1,359	49	1,227	1,491	1,454	16	1,411	1,497
		2	11,563	318	10,704	12,423	21,895	2,008	16,465	27,324
		3	28,897	624	27,209	30,585	31,469	744	29,456	33,482
		4	30,035	4,516	17,821	42,249	31,459	703	29,558	33,360
		5	N/A	N/A	N/A	N/A	31,815	3,513	22,314	41,316
Upper extremity	No	1	545	9	520	569	606	7	587	626
		2	1,916	150	1,511	2,322	6,064	1,492	2,028	10,099
	Yes	3	24,872	2,433	18,293	31,451	24,402	4,965	10,975	37,828
		1	2,207	31	2,122	2,291	2,416	74	2,217	2,614
		2	8,396	685	6,544	10,247	6,928	924	4,430	9,425

Table 2-6. (Continued)

Body region	Fracture/ dislocation	MAIS	Non-CDS strata (in AIS-85 scale)				CDS stratum (in AIS-90 scale)			
			Mean	Implied std.error	99% confidence interval		Mean	Implied std.error	99% confidence interval	
		3	25,278	1,345	21,640	28,916	33,515	1,218	30,220	36,810
Trunk/ abdomen	No	1	2,251	62	2,084	2,417	10,518	1,301	7,000	14,036
		2	10,971	2,354	4,605	17,338	11,260	2,420	4,715	17,804
		3	30,025	7,639	9,364	50,685	33,924	6,305	16,873	50,975
		4	29,590	8,487	6,636	52,544	32,281	3,844	21,885	42,677
		5	57,454	3,171	48,878	66,030	78,169	8,130	56,180	100,158
	Yes	1	1,372	13	1,338	1,406	1,739	48	1,609	1,869
		2	10,509	233	9,879	11,139	12,905	1,304	9,378	16,432
		3	19,398	678	17,564	21,232	45,564	1,627	41,165	49,964
		4	24,771	1,901	19,631	29,912	25,337	1,219	22,039	28,636
		5	N/A	N/A	N/A	N/A	100,540	5,339	86,102	114,978
Other head	No	1	793	6	777	809	1,485	46	1,361	1,608
		2	4,146	1,038	1,338	6,955	10,970	3,586	1,273	20,667
		3	13,287	3,026	5,104	21,470	19,961	566	18,429	21,493
	Yes	1	1,368	11	1,339	1,397	2,246	20	2,191	2,301
		2	5,244	125	4,905	5,583	11,704	1,184	8,502	14,905
		3	37,301	860	34,976	39,626	42,661	1,611	38,306	47,017
		4	49,136	3,424	39,876	58,396	60,468	702	58,570	62,366
Burns	No	1	1,350	45	1,229	1,471	1,385	40	1,276	1,494
		2	3,192	120	2,868	3,517	5,798	306	4,969	6,626
		5	43,723	6,712	25,570	61,876	N/A	N/A	N/A	N/A
Minor external	No	1	757	7	738	776	532	6	516	548

II. Property Damage, Insurance, and Legal Costs

Some crash costs are most easily estimated from insurance data. These include not only insurance claims processing and legal costs but also costs of property damage. Insurance data also is a critical input when analyzing who pays the costs of crashes.

To analyze the insurance-related costs, we purchased data from the Insurance Services Office (ISO). ISO is a data-pooling organization that aggregates claims data from a large cross-section of auto insurers. Data we bought detailed insurance premiums collected and claims paid by selected insurers in 2009. We used those data in conjunction with national insurance statistics and crash data to analyze (1) property damage costs per crash, (2) numbers of people receiving insurance claims payments due to crash injury, and (3) transaction costs of compensation through the insurance and legal systems. This chapter describes the data we purchased, our analyses of them, and what they showed.

Auto Insurance Data Description and Loss Cost Computations

ISO structured its data report around a spreadsheet developed by the Motorcycle Insurance Committee of the National Association of Independent Insurers (NAII, Miller and Lawrence, 2003). ISO was able to break out data only by motorcycle versus other personal auto versus commercial auto, with commercial auto decomposed by vehicle type. They provided data on seven categories of insurance coverage:

- Bodily injury liability (coverage if the policyholder's vehicle injures someone; mandatory in most States; in no-fault insurance States this coverage compensates losses that exceed the no-fault threshold). For motorcycles, some companies separated passenger liability coverage from other bodily injury coverage.
- Property damage liability (coverage if the policyholder's vehicle damages or destroys someone else's property; mandatory in many States).
- Own medical payments (coverage for the policyholder's own injury treatment costs up to a modest ceiling, typically $1,000; often mandatory in States without no-fault insurance).
- Personal injury protection (no-fault coverage for the policyholder's own losses up to a modest ceiling, typically $15,000 to $25,000; mandatory in some States).
- Collision (coverage for damage to the policyholder's vehicle when the policyholder is at fault in the crash or no one is; typically required by the lender if vehicle purchase was financed).
- Comprehensive (coverage for theft or non-crash damage to the policyholder's vehicle; typically required by the lender if vehicle purchase was financed).
- Uninsured and underinsured motorist (coverage for injuries to the policyholder and other occupants of the policyholder's vehicle, as well as the policyholder's property damage when a driver without insurance is at fault or when the at-fault driver has too little insurance to fully compensate the policyholder's losses; mandatory in many States).

For each category, we obtained four data items for policies written in 2009. Coverage in a policy is for a maximum of one year:

- Earned exposure (the number of vehicles covered by insurance for this risk).
- Earned premiums (how much policyholders paid for this coverage, net of any dividends or rebates to policyholders).
- Incurred losses (the amount paid or reserved for future payment of claims against the policies, including amounts that will be paid by reinsurers).
- Incurred claim count (the number of damage claims that the insurance paid for or anticipates paying for as lawsuits and other disputes are resolved).

From the data collected, by vehicle and coverage type, we computed:

- Claims per 1,000 covers (incurred claim count divided by earned exposure, i.e., the number of claims filed per 1,000 policies that offer the specific coverage).
- Claim severity (incurred losses divided by incurred claim count, i.e., the average payments per claim paid).
- Average loss cost (incurred losses divided by earned exposure, a measure influenced by both the frequency of claims and claim severity, i.e., losses per cover).
- Percent of total losses (by vehicle type, incurred losses for each coverage divided by total incurred losses for all coverages).
- Loss ratio (the ratio of incurred losses to earned premiums, i.e., the percentage of premiums that is paid to settle claims).

We used National data on premiums written and loss ratios (Glenn, 2010) to estimate coverage and representativeness of ISO data and to factor up ISO data to National estimates. As Table 7 shows, ISO data include 26.2 percent of private passenger auto premiums and 26.2 percent of commercial auto premiums. Unlike in 1998–1999, when Miller and Lawrence (2000) found losses in ISO data was typical of all auto policies, the loss ratios in 2009 ISO private passenger and commercial auto liability data are lower than the National averages. Loss ratios still are comparable to National averages for other coverages.

Table 2-8 summarizes premiums and exposures earned and policy results.

Property Damage Costs

Across commercial and personal lines, property damage averaged $2,547 per liability claim for damage to other vehicles and $3,122 per collision claim for damage to the insured's vehicle, with an overall average of $2,840. In general, own collision is subject to a deductible but liability is first-dollar coverage. So why is the average insurance payment higher under own collision? Because people do not file small claims below their deductible from fear of increased insurance rates, or with a claim amount below their deductible.

We estimated how often people do not claim for property damage. Ratioing the number of earned exposures from Table 2-8 indicates that 78.3 percent (47.3 million collision policies/60.4 million liability policies) of insured drivers carry collision coverage. About one-third of crashes are single-vehicle crashes, with the large majority of the rest involving two vehicles. So if all crashes with meaningful damage led to claims, we would expect to see 1.566 (2×0.783) times as many own-vehicle claims as liability claims. Indeed, this multiplier could be even higher since drivers share fault in some crashes. The actual ratio is 1.044. The remaining 0.522 smaller claims are not filed.

Damage costs per own damaged vehicle average $3,122 plus deductible. A Web site that specializes in insurance quotes (www.carinsurance.com/kb/content24628.aspx, accessed July 21, 2012) States that insurance professionals suggest collision policies typically carry a $500 deductible but itself estimates a lower $375 average. With a $375 deductible, costs would average $3,497 per vehicle damaged sufficiently to prompt a collision claim in 2009. We assume the $2,547 average in liability claims should apply to own damage as well. Then for the remaining one-third of damaged vehicles (1−1.044/1.566 = 0.3335), costs would average $648 ([$2,547−$3,497×.6665]/0.3335). Presumably that lower property damage cost is representative of damage in unreported crashes. Across all vehicles with property damage compensated by insurers, damage costs per vehicle would average $3,032 ($2,840+$375× 2,303,872/4,511,166).

Blincoe and Luchter (1983) provide ratios that can be applied to average property damage in reported crashes to estimate damage by crash severity. Table 2-10 shows those factors and applies them to estimate property damage per vehicle by MAIS crash severity. It provides both cost per vehicle and cost per crash, as well as associated incidence estimates. In this table, we computed vehicles in insurer- reported property damage crashes as vehicles with property damage compensated (computed as vehicles with claims in ISO divided by the percentage of property damage claims costs in ISO) minus vehicles where someone had an injury reported in the CDS/GES aggregate file. Multiplying vehicle involvements times cost per damaged vehicle indicates that $59.9 billion in damage occurred in those crashes, generating $54.2 billion in insurance payments. By comparison, the insurance data indicated $53.5 billion was paid. The close agreement of these numbers suggests that the severity allocation factors, although old, still are reasonably accurate.

Property Damage Cost per Crash. Insurance compensated $60 billion in crash damage in 2009 (Table 2-10). Because NHTSA no longer collects this information, Table 2-10 uses number of vehicles per crash by MAIS from 1984–1986 NASS and 2009 FARS to compute property damage costs per crash from costs per vehicle. Table 2-11 shows that 2009 GES estimates of vehicles per crash by police-reported KABCO severity are virtually identical to the 1984–1986 NASS estimates. These ratios have remained remarkably stable over time. For unknown reasons, the GES/NASS ratio for fatal crashes of 1.63–1.66 is much higher than the ratio of 1.48 for 2009 from FARS. We used the FARS ratio.

Number of People Whom Auto Insurance Compensates for Injury

ISO includes 4,511,166 property damage claims (Table 2-8). The percentage of property damage premiums covered by ISO insurers is slightly higher than the percentage of claims paid (Table 2-7). Thus, drivers covered by these insurers either (1) have slightly lower crash risks than other insureds, (2) suffer slightly less damage per crash, or (3) buy slightly more costly insurance. Depending on which of these possibilities is correct, insurers paid for damage to 18.3–18.8 million crashed vehicles in 2009. Similarly, exclusive of uninsured motorist coverage, auto insurers paid 4.62–5.14 million injury claims in 2009. We computed these ranges as ISO claims incurred divided by percentage of premiums or claims payments in ISO.

Some own-medical claims and no-fault claims, however, are for injuries that also generate bodily injury claims. Roughly one-third of crashes involve a single vehicle. Thus at most one-third of drivers (half of drivers in multi-vehicle crashes) might be in crashes where another driver was at fault. Those drivers generally would receive bodily injury compensation as their insurer recovered own-medical losses from the at-fault driver's insurance. Because some bodily injury claims are for recovery above no-fault limits, we assume 10 percent of no-fault claims also involve a bodily injury claim. Reducing own-medical claims by one-third and no-fault claims by 10 percent suggests liability insurance compensated 4.2–4.7 million injured people in 2009. This estimate is incomplete.

Despite preponderant State laws mandating liability coverage, an estimated 13.8 percent of U.S. drivers are uninsured (Insurance Research Council, 2011). Uninsured motorist coverage compensates bodily injury and in some States, either by mandate or at buyer option, property damage. A single claim can capture both categories of losses. This coverage is not mandatory everywhere; only 89 percent of personal auto liability insurance buyers purchased it in 2009. It probably compensated another 0.25 to 0.28 million injury claims for insured drivers (89% with coverage × 13.8% of drivers uninsured × 561,680 ISO bodily injury claims against insured drivers/24.3%–27.1% of all bodily injury claims in the ISO data). That brings the total number of people with auto insurance compensation for injury in 2009 to 4.45 to 5.0 million.

Suppose uninsured drivers had average crash risks and carried auto insurance. Then 4.9 to 5.45 million people would have received auto insurance compensation for medical costs, earnings losses, or lost household production in 2009 (4.2–4.7 million/86.2% insured).

Comparison to Other Crash Injury Counts. How does this number compare with estimates from NHTSA data systems and health care administrative system? At the person level, we reweighted NHTSA's CDS and old NASS data for non-CDS strata using 2007–2010 GES data. The weights were matched by police- reported injury severity, belt use, alcohol-involvement in the crash, single- versus multiple-vehicle crash, and occupant/nonoccupant. We then applied GES and police underreporting rates by MAIS from NHTSA's main report. We added fatalities from FARS. The resulting incidence estimate for people injured in crashes was 3,954,503 (2,391,766 from CDS and FARS, plus 1,562,737 from non-CDS strata), including 3,002,384 injured people included in police-reported crashes. Table 2-9 summarizes the estimates. It also uses the cost estimates in Table 2-2 to estimate total cost of these injuries. Since even our lower bound is far higher, computations in the remainder of this report use the 4.9 million lower bound.

HCUP NIS and NEDS offer a further estimate of nonfatal crash injury incidence. They indicate that 2,735,916 people were treated and released for crash injuries in 2007 and 221,366 crash survivors were admitted to hospital in 2008. Adding survivors treated only in physician offices and clinics based on factors from Finkelstein et al. (2006), we estimate 3.6 million crash survivors were medically treated annually in 2007–2008. The comparable total for 2000 was 4.2 million.

Portion of Injured People Compensated. Blincoe et al. (2002) adopt estimates from Miller et al. (1991) that auto liability policy limits average $148,000 per person injured ($210,000 inflated to 2010 dollars) and that 55 percent of those suffering moderate (MAIS-2) to fatal injuries make a claim. Since States have not been shifting liability regimes (e.g., changing to no-fault insurance or raising minimum liability coverage requirements), these factors probably are unchanged. They suggest $31.225 billion (55% × [($63.787 billion –

$17.220 billion) + ($18.842 billion – $8.637 billion)], from Table 2-9) of the $50.017 billion ($75.612 billion in premiums × .651 loss ratio) in insurance compensation pays for medical and earnings/household production losses of people with MAIS-2+ injuries. The 0.95 million compensated injury claims in excess of the injury count estimated in NHTSA's crash reporting survey (23.9% of claims) represent a combination of (1) liability claims for damages exceeding policy limits for own medical or no- fault claims and (2) fraudulent claims. (That percentage is credible given that an estimated 11% of bodily injury (BI) claims and 6% of no-fault personal injury protection [PIP] claims were fraudulent nationally in 2007, so roughly 91 percent of claims were for actual injuries. Remaining compensation would be sufficient to cover 55 percent of MAIS-1 injury costs [($50.017 billion × 91% – $31.225 billion)/($17.220 billion + $9.011 billion)]. This uniform compensation seems more credible in today's insurance market than the estimated 24.9 percent of MAIS-1 injuries covered in 1990 (Miller et al., 1991).)

This shift to uniformity is not surprising. The average property damage liability claim rose from roughly $1,600 in 1998–1999 to $2,547 in 2009. As property damage amounts rise, people are more likely to claim rather than settle informally outside the insurance system. Minor injuries they otherwise might have ignored probably would be divulged and compensated as part of the process.

Table 2-7. Policyholders in 2009 pooled, multi-insurer Insurance Services Office (ISO) data as a percentage of insured vehicles, and representativeness of loss ratios in ISO data

Coverage	Premiums Written		% of Premiums in ISO Data	Loss Ratio		% of Losses in ISO Data
	Nationally	In ISO Data		Nationally	In ISO Data	
Private Passenger Auto Liability	$96,498,093,000	$26,303,105,214	27.3%	68.9	58.3	23.1%
Private Passenger Property Damage	$64,821,016,000	$15,941,256,094	24.6%	58.3	56.7	23.9%
Total Private Passenger	$161,319,109,000	$42,244,361,308	26.2%	64.6	57.7	23.4%
Commercial Auto Liability	$18,394,752,000	$4,864,048,399	26.4%	53.1	50.4	25.1%
Commercial Property Damage	$5,656,866,000	$1,430,695,559	25.3%	54.0	52.9	24.8%
Total Commercial	$24,051,618,000	$6,294,743,958	26.2%	53.3	50.9	25.0%
All Auto Liability	$114,892,845,000	$31,167,153,613	27.1%	66.4	57.1	23.3%
All Own Property Damage	$70,477,882,000	$17,371,951,653	24.6%	58.0	56.4	24.0%
Grand Total	$185,370,727,000	$48,539,105,266	26.2%	63.2	56.8	23.6%
All Bodily Injury	$76,831,734,421	$20,831,443,918	27.1%	65.1	58.5	24.3%
All Property Damage *	$108,538,992,579	$27,707,661,348	25.5%	62.0	57.4	23.6%

* Includes comprehensive (non-crash) coverage.

Table 2-8. Earned premiums, exposures, claims, and losses by auto insurance line and coverage in 2009 pooled, multi-insurer Insurance Services Office data

Coverage	Earned Premium	Earned Exposure (Car Years)	Incurred Claims	Claims per 1,000 Covers	Incurred Losses	Average Cost per Claim	Average Loss Cost	% of Total Losses	Loss Ratio
PERSONAL LIABILITY									
Bodily Injury	$10,664,642,098	53,294,052	561,680	10.5	$5,933,591,572	$10,564	$111	38.7%	55.6
Property Damage	$7,804,414,085	53,465,036	2,023,212	37.8	$4,986,741,517	$2,465	$93	32.5%	63.9
Personal Injury Protection	$3,414,579,139	23,026,608	429,129	18.6	$2,498,591,166	$5,822	$109	16.3%	73.2
Medical Payments	$736,154,211	26,681,405	200,499	7.5	$522,840,594	$2,608	$20	3.4%	71.0
Uninsured/Under insured Motorist	$3,683,315,681	47,639,390	180,508	3.8	$1,397,283,791	$7,741	$29	9.1%	37.9
Total	$26,303,105,214	53,465,036	3,395,028	63.5	$15,339,048,640	$4,518	$287	100.0%	58.3
PERSONAL AUTO PHYSICAL DAMAGE									
Collision	$11,511,031,389	41,708,624	2,173,779	52.1	$6,634,847,526	$3,052	$159	43.3%	57.6
Comprehensive	$4,430,224,705	42,403,787	2,393,853	56.5	$2,399,668,164	$1,002	$57	15.6%	54.8
Total	$15,941,256,094	42,403,787	4,567,632	107.7	$9,034,515,690	$1,978	$213	58.9%	56.9
COMMERCIAL LIABILITY									
Bodily Injury	$3,502,825,417	6,921,091	53,365	7.7	$1,763,305,023	$33,042	$255	11.5%	50.3
Property Damage	$1,260,456,165	6,921,091	184,082	26.6	$634,507,411	$3,447	$92	4.1%	50.3
No Fault	$100,766,817	2,295,294	7,690	3.4	$51,483,863	$6,695	$22	0.3%	51.1
Total	$4,864,048,399	16,137,476	245,137	15.2	$2,449,296,297	$9,992	$152	16.0%	50.4
COMMERCIAL AUTO PHYSICAL DAMAGE									
Collision	$1,081,294,659	5,573,201	130,093	23.3	$557,378,289	$4,284	$100	3.6%	51.5

Coverage	Earned Premium	Earned Exposure (Car Years)	Incurred Claims	Claims per 1,000 Covers	Incurred Losses	Average Cost per Claim	Average Loss Cost	% of Total Losses	Loss Ratio
Comprehensive	$349,400,900	5,422,795	92,322	17.0	$199,488,341	$2,161	$37	1.3%	57.1
Total	$1,430,695,559	5,573,201	222,415	39.9	$756,866,630	$3,403	$136	4.9%	52.9
ALL POLICIES EXCEPT UNINSURED MOTORIST AND COMPREHENSIVE									
Property Damage	$21,657,196,298	107,667,952	4,511,166	41.9	$12,813,474,743	$2,840	$119	83.5%	59.2
Liability	$9,064,870,250	60,386,127	2,207,294	36.6	$5,621,248,928	$2,547	$93	36.6%	62.0
Own (Deductible)	$12,592,326,048	47,281,825	2,303,872	48.7	$7,192,225,815	$3,122	$152	46.9%	57.1
Bodily Injury	$18,418,967,682	112,218,450	1,252,363	11.2	$10,769,812,218	$8,600	$96	70.2%	58.5

Personal lines coverage includes private passenger vehicles and motorcycles.

Table 2-9. Estimated people injured in crashes and compensated losses (2010 dollars) by injury severity and crash strata

MAIS	CDS Cases	Cases Adjusted for Under- reporting	Paid/ Person*	Cost (Millions)	Non-CDS Cases	Cases Adjusted for Under- reporting	Paid/ Person*	Cost (Millions)
1	1,364,841	2,026,599	$8,497	$17,220	964,805	1,432,601	$6,029	$8,637
2	179,341	248,009	$88,563	$21,964	65,602	90,720	$40,565	$3,680
3	56,419	65,272	$210,000	$13,707	30,658	35,468	$163,258	$5,790
4	13,129	14,534	$210,000	$3,052	2,305	2,552	$210,000	$536
5 & 6	3,932	4,353	$210,000	$914	1,261	1,396	$210,000	$293
Fatal	32,999	32,999	$210,000	$6,930			$210,000	
All	1,650,661	2,391,766		$63,787	1,064,631	1,562,737		$18,936

* If 100% claimed; in reality 55% of AIS 2–5 claim. Totals were computed before rounding.

Table 2-10. Property damage in crashes reported to insurers, 2009*

MAIS	Fraction of Mean Cost	Property Damage/ Vehicle	Vehicles	Crashes	Property Damage/ Crash	Total Property Damage
0	0.8981	$2,723	16,936,098	9,733,390	$4,738	$46,117,102,074
1	1.9172	$5,813	1,784,195	969,671	$10,696	$10,371,421,164
2	2.2420	$6,798	230,173	133,048	$11,760	$1,564,684,072
3	3.5032	$10,622	78,673	46,278	$18,057	$835,637,812
4	4.7898	$14,523	18,799	11,463	$23,817	$273,017,849
5	4.7898	$14,523	7,734	4,834	$23,236	$112,314,709
6	4.7898	$14,523	45,632	30,797	$21,518	$662,699,017
1-6	2.1051	$6,383	2,165,206	1,196,091	$11,554	$13,819,774,624
All	1.0000	$3,032	19,101,304	10,929,481	$5,299	$59,936,876,698

* Excludes comprehensive (non-crash) coverage

Table 2-11. Vehicles per crash by police-reported crash severity

Crash Severity	GES 2009	NASS 1982-86
O – Property Damage Only	1.75	1.75
C – Possible Injury	1.94	1.93
B – Non-incapacitating Injury	1.76	1.75
A – Incapacitating Injury	1.71	1.74
K – Fatal Injury	1.66	1.63
All	1.78	1.76

Table 2-12. Insurance administration and legal costs per person by MAIS severity (2010 dollars, computed at a 3% discount rate)

MAIS	Insurance Administration	Legal
0	$143	$0
1	$3,935	$1,410
2	$9,370	$6,739
3	$24,348	$19,645
4	$37,372	$35,307
5	$79,967	$91,197
Fatal	$28,230	$106,488

Insurance Administration and Legal Costs per Person

We estimated insurance administration and legal costs from medical, work loss, and property damage costs using equations from Blincoe et al. (2000). We modified the equations to incorporate the 55 percent estimate of people with MAIS-1 injuries compensated. In addition, we adopted a more updated policy limit of $210,000 and expected life insurance payout of $59,680 and an expected Worker's Compensation payout of $100,000 for fatalities that occur on the job. Insurance Administration costs reflect administrative costs from a variety of insurance coverages. These include medical expense claims, liability claims, disability claims, Worker's Compensation, welfare payments, sick leave, property damage claims, and life insurance. The derivation and sources for administrative cost rates are detailed in Miller, Viner, et al., 1991. Table 2-12 shows the resulting administrative cost estimates for each injury severity level.

III. Miscellaneous Costs

In this brief chapter we examine various costs not covered in the previous chapters, including those incurred by State and local governments, such as crash-related damage to public property and public services like police and fire department attendance at crash sites.

Adding Roadside Furniture Damage to Property Damage

The insurance data suggest property damage averages $3,032 per crash-involved vehicle damaged seriously enough to prompt an insurance claim, with 18.3 to 18.8 million vehicles damaged that extensively in 2009. These estimates exclude most costs of damage to signs, lampposts, guardrails, and other roadside furniture. State and local governments absorb the roadside furniture costs not covered by insurance.

Estimated costs of roadside furniture damage by crash severity came from 1,462 crashes in 2008 tracked by the Missouri Claims Recovery Department. The data excluded costs not recovered from at-fault drivers and their insurers. As Table 2-13 shows, the costs average $59 per fatal crash and $47 per injury crash. These results, based on a single year in a single State, should be treated with caution.

Public Services

Public services costs are paid almost entirely by State and local government. Using the data underlying the crash cost estimates (Miller et al., 1991), we separated out EMS, police, fire, vocational rehabilitation, and court costs.

The States of Missouri and Washington provided average incident management costs. In 2009 dollars, the estimated mean cost per crash attendance was $82 for 315 crashes in Missouri and $125 for 3,880 crashes in Washington (assuming the response rate to serious injury [A] crashes was 60% of the response rate to fatal [K] crashes). We adopted Washington State's estimate because the data was much more complete than the Missouri data. Using data on the percentage of crashes attended, we broke the estimate down by police-reported crash severity.

To break the costs of incident management (and vehicle and roadside furniture damage) down into cost per person involved in a crash by injury severity, we followed the method used by Miller, Viner, Rossman, et al. (1991). We first cross-tabulated the number of people in a crash by the Abbreviated Injury Scale (AIS) severity of their maximum injury (MAIS), and by the maximum MAIS of anyone in the crash (AAIS). Second, we used that cross-tabulation to iteratively estimate costs by MAIS. We first divided the cost for a property damage only (PDO) crash by the uninjured people involved in a PDO crash to get a cost per uninjured person. Next, we used that cost per uninjured person to compute the cost of an MAIS-1 crash net of the costs associated with uninjured people. Dividing by the number of MAIS-1 injury victims in a crash then yields the cost per MAIS-1 victim. This process was repeated sequentially to compute costs for all MAIS levels. We also counted the number of vehicles per crash by MAIS.

Table 2-13. Crashes by severity, portion involving roadside furniture damage, costs per crash with costs and cost per crash, Missouri, 2008

Severity	Crashes	With Furniture Damage	$/Crash with Costs	Cost/Crash
Fatal	619	102	$356	$59
Injury	21,055	2,178	$452	$47

Table 2-14. Selected ancillary crash costs per person by MAIS (2010 dollars)

MAIS	Vehicle Damage	Roadside Furniture	Incident Management	Vocational Rehabilitation	Fire Department	Police
0	$1,816	$12	$1.60	$0	$7	$12
1	$5,382	$22	$0.60	$17	$9	$79
2	$5,756	$22	$0.30	$106	$95	$99
3	$10,860	$22	$81	$230	$227	$108
4	$16,306	$22	$81	$282	$639	$118
5	$15,070	$22	$78	$262	$651	$126
Fatal	$11,180	$32	$112	$0	$543	$247

Table 2-14 shows the resulting estimated costs per person injured by MAIS severity, as well as estimates for police, fire department, and vocational rehabilitation inflated from prior NHTSA crash cost studies. These factors are small, but the limited geographic coverage of the data underpinning them and the age of some of them mean their uncertainty is wide. A recent National Cooperative Highway Research Program project charged with updating most of these costs was unable to obtain data from additional jurisdictions.

Motor vehicle crashes also result in added societal costs due to congestion and workplace disruption. Congestion costs, which include travel delay, excess fuel consumption, and added greenhouse gases and criteria pollutants are examined in a separate chapter of this report. Workplace costs were estimated by adjusting the workplace costs from Blincoe et al., 2002 to 2010 levels using the employment cost index for total compensation published by the Bureau of Labor Statistics. Table 2-15 summarizes the unit costs by injury severity and cost component for 2010. All injury unit costs are expressed on a per person injured basis. The costs for PDO's are expressed on a per-damaged-vehicle basis. Medical costs include both medical care from Table 3 and vocational rehabilitation costs from Table 2-14. Property damage costs include both vehicle damage and roadside furniture from Table 2-14. Emergency Services includes Incident management, Fire Department, and Police from Table 2-14. Market and Household Productivity are from Table 3. Legal and Insurance Administration costs are from Table 2-12.

Each fatality results in economic impacts of roughly $1.4 million, due primarily to lost productivity and legal costs. MAIS 5 injuries are almost as costly at $1.1 million. The most costly impact for these most serious of survivor injuries is the cost of medical care, but there are also significant costs from lost productivity, legal costs, and insurance administrative costs. For all cost categories, injury costs gradually decline as severity decreases.

IV. Police-Reported Versus Unreported Crash Costs

As noted in Chapter 5, over half of all PDO crashes and about a quarter of all non-fatal injury crashes are not reported to police. However, analyses of safety countermeasures frequently rely only on police- reported incidence data. Crashes that get reported to police are likely to be more severe than unreported crashes because vehicles are more likely to require towing and occupants are more likely to require hospitalization or emergency services. These crashes are typically also likely to require more time to investigate and clear from roadways than unreported crashes. Analysis based solely on police- reported crashes should thus be based on unit costs that are specific to police-reported crashes. For injury related costs, this is more or less automatically accounted for by the shift in the injury severity profile. Unreported crashes have a lower average severity profile than do reported crashes. However, for non-injury related cost components – property damage and congestion costs – there is no profile to shift. In addition, emergency services have higher involvement rates for police-reported crashes.

A separate set of costs was developed in Chapter 3 for police-reported and unreported congestion costs. To estimate separate costs for property damage, we used property damage cost data from the MDAC survey. Data was derived separately for reported and unreported crashes. Table 2-16 lists the results. The mean property damage cost of a crashed vehicle in the MDAC survey was $4,476. However, the mean property damage cost for vehicles in crashes reported to the police was $5,607, and the mean cost for a vehicle in crashes not reported to the police was $1,907. To estimate separate unit costs for vehicles in each crash type, we took the ratio of each crash type to the mean overall cost and applied these factors to the average property damage cost previously derived from insurance data. Since these ratios were derived independently from both the main incidence and property damage analyses, a further adjustment was made to normalize the unit costs so that the sum of reported and unreported crashes matched the overall totals.[6] A similar approach was used for emergency services. Emergency Services consists of separate police, fire, and incident management components. Each component was distributed assuming that unit costs per case were identical for both reported and unreported cases of a specific severity for any case for which police, fire, or incident management teams actually responded. The difference in unit costs for reported and unreported cases is thus a function of differing response rates. For police-reported cases, response rates are assumed to be 100 percent by definition. This is confirmed by the 100 percent rates reported in the MDAC survey for police-reported cases. For unreported cases, MDAC survey police response rates were reported to be 100 percent for all injury cases MAIS 3 and greater. Police response rates for unreported MAIS 0, MAIS1, MAIS2, and PDO cases were reported to be 17.1 percent, 29.2 percent, 37.8 percent, 11.5 percent respectively.

Fire response is assumed to be a subset of police response cases. Fire response rates derived from Blincoe et al. (1992) were thus assumed for police-reported cases, and were further modified by the relative unreported/reported police response rate in the MDAC survey for unreported cases.

Table 2-15. Summary of Unit Costs, Police-Reported and Unreported Crashes, 2010, 3-Percent Discount Rate, 2010 Dollars

	PDO	MAIS0	MAIS1	MAIS2	MAIS3	MAIS4	MAIS5	Fatal
Medical	$0	$0	$3,801	$24,481	$70,902	$196,911	$441,618	$11,317
EMS	$28	$21	$89	$194	$416	$838	$855	$902
Market	$0	$0	$2,858	$39,637	$110,088	$174,728	$352,178	$933,262
Household	$60	$45	$941	$11,921	$33,505	$44,593	$107,070	$289,910
Insurance	$191	$143	$3,935	$9,370	$24,348	$37,372	$79,967	$28,322
Workplace	$62	$46	$341	$2,644	$5,776	$6,361	$11,091	$11,783
Legal Costs	$0	$0	$1,410	$6,739	$19,645	$35,307	$91,197	$106,488
Subtotal	$341	$255	$13,375	$94,986	$264,680	$496,110	$1,083,976	$1,381,984
Congestion	$1,077	$760	$1,109	$1,197	$1,434	$1,511	$1,529	$5,720
Prop. Damage	$2,444	$1,828	$5,404	$5,778	$10,882	$16,328	$15,092	$11,212
Subtotal	$3,521	$2,588	$6,513	$6,975	$12,316	$17,839	$16,621	$16,932
Total	$3,862	$2,843	$19,888	$101,961	$276,996	$513,949	$1,100,597	$1,398,916

Note: Unit costs are expressed on a per-person basis for all injury levels. PDO costs are expressed on a per-damaged-vehicle basis.

Table 2-16. Per Vehicle Property Damage in MDAC Survey

Statistic	All crashes		
	All	Reported	Unreported
Number	1847	1256	591
Mean	$4,476	$5,607	$1,907
Median	$1,698	$2,000	$762
SE of Mean	$846	$1,200	$408
95% LCL of Mean	$2,816	$3,251	$1,107
95% UCL of Mean	$6,136	$7,962	$2,708
Minimum	$0	$0	$0
25th Percentile	$576	$884	$241
75th Percentile	$3,685	$4,265	$1,755
Maximum	$310,000	$310,000	$300,000
Mean Ratio to All	1.000	1.253	0.426

Incident management response rates were estimated based on data from Washington State cited in NCHRP Working Paper 4 (Bahar & Miller, 2010), which indicate response rates of 23.2 percent for K and A injuries, 2.3 percent for B and C injuries, and 5.9 percent for O injuries. In order to translate these into equivalent MAIS levels, a KABCO/MAIS injury matrix was established. In order to reflect the fact that within each KABCO level, incident response rates were likely to be more heavily weighted towards the more serious crashes, the initial incidence matrix was modified by applying the relative Fire Department response rates across MAIS severities as a model proxy. For each MAIS category, relative weights were then computed across the 5 KABCO categories, and these weights were applied to the corresponding average incident management response rates and then summed to calculate an average response rate by MAIS severity level. These rates were assumed to represent police-reported cases. As with Fire response, they were further modified using the relative unreported/reported police response rate from the MDAC survey to estimate incident management response rates for unreported cases. Table 2-17 summarizes the inputs and results of this process for each EMS component.

Table 2-17. Summary of Police-Reported and Unreported Emergency Services Unit Costs

	Response Rates				Unit Costs	
	Reported Crashes	Unreported Crashes	Average Unit Cost	Percent Unreported	Reported Crashes	Unreported Crashes
Police Response						
Fatal	100.00%	100.00%	$247.00	0.00%	$247.00	$247.00
MAIS 0	100.00%	17.16%	$12.00	53.14%	$21.44	$3.68
MAIS 1	100.00%	29.17%	$79.00	25.45%	$96.37	$28.11
MAIS 2	100.00%	37.84%	$99.00	19.95%	$113.01	$42.76
MAIS 3	100.00%	100.00%	$108.00	4.31%	$108.00	$108.00
MAIS 4	100.00%	100.00%	$118.00	0.00%	$118.00	$118.00
MAIS 5	100.00%	100.00%	$126.00	0.00%	$126.00	$126.00
PDO	100.00%	11.54%	$17.00	59.72%	$36.04	$4.16
Fire Department Response						
Fatal	95.00%	95.00%	$543.00	0.00%	$543.00	$543.00
MAIS 0	1.00%	0.17%	$7.00	53.14%	$12.50	$2.15
MAIS 1	1.00%	0.29%	$9.00	25.45%	$10.98	$3.20
MAIS 2	15.00%	5.68%	$95.00	19.95%	$108.45	$41.04
MAIS 3	35.00%	35.00%	$227.00	4.31%	$227.00	$227.00
MAIS 4	90.00%	90.00%	$639.00	0.00%	$639.00	$639.00
MAIS 5	95.00%	95.00%	$651.00	0.00%	$651.00	$651.00
PDO	1.00%	0.12%	$9.00	59.72%	$19.08	$2.20
Incident Management Response						
Fatal	22.45%	22.45%	$112.00	0.00%	$112.00	$112.00
MAIS 0	5.80%	1.00%	$2.00	53.14%	$3.57	$0.61
MAIS 1	5.65%	1.65%	$1.00	25.45%	$1.22	$0.36
MAIS 2	9.78%	3.70%	$0.00	19.95%	$0.00	$0.00
MAIS 3	15.67%	15.67%	$81.00	4.31%	$81.00	$81.00
MAIS 4	17.85%	17.85%	$81.00	0.00%	$81.00	$81.00
MAIS 5	20.49%	20.49%	$78.00	0.00%	$78.00	$78.00
PDO	5.80%	0.67%	$2.00	59.72%	$4.24	$0.49
Total Emergency Services						
Fatal			$902.00	0.00%	$902.00	$902.00
MAIS 0			$21.00	53.14%	$37.51	$6.44
MAIS 1			$89.00	25.45%	$108.57	$31.67
MAIS 2			$194.00	19.95%	$221.46	$83.80
MAIS 3			$416.00	4.31%	$416.00	$416.00
MAIS 4			$838.00	0.00%	$838.00	$838.00
MAIS 5			$855.00	0.00%	$855.00	$855.00
PDO			$28.00	59.72%	$59.36	$6.85

The results of this analysis for congestion, property damage, and emergency services are presented in Tables 2-18 and 2-19 together with the other cost components that did not vary by reporting status. The differences seem negligible at the more severe injury levels due to the overwhelming costs of factors such as lost productivity and medical care which do not vary by reporting status, except through the shift in injury profiles. However, at lower severity levels the unit costs are significant. For PDO vehicles and MAIS0s, police-reported crashes have costs that are three times those of unreported crashes. For minor (MAIS1) injuries, reported crashes cost 40 percent more than unreported crashes. These ratios decline as injury severity increases. Note that for MAIS4s, MAIS5s, and Fatalities, property damage costs are identical under both reported and unreported cases. All injuries at these levels are believed to be reported to police, and the original property damage cost estimate is thus assumed to represent police- reported cases. These same costs are thus listed under both scenarios

Table 2-18. Summary of Unit Costs, Police-Reported Crashes, 3-percent discount Rate (2010 Dollars)

	PDO	MAIS0	MAIS1	MAIS2	MAIS3	MAIS4	MAIS5	Fatal
Medical	$0	$0	$3,801	$24,481	$70,902	$196,911	$441,618	$11,317
EMS	$59	$38	$109	$221	$416	$838	$855	$902
Market	$0	$0	$2,858	$39,637	$110,088	$174,728	$352,178	$933,262
Household	$60	$45	$941	$11,921	$33,505	$44,593	$107,070	$289,910
Insurance	$191	$143	$3,935	$9,370	$24,348	$37,372	$79,967	$28,322
Workplace	$62	$46	$341	$2,644	$5,776	$6,361	$11,091	$11,783
Legal Costs	$0	$0	$1,410	$6,739	$19,645	$35,307	$91,197	$106,488
Subtotal	$372	$272	$13,395	$95,013	$264,680	$496,110	$1,083,976	$1,381,984
Congestion	$2,104	$1,416	$1,426	$1,450	$1,490	$1,511	$1,529	$5,720
Prop. Damage	$3,599	$2,692	$7,959	$8,510	$16,027	$16,328	$15,092	$11,212
Subtotal	$5,704	$4,108	$9,385	$9,960	$17,517	$17,839	$16,621	$16,932
Total	$6,076	$4,380	$22,779	$104,974	$282,197	$513,949	$1,100,597	$1,398,916

Note: Unit costs are expressed on a per-person basis for all injury levels. PDO costs are expressed on a per-damaged-vehicle basis.

Table 2-19. Summary of Unit Costs, Unreported Crashes, 3-percent discount Rate (2010 Dollars)

	PDO	MAIS0	MAIS1	MAIS2	MAIS3	MAIS4	MAIS5	Fatal
Medical	$0	$0	$3,801	$24,481	$70,902	$196,911	$441,618	$11,317
EMS	$7	$6	$32	$84	$416	$838	$855	$902
Market	$0	$0	$2,858	$39,637	$110,088	$174,728	$352,178	$933,262
Household	$60	$45	$941	$11,921	$33,505	$44,593	$107,070	$289,910
Insurance	$191	$143	$3,935	$9,370	$24,348	$37,372	$79,967	$28,322
Workplace	$62	$46	$341	$2,644	$5,776	$6,361	$11,091	$11,783
Legal Costs	$0	$0	$1,410	$6,739	$19,645	$35,307	$91,197	$106,488
Subtotal	$320	$240	$13,318	$94,876	$264,680	$496,110	$1,083,976	$1,381,984
Congestion	$384	$180	$180	$180	$180	$180	$180	$458
Prop. Damage	$1,224	$916	$2,707	$2,894	$5,451	$16,328	$15,092	$11,212
Subtotal	$1,609	$1,096	$2,888	$3,075	$5,632	$16,508	$15,272	$11,670
Total	$1,928	$1,337	$16,205	$97,951	$270,312	$512,618	$1,099,248	$1,393,654

Note: Unit costs are expressed on a per-person basis for all injury levels. PDO costs are expressed on a per-damaged-vehicle basis.

3. CONGESTION IMPACTS

Motor vehicle crashes result in significant time delays to other motorists who are inconvenienced by lane closures, police, fire, or emergency services activity, detours, and general traffic slowdowns resulting from rubbernecking and chain reaction braking. This results in a significant time penalty for those affected, which can be valued based on wage rates and the value people place on their free time. It also results in wasted fuel, increased greenhouse gas production, and increased pollution as engines idle while drivers are caught in traffic jams and slowdowns. These impacts affect drivers' transportation costs and negatively impact the health and economic welfare of the Nation.

Assessing congestion costs is difficult because virtually every crash occurs under unique circumstances. Differences in crash severity, vehicle involvement, roadway type, time of day, traffic density, emergency services response time, weather, hazardous material spillage, lane configurations, driver behavior, and other variables can influence the extent of congestion and the resulting societal impacts. While there are a number of studies that document the impact of crashes on roadway congestion, most of these focus very narrowly on impacts for a specific roadway, and in most cases, these roadways are urban interstates.

A few studies have attempted to project congestion impacts from crashes at a higher level. Chin, Franzese, Greene, Hwang, & Gibson (2004), used traffic engineering modeling methods to derive estimates of delay impacts Nationally for freeways and principal arterials. Zaloshnja, Miller, and Spicer (2000) used relative traffic density data to scale results from a study of urban interstates in Minneapolis- St.Paul to estimate the delay hours for police-reported crashes involving trucks and buses with a gross vehicle weight rating over 10,000 pounds across six different urban and rural roadway categories.

More recently, the Federal Motor Carrier Safety Administration (FMCSA) contracted with the US DOT/Volpe Center in Cambridge, MA to produce a simulation based estimate of the per-crash impacts of congestion from commercial vehicle crashes (Hagemann et al., 2013). This study involved traffic simulation measurements using TSIS-CORSIM, a micro-simulation tool developed by the University of Florida McTrans Center. TSIS-CORSIM simulates traffic responses to specific roadway and crash scenarios and produces estimates of aggregate vehicle delay hours and added fuel consumption. The authors of the study then linked the TSIS-CORSIM results to the Environmental Protection Agency's Motor Vehicle Emission Simulator (MOVES) model to produce estimates of greenhouse gas and criteria pollutant emissions. The estimation process involved Monte Carlo simulations of 77 different crash scenarios in order to capture the variety of possible outcomes across numerous sets of crash circumstances. These results were then weighted based on nationwide crash incidence, producing average impacts for crashes on 5 different categories of roadways varying by three different crash severities (fatal crashes, injury crashes, property-damage-only crashes). While any simulation process is subject to uncertainty, the FMCSA study is arguably the most sophisticated attempt thus far to estimate nationwide congestion costs from crashes. However, the FMCSA study's focus on commercial vehicle crashes limits its applicability to this current effort, which examines all motor vehicle crashes. Commercial vehicle crashes make up only about 5 percent of all police-reported crashes nationwide. More importantly, they typically have more serious congestion consequences than other crashes. This results from several factors, most notably, that they are more likely to involve lane closings, that they take longer to clear from the roadway (especially in the case of hazardous waste or cargo spillage), and that they are more likely to occur during normal weekday hours, when traffic density is highest, and less likely to occur on weekends and at night when traffic density is lighter.

The approach taken in this study involves a synthesis of past approaches. It uses empirical data derived from both current data sources and previous literature to develop a basic congestion model. This model will estimate the congestion impacts from lane closings, rubbernecking, and subsequent traffic dispersal across the same roadway categories examined in the FMCSA study. The model is run once with data and assumptions appropriate for the universe of all crashes, and then again with data appropriate for commercial vehicle crashes. The results of these two sets of outputs are then used to compute normalizing factors that can be applied to the FMCSA results for commercial vehicle crashes, to derive an estimate that is more representative of the overall universe of traffic crashes. This linkage to the FMCSA report is motivated by the ability of its simulation methods to capture several aspects that are not easily estimated using more conventional approaches. These include the impact of detours, and more importantly, the ability to capture non-linear impacts that cause disproportionate congestion under extreme circumstances that cannot be reflected using average input values.

Methodology

The primary measure of lost capacity due to congestion is vehicle hours. In its most literal sense "Vehicle Hours" represents the sum of the net delay encountered by each individual affected motorist. In practice, it is measured more broadly by determining the change in capacity over specific roadway segments during the course of a crash event and its aftermath. Estimating lost vehicle hours nationwide thus requires measures of crash event durations, roadway capacity, lane closures, and capacity losses. These factors influence various aspects of congestion delay. These include delay caused by lane closings, delay caused by rubbernecking in lanes travelling in the direction of the crash, delay caused by rubbernecking in lanes travelling in the opposite direction of the crash, delay caused by dispersal of the remaining traffic queue after the crash is cleared from the road, and delay caused by detours. Each of these factors will be addressed separately.

Lane Closings

Capacity losses due to lane closure is a function of traffic density on the affected roadway at the time of the crash, the duration of the crash event, the probability of and extent of lane closures, and proportional reduction in travel capacity through the roadway that results from those lane closures. These factors combine in a direct multiplicative relationship to determine lost capacity due to lane closures as follows:

$$VC = AAHT*CD*PLC*RCL$$

where:

VC = Vehicle capacity lost due to lane closure

AAHT = Average annual hourly traffic (vehicles) past the crash site during the time affected by the crash

CD = crash duration[7]

PLC = Probability-of-lane closure

RCL = Reduced capacity (%) in the direction of the crash given lane closure.

Annual Average Hourly Traffic

Annual average hourly traffic (AAHT) is the average number of vehicles that would pass the crash site during the time affected by the crash. AAHT can be derived by merging together data regarding roadway capacity, travel profiles, and crash occurrence profiles. The initial basis for computing AAHT is the annual average daily traffic (AADT)[8] data collected by the Federal Highway Administration through their Highway Performance Monitoring System (HPMS). The HPMS is the most comprehensive source of data about the use of the Nation's roadways. It currently collects data from over 110,000 highway sample segments. HPMS monitors roadway traffic to produce AADT statistics for the various roadway types across the country.

The impact of congestion is largely a function of traffic density, and this can vary significantly by road type. For this analysis, AADT estimates were collected for the following roadway types:

Urban Interstate/Expressway, Urban Arterial,
Urban Other,
Rural Interstate/Principal Arterial, and
Rural Other.

All U.S. roadway types are collapsed into these five broader categories, which match those used in the FMCSA heavy truck crash report. They were chosen for this report to facilitate the utilization of some data elements in the FMCSA report, which were stratified using these roadway categories.

The AADT data obtained from FHWA for these categories are summarized in the first column of Table 3-1.[9] These represent the mean VMT weighted AADT counts across all roadways within the five basic roadway categories. These five categories collapse nine categories that are collected under the HPMS system. The groupings are combined as follows:

Urban Interstate/Expressway includes the two roadway functional categories "**urban principal arterial - urban interstate**" and "**urban principal arterial- other freeways and expressways.**."

Urban Arterial includes the two roadway functional categories "**urban principal arterial – other**" and "**urban minor arterial.**."

Urban Other includes the two roadway functional categories "**urban collector" and "urban local.**."

Rural Interstate/Principal Arterial includes the two roadway functional categories "**rural principal arterial - rural interstate**" and "**rural other principal arterial.**."

Rural Other includes "**rural minor arterials,**," "**rural major collectors,**," "**rural local,**" and "**rural minor collectors.**."

The mean VMT weighted AADTs were calculated using the following formula:

$$\frac{\sum AADT * AADT * SegmentLength}{\sum AADT * SegmentLength}$$

However, data for three of the HPMS categories, *"urban local,"* *"rural local,"* and "*rural minor collectors*" were only collected in summary format from the States and could not be directly weighted using this method. HPMS does calculate AADT values that include these 3 roadway functional categories using a standard method that does not weight by VMT.[10] The impact of these segments was therefore estimated by comparing the ratio of the Urban Other and Rural Other AADTs both including and excluding these 3 categories. This ratio was applied to the mean VMT weighted AADTs for Urban Other and Rural Other that excluded these 3 categories to produce the estimates noted in Table 3-1.

As would be expected, urban roadways experience far greater travel densities than rural roadways do, and interstates and freeways are more travelled than other roadway categories. The obvious implication of this is that crashes that occur on urban roadways will affect more vehicles than those that occur on rural roadways, as will those that occur on interstates and freeways.

AADT statistics represent the average number of vehicles that travel a roadway segment over the course of a day. However, travel within a given day is not evenly distributed, and travel patterns vary on weekends from those that occur on weekends. Festin (1996) documented daily travel patterns over the 24-hour cycle on both weekdays and weekends using data obtained from 5,000 Automatic Traffic Recorder (ATR) sites nationwide. Figures 3-A and 3-B illustrate the average traffic density patterns for weekday and weekend travel.

A typical weekday pattern starts at a low point between 2 and 4 a.m., and then starts a rapid acceleration through about 8 a.m. as workers commute to jobs. It then drops into a mid-morning lull, before picking up in the afternoon and peaking during the evening rush hour, after which it rapidly declines. By contrast, on weekends traffic displays a gradual rise from about 4 a.m. until it peaks between noon and 4 p.m., after which it declines steadily into the evening. These patterns are important because they determine the density of traffic that will be present at different times of day or days of the week, and crashes that occur during peak hours will cause considerably more congestion than those that occur during off-peak hours.

Crash frequencies also vary by time of day and day of week, as well as by crash type. Table 3-2 shows the frequency of crashes by time of day for weekends and weekdays, for crashes of different severity compiled from NHTSA's FARS and GES databases for the years 2005-2010. In Figures 3-C and 3-D this data is charted for each crash severity level against the traffic density profiles illustrated in Figures 3-A and 3-B.

It is apparent from these figures that while crash frequencies may follow a somewhat similar pattern to traffic density, they diverge in ways that have implications for congestion impacts. Moreover, this divergence is more pronounced on weekends.

During weekdays, crash patterns follow the general shape of traffic density patterns. This is expected since traffic density is a measure of exposure. In the early morning hours, injury and PDO crashes follow traffic density patterns very closely. The highest number of these crashes occur during the mid-day timeframe up to the end of rush hour, when traffic density is highest. After peak rush hour, they occur at a rate that is above traffic density. Fatal crashes follow a similar pattern, but generally occur at rates below traffic density during the day but above density in the early morning hours and after rush hour when traffic density is relatively light.

Table 3-1. Annual Average Daily Traffic by Roadway Category 2005-2008

Roadway Functional Class	AADT
Urban Interstate/ Expressway	113,814
Urban Arterial	23,996
Urban Other	2,908
Rural Interstate/Principal Arterial	25,579
Rural Other	1,502

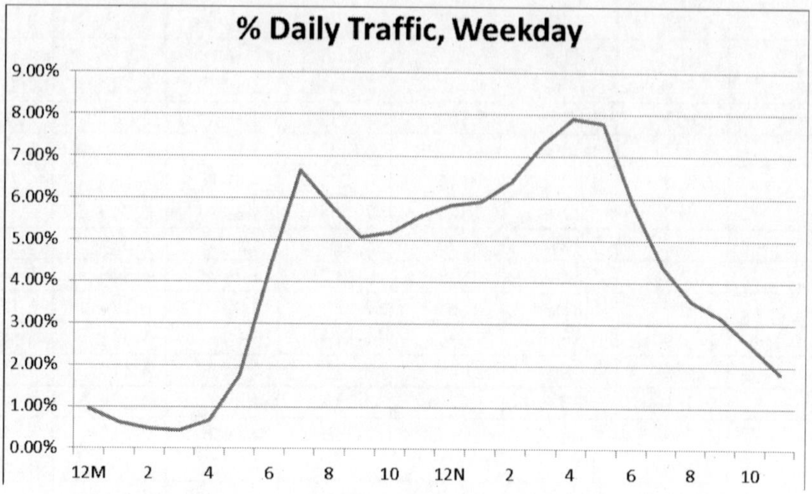

Figure 3-A. % Daily Traffic, Weekday.

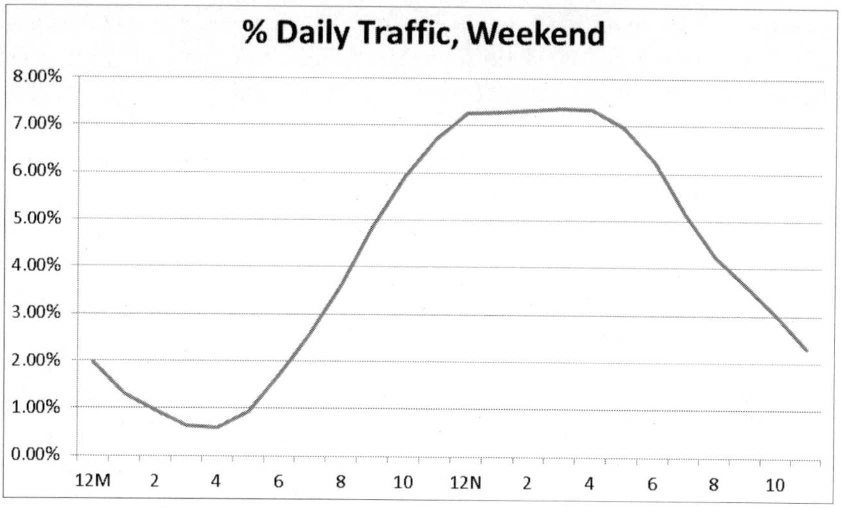

Figure 3-B. % Daily Traffic, Weekend.

Table 3-2. Distribution of Daily Traffic and Crash Occurrences by Crash Severity

Time	% Daily Traffic	Weekday Crashes			Weekend Crashes		
		Fatal Crash	Injury Crash	PDO Crash	Fatal Crash	Injury Crash	PDO Crash
12 Midnight	0.97%	3.91%	1.57%	1.50%	5.04%	3.22%	3.47%
1 a.m.	0.62%	3.15%	1.13%	1.05%	5.94%	3.08%	2.98%
2 a.m.	0.47%	3.22%	1.06%	0.95%	7.50%	3.78%	3.37%
3 a.m.	0.44%	2.48%	0.88%	0.76%	6.28%	3.04%	2.94%
4 a.m.	0.67%	1.84%	0.59%	0.63%	4.17%	2.09%	2.01%
5 a.m.	1.76%	2.13%	0.78%	0.83%	3.05%	1.71%	1.64%
6 a.m.	4.33%	3.44%	1.77%	1.92%	2.67%	1.75%	1.68%
7 a.m.	6.66%	4.08%	3.78%	4.35%	2.39%	1.68%	2.09%
8 a.m.	5.85%	3.77%	6.46%	6.97%	2.07%	2.32%	2.57%
9 a.m.	5.07%	3.18%	4.78%	4.83%	2.22%	3.02%	3.17%
10 a.m.	5.18%	3.31%	3.86%	4.24%	2.45%	3.81%	4.12%
11 a.m.	5.58%	3.68%	4.40%	4.27%	2.82%	4.79%	4.77%
12 Noon	5.85%	4.03%	5.36%	5.65%	3.31%	5.68%	5.78%
1 p.m.	5.92%	4.28%	6.01%	6.00%	3.71%	6.26%	6.33%
2 p.m.	6.39%	4.82%	6.14%	6.02%	3.99%	6.53%	6.32%
3 p.m.	7.28%	5.47%	7.82%	7.57%	4.31%	6.74%	6.34%
4 p.m.	7.90%	5.77%	8.59%	8.52%	4.55%	7.05%	6.06%
5 p.m.	7.79%	5.68%	8.87%	9.23%	4.74%	6.34%	6.55%
6 p.m.	5.88%	5.81%	7.86%	7.91%	5.08%	6.28%	6.15%
7 p.m.	4.38%	5.55%	5.37%	5.14%	5.10%	5.38%	5.58%
8 p.m.	3.54%	5.28%	4.05%	3.57%	4.94%	4.47%	4.65%
9 p.m.	3.16%	5.30%	3.47%	3.23%	4.89%	4.07%	4.31%
10 p.m.	2.49%	5.08%	3.01%	2.72%	4.54%	3.71%	3.89%
11 p.m.	1.82%	4.72%	2.37%	2.13%	4.23%	3.19%	3.22%
Total	100.00%	100.00%	100.00%	100.00%	100.00%	100.00%	100.00%

On weekends, a significant proportion of all three crash types occur between midnight and 8 a.m., when traffic congestion is relatively light. This is especially true for fatal crashes, with a particular concentration of fatalities in the midnight to 6 a.m. timeframe. This likely reflects the increased consumption of alcohol on weekend evenings, with its resulting toll on driver skills and judgment. For the remaining hours of the weekend, crash frequencies are lower than would be expected based solely on exposure, but when they occur, they occur during times of relatively high traffic density.

Given these different relationships between traffic density and crash occurrence on weekdays versus weekends, the two time frames must be analyzed separately. Therefore, to derive AAHT for traffic crashes, AADT was first split into weekday and weekend values using factors developed in the FMCSA study. HPMS does not publish separate AADTs for weekdays and weekends, so FMCSA analyzed ATR data to derive ratios between the weekend, weekday, and total sample means. These provided scaling factors that were used to transform the overall AADT number into separate weekend and weekday AADTs. These scaled AADT values were then weighted to reflect the relative frequency of crashes on weekdays and weekends[11], and the resulting AADT totals were used to derive overall ratio weights to apply to the AAHT results.

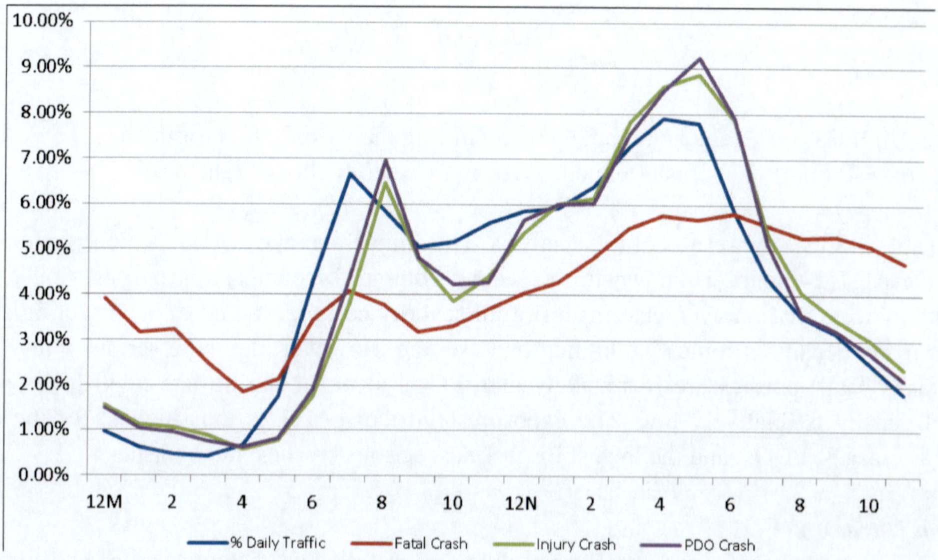

Figure 3-C. Weekday Crash Frequencies versus Congestion.

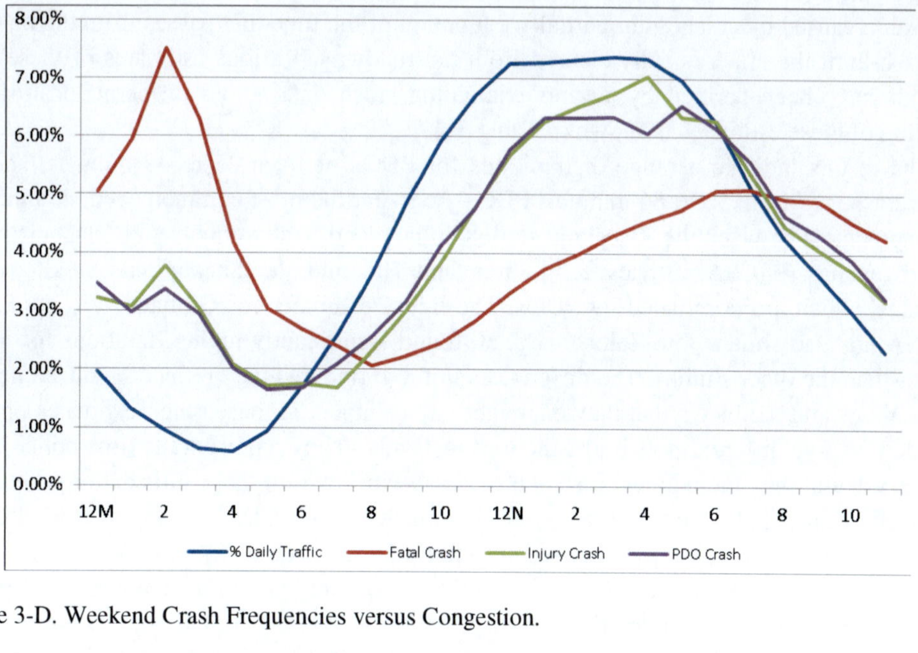

Figure 3-D. Weekend Crash Frequencies versus Congestion.

AAHT was then calculated for each crash severity and roadway type by weighting the hourly distribution of travel density by the portion of crashes that occur during each specific hour. For example, about 1% of average weekday travel occurs around midnight on weekdays, when the weighted AAHT on urban interstates/freeways is 1166 vehicles, and about 4 percent of weekday fatal crashes occur; but on weekends about 2 percent of fatal crashes occur around midnight, when AAHT is 1925 and about 5 percent of weekend fatal crashes occur. Summing these aggregates across the full 24-hour period gives an exposure adjusted AAHT value as follows:

$$\sum_{h=1}^{24} AADT(h) * Cf(h)$$

where *AADT(h)* = Average annual daily travel during the specific hour of the day
Cf(h) = Proportional crash frequency during the specific hour of the day.

Table 3-3 lists the results of this analysis. The simple average AAHT, which is computed as the AADT/24 hours, is shown in the second column for comparison purposes only. The results indicate that on average, crashes of all severity categories tend to occur during times when travel density is somewhat higher than average. However, the more serious injury (and especially fatal) crashes are more likely than PDOs to occur during late night hours when travel density is relatively low. The exposure adjusted AAHT is thus highest for the least serious crashes, PDOs, and the lowest for the most serious crashes, fatal crashes.

Crash Duration (CD)

Comparisons across studies of travel delay often focus on different issues and produce conflicting results. One metric that is common to many studies is crash duration – the time during which the crash affects travel on the roadway. Estimates of vehicle hours spent in congestion caused by crashes are partially a function of the time that passes from the onset of the crash until the crash is fully cleared from the roadway. Various estimates of these time intervals have been derived by authors examining crash data across different locations. A sampling of these estimates is shown in Table 3-4.

The results indicate a range of durations for PDOs of from 27 to 49 minutes, and for injury crashes of from 50 to 86 minutes. PDOs are by far the most common event and there is close agreement between the Zaloshnja et al., Giuliano and Skabardonis, et al. PDO estimates as well as the FMCSA, Boyles et al., Lan and Hu, and the Skabardonis et al. overall estimates, which are dominated by PDOs. The injury estimates for Giuliano, Wu et al., and FMCSA are also similar, but Zaloshnja et al. found significantly higher durations for injury crashes than the other studies. The differences in these findings may reflect the differences in the localities and roadways that they examined. Zaloshnja et al. based their estimates on data provided for five different police jurisdictions in 1989.[13] They represent the time police spent at the crash and they include all crash and roadway types within those jurisdictions. Giuliano examined crashes that occurred in a section of a major Los Angeles Freeway in 1983-85. The Giuliano values were subsequently used in reports by Chin et al. (2002). Wu and Kachroo examined crashes in Northern Virginia in 1997, and Boyles, Fjardo, and Waller used police logs from the Georgia Department of Transportation in 2004. Lan and Hu examined urban freeways in the Twin Cities area of Minnesota during 1994-95. Skabardonis, Chia-Chavala, and Rydzewski (1998) examined crashes on California's I-880 freeway in 1995.

There are only two studies that estimate durations for fatal crashes, but agreement is fairly close. We note that the FMCSA study specifically examined crash data for commercial vehicles while the Zaloshnja et al. study collected data for all crash types. One might expect that time intervals to clear heavy truck crashes might be longer, but this is not apparent from these two samples. Further, the FMCSA study time intervals, which represent data from 23,388 crashes in Pennsylvania from 2006 to 2008, represent time intervals during which the road was closed due to the crash, whereas the Zaloshnja estimates represent the time interval during which police were present at the crash. Other factors such as local differences and reporting parameters may obscure such differences. However, it is also possible that the time difference may not be significant. Studies that examine crashes both with and without trucks have generally found more vehicle hours of delay in heavy truck crashes.[14] However, this is often a function of more lane closures due to the size of the vehicle as well as more equipment needed to remove them. In addition, in the rare case of hazardous waste or other cargo spillage, multiple lanes may be blocked by spilled cargo or fuel.

Table 3-3. Crash Exposure Adjusted AAHT by Roadway Class and Crash Severity

Roadway Category	AADT	Simple Average AAHT	Crash Exposure Adjusted AAHT		
			Fatal	Injury	PDO
Urban Interstate Expressway	113814	4742	4934	6218	6300
Urban Arterial	23996	1000	1040	1311	1328
Urban Other	2908	121	126	159	161
Rural Interstate/Principal Arterial	25579	1066	1114	1392	1407
Rural Other	1502	63	65	82	83

Table 3-4. Crash Duration Estimates (Minutes)

Study	Crash Type	PDO	Injury	Fatal	All
Zaloshnja, Miller, and Spicer (5 U.S. urban freeways, 1989 -Unk # observations)	All	49	86	233	60
FMCSA (2012) (Pennsylvania, 2006-2008 -23,388 observations)	Truck	35	55	216	44
Giuliano (1988) (LA I-10 freeway, 1983-85 -270 obs)		All	44	56	49
Wu, Kachroo, and Ozbay (1998) (Northern VA, 1997 -33 observations)	All	27	50		34
Boyles, Fajardo, and Waller (2006) (Atlanta, 2004 -2,970 observations)	All				42
Lan and Hu (1999)[12] (3,877 observations) (Minnesota Urban freeways, 1994-95)	All				39
Skabardonis, Chira-Chavala, and Rydzewski (1998) (CA I-880 freeway 1995 -92 observations)	All	40	63		43
Simple Average (minutes)=		39	62	224	44
Simple Average (hours)=		0.65	1.04	3.74	0.74

Each of these studies focused on specific locations and since they do not overlap, it is not surprising that there is variation in their results. These results reflect differences in traffic density, which is specific to different roadway types and cities, and which impacts police, rescue, and cleanup operation response times. It also reflects differences in the metric used, which varies from empirical observations to police logs. It's unclear whether police presence on the scene would exceed the timeframe when the crash has been cleared. There is likely to be a delay between the crash occurrence and the police response. Conversely, in some cases police may remain at a crash site to complete crash reports after the vehicles have been cleared from the road.

The definition of crash duration can vary by study. The universe of events that reflect the full impacts of a crash begins with the crash incident and ends when traffic patterns return to normal. Within this interval, a number of events occur that influence the impact on traffic congestion. The first of these is the arrival of law enforcement or emergency service personnel at the scene. This usually occurs a short period of time after the incident is detected by State or local traffic or emergency systems. While detection usually involves some level of delay from the time of the incident, with the proliferation of cell phones, this interval is becoming increasingly shorter. Arrival of emergency service or police can result in an increase in congestion since these vehicles may block additional lanes while at the crash site. Subsequently, police may formally close down travel lanes and/or move crashed vehicles off the roadway to relieve congestion. Once the vehicle is removed from the scene and emergency personnel and police have left, traffic will gradually return to normal.

Figure 3-E illustrates the general impact on traffic flow over time from a crash event.[15]

The results indicate a range of durations for PDOs of from 27 to 49 minutes, and for injury crashes of from 50 to 86 minutes. PDOs are by far the most common event and there is close agreement between the Zaloshnja et al., Giuliano and Skabardonis, et al. PDO estimates as well as the FMCSA, Boyles et al., Lan and Hu, and the Skabardonis et al. overall estimates, which are dominated by PDOs. The injury estimates for Giuliano, Wu et al., and FMCSA are also similar, but Zaloshnja et al. found significantly higher durations for injury crashes than the other studies. The differences in these findings may reflect the differences in the localities and roadways that they examined. Zaloshnja et al. based their estimates on data provided for five different police jurisdictions in 1989.[13] They represent the time police spent at the crash and they include all crash and roadway types within those jurisdictions. Giuliano examined crashes that occurred in a section of a major Los Angeles Freeway in 1983-85. The Giuliano values were subsequently used in reports by Chin et al. (2002). Wu and Kachroo examined crashes in Northern Virginia in 1997, and Boyles, Fjardo, and Waller used police logs from the Georgia Department of Transportation in 2004. Lan and Hu examined urban freeways in the Twin Cities area of Minnesota during 1994-95. Skabardonis, Chia-Chavala, and Rydzewski (1998) examined crashes on California's I-880 freeway in 1995. estimate of this queue dispersal impact will thus be made separately based on the results of the following analysis of impacts during the accident duration phase. See *Post-Crash Duration Queue Dispersal* later in this report for further discussion and analysis of the impacts of this phase of crash impacts.

Most of the studies noted in Table 3-4 share the basic limitation of being conducted only on specific urban freeways during relatively short timeframes. The exception is the FMCSA report, which examined crashes on all roadway types in Pennsylvania from 2006 to 2008. However, the FMCSA report was confined to heavy truck crashes, and wasn't aggregated by lane blockage. The framework for this current analysis is specific to five different roadway types, and is intended to be representative of all roadway crashes. To accommodate this structure, we obtained data for all crash types from the Pennsylvania DOT traffic records database for the years 2006-2010. The Pennsylvania DOT data represent a complete census of police-reported crashes for this time period, broken out by roadway type, roadway blockage status, crash severity, and crash duration. This data describes the characteristics of roughly 418,000 crashes that occurred in Pennsylvania during this timeframe. From this data, we obtained counts of crashes stratified into the 7 duration categories coded in the Pennsylvania DOT database. An example of this output is shown in Table 3-5.

From this data, we computed the proportion of crashes that occurred under each duration category, and applied that proportion to the average duration of crashes in that category. An example of this process is shown in Table 6 for crashes with full road closure.

Table 3-5. 2006-2010 PA Statewide Crashes by Roadway Closure and Severity on Urban Interstate Expressways

Closure	Severity	Unknown Time/ None	< 30 Minutes	30-60 Minutes	1-3 Hours	3-6 Hours	6-9 Hours	> 9 Hours	Total
Fully	Fatal	4	7	28	131	100	24	7	301
	Injury	27	1,667	1,850	685	146	26	20	4,421
	PDO	19	1,303	1,036	351	68	17	11	2,805
	All	50	2,977	2,914	1,167	314	67	38	7,527
None	Fatal	36	0	0	0	0	0	0	36
	Injury	9,195	0	0	0	0	0	0	9,195
	PDO	16,291	0	0	0	0	0	0	16,291
	All	25,522	0	0	0	0	0	0	25,522
Partially	Fatal	2	3	12	54	24	3	1	99
	Injury	121	3,606	3,064	555	32	6	2	7,386
	PDO	101	3,754	2,295	347	35	5	5	6,542
	All	224	7,363	5,371	956	91	14	8	14,027
Unknown	Fatal	0	0	0	0	0	0	0	0
	Injury	4	2	1	0	0	0	0	7
	PDO	4	0	0	0	0	0	0	4
	All	8	2	1	0	0	0	0	11
Totals	Fatal	42	10	40	185	124	27	8	436
	Injury	9,347	5,275	4,915	1,240	178	32	22	21,009
	PDO	16,416	5,057	3,331	698	103	22	16	25,643
	All	25,805	10,342	8,286	2,123	405	81	46	47,088

Table 3-6. PA Crash Duration, Urban Interstate Expressways, 2006-2010, Full Roadway Closure (Minutes)

Severity	< 30 minutes	30-60 minutes	1-3 Hours	3-6 Hours	6-9 Hours	>9 Hours	Total
Fatal	2.36%	9.43%	44.11%	33.67%	8.08%	2.36%	100.00%
Injury	37.94%	42.10%	15.59%	3.32%	0.59%	0.46%	100.00%
PDO	46.77%	37.19%	12.60%	2.44%	0.61%	0.39%	100.00%
All	39.82%	38.97%	15.61%	4.20%	0.90%	0.51%	100.00%
Severity	**Minutes**						
	15	**45**	**90**	**240**	**420**	**540**	**Total**
Fatal	0.35	4.24	39.70	80.81	33.94	12.73	171.77
Injury	5.69	18.95	14.03	7.97	2.49	2.46	51.59
PDO	7.02	16.73	11.34	5.86	2.56	2.13	45.64
All	5.97	17.54	14.05	10.08	3.76	2.74	54.14

The table indicates that, for fatal crashes on urban interstate expressways where there was full roadway closure, the average incident duration was 172 minutes. For injury crashes, the average duration was 52 minutes, and for PDO crashes, the average duration was 46 minutes. Note that the averages from some duration categories are skewed within the duration. This reflects weighting of cases within each category. The weighted time values chosen to represent each category were adopted from FMCSA, based on data from crashes in Kentucky. Table 3-7 summarizes the results for each roadway type and crash severity.

Table 3-7. Average Roadway Closure Times by Crash Severity and Roadway Type

Fatal Crashes						
Roadway Type	**Closure (minutes)**			**Closure (hours)**		
	Full	**Partial**	**Combined**	**Full**	**Partial**	**Combined**
Urban Int/Expressway	171.77	134.07	162.44	2.86	2.23	2.71
Urban Arterial	161.01	95.33	151.80	2.68	1.59	2.53
Urban Other	159.99	56.25	152.80	2.67	0.94	2.55
Rural Int/Arterial	175.84	118.05	161.74	2.93	1.97	2.70
Rural Other	146.89	83.40	139.14	2.45	1.39	2.32
Injury Crashes						
Roadway Type	**Closure (minutes)**			**Closure (hours)**		
	Full	**Partial**	**Combined**	**Full**	**Partial**	**Combined**
Urban Int/Expressway	51.59	34.85	41.12	0.86	0.58	0.69
Urban Arterial	47.94	29.51	37.18	0.80	0.49	0.62
Urban Other	52.89	31.54	43.31	0.88	0.53	0.72
Rural Int/Arterial	74.22	47.72	57.71	1.24	0.80	0.96
Rural Other	60.58	41.79	51.27	1.01	0.70	0.85
PDO Crashes						
Roadway Type	**Closure (minutes)**			**Closure (hours)**		
	Full	**Partial**	**Combined**	**Full**	**Partial**	**Combined**
Urban Int/Expressway	45.64	31.67	35.87	0.76	0.53	0.60

PDO Crashes						
Roadway Type	Closure (minutes)			Closure (hours)		
	Full	Partial	Combined	Full	Partial	Combined
Urban Arterial	43.48	27.51	32.26	0.72	0.46	0.54
Urban Other	50.26	28.94	37.39	0.84	0.48	0.62
Rural Int/Arterial	74.86	44.00	52.26	1.25	0.73	0.87
Rural Other	58.63	38.57	45.99	0.98	0.64	0.77

Table 3-8. Average Roadway Closure Times by Crash Severity and Roadway Type, Adjusted for Response Lag

Fatal Crashes						
Roadway Type	Closure (minutes)			Closure (hours)		
	Full	Partial	Combined	Full	Partial	Combined
Urban Int/Expressway	176.77	139.07	167.44	2.95	2.32	2.79
Urban Arterial	166.01	100.33	156.80	2.77	1.67	2.61
Urban Other	164.99	61.25	157.80	2.75	1.02	2.63
Rural Int/Arterial	185.84	128.05	171.74	3.10	2.13	2.86
Rural Other	156.89	93.40	149.14	2.61	1.56	2.49
Injury Crashes						
Roadway Type	Closure (minutes)			Closure (hours)		
	Full	Partial	Combined	Full	Partial	Combined
Urban Int/Expressway	56.59	39.85	46.12	0.94	0.66	0.77
Urban Arterial	52.94	34.51	42.18	0.88	0.58	0.70
Urban Other	57.89	36.54	48.31	0.96	0.61	0.81
Rural Int/Arterial	84.22	57.72	67.71	1.40	0.96	1.13
Rural Other	70.58	51.79	61.27	1.18	0.86	1.02
PDO Crashes						
Roadway Type	Closure (minutes)			Closure (hours)		
	Full	Partial	Combined	Full	Partial	Combined
Urban Int/Expressway	50.64	36.67	40.87	0.84	0.61	0.68
Urban Arterial	48.48	32.51	37.26	0.81	0.54	0.62
Urban Other	55.26	33.94	42.39	0.92	0.57	0.71
Rural Int/Arterial	84.86	54.00	62.26	1.41	0.90	1.04
Rural Other	68.63	48.57	55.99	1.14	0.81	0.93

As previously noted, the duration times in Tables 3-5, 3-6, and 3-7 represent the time spent by police at the crash site. However, crashes will begin to influence congestion from the time of their occurrence. There is typically a delay between the crash occurrence and the arrival of emergency personnel such as police or ambulance. A review of online studies or articles focused on local police or EMS response times indicates that responses can typically take 5 to 20 minutes, depending on the nature of the jurisdiction, with longer response times generally occurring in rural areas. We did not find studies that examine National average response times. To account for this time lag, we added 5 minutes to the duration times for urban roadway crashes and 10 minutes to rural roadway crashes. We believe these may be conservative estimates, since 5 minutes appears to be a minimum response time except in very small jurisdictions, and many rural response times exceeded 10 minutes. In addition, published response times are typically based on the elapsed time between emergency notification or dispatch and EMS or police arrival at the scene. The lag between the crash and such notification is thus not accounted for in these studies. As noted previously, the proliferation of cell phones and other portable communication devices likely minimizes the time between crash events and emergency notification. Table 3-8 modifies Table 3-7 to include these average response time lag estimates.

The average crash duration times in Table 3-8 served as the basis for each congestion aspect in this analysis. For crashes where there was lane closure, the weighted combined duration for both full and partial closure is applicable. It seems likely that crashes that do not require lane closure would disrupt traffic for a shorter duration than those that do. Directionally, this is confirmed by the significant decrease in duration found across all roadway types when comparing cases that involved full closure to those that involved only partial closure. The PA database does not include duration information for crashes that did not involve road closure, and the other studies cited in Table 4 did not discriminate between crashes with lane closure and those without. Since we lack an estimate for these crashes, for this analysis, it will be assumed that crashes that did not require road closure have similar durations to those that required only partial closure. We note that this may produce an upward bias for this segment of the analysis, but this data is the closest match available. However, the overall average crash duration (across all crash types) resulting from this assumption fits well within the overall averages found in other studies. Within individual categories, for fatal crashes the current study estimates a shorter duration than the Miller and FMCSA studies. However, the Miller study reflects only urban freeways whereas this study represents all roadway types, so this might be expected. Likewise, the FMCSA study represents only heavy truck crashes, whereas this study represents all crash types, so a shorter duration might be expected. The current study's injury estimate is lower than those found in previous studies, but reasonably close to three of the five studies that examined injury crashes. Again, the biggest difference is with the Miller study that examined only urban crashes, but most other studies were also based primarily on urban freeways. For PDO crashes, this study's duration estimate fits well within the range of other studies findings, as does the overall average for all crash types. Overall, the PA data used for this study is the most recent data source, represents the broadest group of roadways and crash types, and is based on the largest number of observations. Table 9 summarizes the various duration estimates from these studies and those used in this study.

The final crash duration times are summarized in Table 10. The closure values will be used for calculating the impacts of crashes with lane closings. The no-closure values will be used for impacts resulting from crashes without lane closings. The combined values, which are derived using PA frequency weights, will be used for calculating opposite direction rubbernecking estimates, which can occur for all crash types.

Probability of Lane Closure

When crashes occur, some vehicles will remain in the roadway while others will end up on the side of the roadway where they won't directly block traffic. Roadways can be blocked by vehicles that come to rest either entirely or partially within the roadway travel lanes, or by debris from the crash (vehicle parts, cargo, damaged roadway structures, etc.). Roadway blockage can also result when police or emergency equipment responds to the crash. Lane blockage can result from formal lane closures set up by police or from de facto closures due to the presence of crashed vehicles. In many crashes, both will occur. Nearly all crashes result in some sort of delay, either through blockage or due to rubbernecking, but a closed lane clearly has a bigger impact. The extent to which lanes are blocked thus has a direct impact on the amount of congestion that results from a crash.

Table 3-9. Summary of Crash Duration Estimates (Minutes)

Study	Crash Type	No. Obs.	PDO	Injury	Fatal	All
Zaloshnja, Miller, and Spicer (5 U.S. Urban freeways, 1989)	All	Unk	49	86	233	60
FMCSA (2012) (Pennsylvania, 2006-2008)	Truck	23,388	35	55	216	44
Giuliano (1988) (LA I-10 freeway, 1983-85)	All	270	44	56		49
Wu, Kachroo, and Ozbay (1998) (Northern VA, 1997)	All	33	27	50		34
Boyles, Fajardo, and Waller (2006) (Atlanta, 2004)	All	2,970				42
Lan and Hu (1999) (Minnesota Urban freeways, 1994-95)	All	3877				39
Skabardonis, Chira-Chavala, and Rydzewski (1998) (CA I-880 frwy 1995)	All	92	40	63		43
NHTSA 2013 (this study) (Pennsylvania 2006-2010)	All	418,000	41	45	151	43

Table 3-10. Crash Duration by Roadway Type and Crash Severity (Hours)

	Fatal Crashes		Injury Crashes		PDO Crashes	
Roadway Type	**Closure**	**No Closure**	**Closure**	**No Closure**	**Closure**	**No Closure**
Urban Int/Express	2.79	2.32	0.77	0.66	0.68	0.61
Urban Arterial	2.61	1.67	0.70	0.58	0.62	0.54
Urban Other	2.63	1.02	0.81	0.61	0.71	0.57
Rural Int/Arterial	2.86	2.13	1.13	0.96	1.04	0.90
Rural Other	2.49	1.56	1.02	0.86	0.93	0.81
All Roadways	2.71	1.85	0.97	0.81	0.86	0.74
Roadway Type	**Combined**		**Combined**		**Combined**	
Urban Int/Express	2.75		0.72		0.64	
Urban Arterial	2.48		0.64		0.58	
Urban Other	2.50		0.73		0.63	
Rural Int/Arterial	2.79		1.05		0.94	
Rural Other	2.39		0.95		0.86	
All Roadways	2.61		0.90		0.78	

Chin et al. (2004), developed estimates of the probability-of-lane closure (PLC) from 1998 FARS data. For crashes involving single vehicles, the probability that a fatal crash would not cause a lane closure was derived assuming that any crash that was located on the roadway facility would cause lane closure, but those that were located outside the facility right of way or classified as off-road crashes would not. The probability of a fatal crash not closing a lane was computed as the ratio of these off roadway crashes to the total number of crashes for which the location was known. This resulted in an estimated probability that lanes would not be closed of 10.8 percent, leaving a probability of closure of 89.2 percent. Chin adopted this same rate for non-fatal injuries as well.[16] Chin assumed that fatal or injury crashes involving two or more vehicles would always involve lane closure. For PDO crashes, Chin used data derived from Giuliano (1989) in her study of incident data on LA freeways. This data indicated a 60 percent probability-of-lane closure in non-injury crashes. Chin overrode these probabilities if more than three cars or more than one truck was involved in the crash. Under these circumstances, Chin assumed that there would be lane closures even if there was no injury.

For this study, we use the actual crash records from the same Pennsylvania database that produced estimates of crash duration. Based on this data, the proportion of police-reported crashes that involved lane closings is summarized in Table 3-11. The fatal crash estimates show close agreement with Chin's 89 percent closure rate estimate based on roadway and crash involvement characteristics. However, they indicate a lower closure rate for injury crashes, which differs from Chin's assumption that the rates would be similar for both fatal and injury crashes. Likewise, the rates of lane closure for PDOs are lower than the 60 percent rate that Giuliano derived in 1989 for LA freeways. This difference likely reflects the broader scope of the Pennsylvania, all-roadway data used in the current analysis.

Reduced Capacity

Given that some level of closure occurs, how does this affect roadway capacity? Logically, this is a function of both the number of blocked lanes and the number of lanes available. This issue was originally examined by Blumentritt, Ross, Glazer, Pinnell, and McCasland (1981). Blumentritt and colleagues obtained data from 10 agencies that operate freeway ramp metering installations to estimate the capacity reduction that results from lane closures as a function of the number of lanes on the roadway and the number of lanes closed. Blumentritt and his group limited their study to up to two lane closures. Chin et al. (2004) expanded his estimates to include up to four lanes. Table 3-12 below summarizes the values derived by Chin.

Table 3-11. Portion of Police-Reported Crashes Involving Lane Closure, PA Crashes 2006-2010

Roadway Type	Fatal	Injury	PDO
Urban Interstate/ Expressway	92%	56%	36%
Urban Arterial	86%	52%	42%
Urban Other	92%	60%	47%
Rural Interstate/Principal Arterial	90%	55%	29%
Rural Other	90%	57%	39%

Table 3-12. Reduced Capacity Due to Freeway Crashes (Normal Capacity = 1.0)

Effect of Crash	Number of Freeway Lanes (One Direction)				
	1	2	3	4	5+
Vehicle on Shoulder	0.45	0.75	0.84	0.89	0.93
1 lane blocked	0.00	0.32	0.53	0.56	0.75
2 lanes blocked	NA	0.00	0.22	0.34	0.50
3 lanes blocked	NA	NA	0.00	0.15	0.20
4 lanes blocked	NA	NA	NA	0.00	0.10

Source: Chin et al., 2004, Table 3-7.

This analysis adopts Chin's estimates of reduced capacity. To use this data, frequency weights must be derived that reflect the lane profile of motor vehicle crashes. For this purpose 2008-2009 FARS data was used. FARS is the only nationwide database containing information on the number of lanes in the roadway by roadway type. However, FARS codes lane data differently for divided and undivided highways. On undivided highways, the FARS lane count represents lanes going in both directions. On divided highways, FARS lane counts represent only the lanes going in the direction of travel that the crash occurred in. Table 3-12 uses unidirectional stratification, i.e., it expresses capacity reduction based on lanes going in one direction only. FARS data for undivided highways must thus be adjusted to normalize them to a single direction basis. This was done by dividing the lane count for undivided highways by two and re-assigning them accordingly. While there are many undivided roadway segments where there are odd numbers of lanes, such as two lanes in one direction and one in the other, data is not available to determine which side the crash occurred in on these odd number of lanes segments. While not precise, halving the two-way lane count assumes that over the universe of crashes on such roads, about half would occur in either direction. The practical application of this would result, for example, in assigning all fatalities coded in FARS as occurring in two-lane, undivided expressways as occurring in roadways where there was a single lane in each direction. In addition, half the fatalities that were coded as occurring in three-lane undivided expressways would be assigned to roadways where there was a single lane in each direction.

Chin also derived probability distributions of the number of lanes closed, given that there was some level of lane closure. These distributions are shown in Table 3-13.

Table 3-13. Probability Distribution of the Number of Lanes Closed Given Lane Closure, by Vehicle Involvement

Number of Vehicles Involved	Type of Vehicles Involved	Lanes Closed			
		1	2	3	4+
I Vehicle	Any Type	0.997	0.001	0.001	0.001
2 Vehicles	2 Cars, or 1 car and 1 truck	0.950	0.048	0.001	0.001
	2 trucks	0.001	0.997	0.001	0.001
3 Vehicles	3 cars, or 2 cars and 1 truck	0.500	0.450	0.049	0.001
	1 car and 2 trucks or 3 trucks	0.001	0.600	0.300	0.099
More Than 3 Vehicles	Any type	0.001	0.099	0.800	0.100

In order to determine incidence of lane closure, FARS data was examined consistent with the vehicles involved categories used in Table 3-13. Data was collected separately for divided roadways, two direction undivided roadways, and single direction undivided roadways. Counts for the number of crashes that occurred under each lane category in one direction were determined by adding the totals for divided roadways, single direction undivided roadways, and half of the crashes for two directional undivided roadways. This produced total counts of crashes by lane counts for each vehicle involvement category, which were in turn used to determine the relative frequency of crashes for each cell. An example of these results for Urban Interstates/Expressways[17] is shown in Table 3-14.

The probability distributions from Table 3-13 were combined with the crash involvement frequencies derived in Table 3-14 to determine the distribution of lane blockage for roadways of various lane counts.

The results are shown in Table 3-15.

So, for example, based on the relative frequency of crashes by roadway width (lane count) on a roadway with two lanes in the direction of travel of the crashed vehicle, 91 percent of the time only one lane is blocked, but 9 percent of the time two lanes are blocked.

The results from Table 3-15 were then combined with the data from Table 3-12 and the lane frequency counts that were behind the proportions in Table 3-14 to determine the weighted average capacity reduction percentage for each roadway category. As previously mentioned, lane information by roadway type is not available in GES or CDS data bases. Therefore, the blockage estimates derived from FARS will be used for both nonfatal injury and PDO crashes as well. The results are shown in Table 3-16.

Table 3-14. Crash Distribution within Lane Categories, Urban Interstate/Expressway

Number of Vehicles Involved	Type of Vehicles Involved	Lanes in Direction of Crash			
		1	2	3	4+
I Vehicle	Any Type	0.700	0.594	0.562	0.489
2 Vehicles	2 Cars, or 1 car and 1 truck	0.253	0.308	0.305	0.345
	2 trucks	0.002	0.007	0.006	0.005
3 Vehicles	3 cars, or 2 cars and 1 truck	0.034	0.057	0.071	0.095
	1 car and 2 trucks or 3 trucks	0.001	0.005	0.006	0.007
More than 3 Vehicles	Any type	0.009	0.029	0.051	0.058
All		1.000	1.000	1.000	1.000

Table 3-15. Distribution of Lane Blockage, Given at Least One Lane Blocked, Urban Interstate/Expressway

Effect of Crash	Number of Freeway Lanes (One Direction)				
	1	2	3	4	5+
1 lane blocked	1.00	0.91	0.89	0.86	0.86
2 lanes blocked	NA	0.09	0.06	0.08	0.08
3 lanes blocked	NA	NA	0.05	0.05	0.05
4 lanes blocked	NA	NA	NA	0.01	0.01
All	1.00	1.00	1.00	1.00	1.00

Table 3-16. Derivation of Average Vehicle Capacity Reduction,[18] Crashes with at Least One Lane Closed

	Urban Interstates/ Expressways	Urban Arterials	Urban Other	Rural Interstate/ Principal Arterials	Rural Other
Fatal Crashes					
AADT	113,814	23,996	2,908	25,579	1,502
AAHT	4,934	1,040	126	1,114	65
AAHT - one way	2,467	520	63	557	33
Crash duration (hours)	2.79	2.61	2.63	2.86	2.49
% w/at least 1 lane closed	91.7%	86.1%	91.9%	90.3%	89.8%
% Blockage	62.0%	75.8%	89.1%	78.9%	97.2%
Avg. vehicle capacity reduction	3,913	887	136	1,135	71
Injury Crashes					
AADT	113,814	23,996	2,908	25,579	1,502
AAHT	6,218	1,311	159	1,392	82
AAHT - one way	3,109	655	79	696	41
Crash duration (hours)	0.77	0.70	0.81	1.13	1.02
% w/at least 1 lane closed	56.2%	51.6%	60.3%	54.9%	57.4%
% Blockage	62.0%	75.8%	89.1%	78.9%	97.2%
Avg. vehicle capacity reduction	832	180	34	340	23
PDO Crashes					
AADT	113,814	23,996	2,908	25,579	1,502
AAHT	6,300	1,328	161	1,407	83
AAHT - one way	3,150	664	80	703	41
Crash duration (hours)	0.68	0.62	0.71	1.04	0.93
% w/at least 1 lane closed	36.5%	42.3%	47.3%	29.4%	39.4%
% Blockage	62.0%	75.8%	89.1%	78.9%	97.2%
Avg. vehicle capacity reduction	485	132	24	169	15

Table 3-16 summarizes the inputs and results of this analysis of the congestion impacts of lane closure. The results indicate relative impacts that reflect differences in both traffic density and crash severity. Fatal crashes have the greatest impact on congestion followed by nonfatal injury crashes and PDO crashes. This reflects the added duration of the crash events as well as the added rate of lane closure that results from emergency and police response to the more serious crashes involving death or injury. Within each severity category, urban interstates/expressway experience the highest per-crash congestion impact, followed by urban arterials and rural interstates/principal arterials. These impacts reflect the much higher traffic density found in urban roadways, and to a lesser extent on rural interstates.

Rubbernecking

Rubbernecking occurs when drivers slow down as they pass an unexpected highway incident such as an accident or broken down vehicle. The term is most commonly used to describe unnecessary slowing on the part of drivers, but as used here it encompasses any slowing that affects capacity when a lane isn't blocked, regardless of specific motivation. This slowing can occur because the drivers of passing vehicles are curious, or because the presence of the crashed vehicle makes them cautious, or because they must respond to braking of other vehicles ahead of them. Under moderate or heavy congestion levels, once an initial vehicle brakes, it can cause a corresponding wave of reactive braking to ripple back through traffic, slowing all subsequent drivers until they pass the crash site. Within the context of crashes, rubbernecking is caused by the presence of crashed vehicles or police or emergency equipment at the scene that is not directly blocking the travel lane. Rubbernecking can occur in the direction of travel for the crashed vehicle when vehicles are removed to the side of the road, or in open lanes when vehicles are blocking only a portion of the roadway. Rubbernecking can also affect traffic going in the opposite direction from the crash affected lanes.

The aggregate reduced capacity estimates used in the previous analysis of lane closures essentially include the impact of both rubbernecking and slowed traffic in cases where lanes are closed. Rubbernecking within same direction lanes for crashes with lane closure thus does not have to be separately estimated. This section will therefore focus on rubbernecking in same direction lanes for cases where there is no lane closure, as well as rubbernecking impacts in opposite direction lanes. These estimates will be calculated as follows.

$$VCL = (AAHT*CD*(1\text{-}PLC)*RCR) + (AAHT*CD*RO*RCO)$$

where:

VCL = Vehicle capacity lost due to rubbernecking when no lane closure and in opposite direction of travel

AAHT = Average annual hourly travel (vehicles) past the crash site during the time affected by the crash

CD = crash duration

PLC = Probability-of-lane closure

RCR = Reduced capacity when no lanes blocked

RO = Rate of rubbernecking in opposite lane

RCO = Reduced capacity due to rubbernecking in opposite lane

Frequency of Crashes With No Lane Closure (1-PLC)

AAHT, CD, and PLC were derived in the previous discussion of congestion due to lane closures. The frequency of crashes with no lane closures is derived as 1-PLC. These represent crashes that occur far enough outside the travelled roadway, or are of a minor enough severity, that neither the involved vehicles nor police or emergency response vehicles and operations will result in lane blockage or closure.

Reduced Capacity When No Lanes Blocked (RCR)

RCR is derived from the "Vehicle on Shoulder" row in Table 3-12. This represents the reduced capacity that results when there is no lane closure. Essentially, this represents the impact of rubbernecking as vehicles slow down to pass the crashed vehicles on the side of the road. For each roadway width (number of lanes), RCR is calculated as the product of the reduced capacity when there are vehicles on the shoulder and the relative frequency of crashes on roadways with the specified width for the lane (originally developed as described under the previous lane closure discussion). So, for example, for urban interstate/ expressways, a vehicle on the shoulder of a two-lane (in one direction) roadway will reduce roadway capacity by 25 percent (1- the value in Table 12). 34.7 percent of all urban interstate/ expressway crashes occur under these two lane conditions. The product of these numbers added to these products for each lane count determines the average reduced capacity across all roadway widths that are urban interstates/expressways. The results are shown for all roadway categories in Table 3-17.

Rate of Rubbernecking in Opposite Lane (RO)

There are very little data available regarding the incidence of rubbernecking in opposite direction lanes. This analysis is based on a study conducted by Masinick and Teng in 2004. That study was the first to evaluate the rubbernecking impact of accidents in traffic in the opposite direction based on archived traffic and accident data. Masinick and Teng examined archived data from the year 2000 on roadway occupancy behavior on a 10-mile section of freeway in the Hampton Roads, Virginia, area. They judged each crash based on particular criteria to determine whether it would be considered to have a significant impact on roadway capacity during the crash. They defined a significant impact as one in which they observed a sharp increase in segment occupancy soon after the crash occurred, with this increase held fairly constant for the duration of the incident, after which occupancy rates returned to normal. On this basis, they determined that there were 201 crashes that had a significant impact on traffic flow in the direction of the accident, 102 that had a significant impact in the opposite direction, and 84 that had this impact in both directions. The discrepancy between the 84 and 102 opposite direction crashes was a function of greater volume in the opposite direction (e.g., a rush hour crash that occurred in the opposite direction of rush hour traffic), or possibly of roadway design that produced more significant impacts in one direction. The ratio of opposite direction rubbernecking was thus either 51 percent or 42 percent of the rate in the crash direction, depending on whether these factors are accounted for. This study uses the .51 ratio that includes all cases, since both factors are legitimate causes of rubbernecking attributable to the crash.

A more recent study of rubbernecking by Knoop, Hoogendoorn, and van Zuylen (2008) found reverse direction rubbernecking in one of two crashes that they studied in detail (a 50% ratio), but this is far too small a sample to draw conclusions from. Note that Masinick and Teng (2004) found only 24 percent of the crashes they observed involved a significant impact on traffic flow in the direction of the crash. This contrasts with the data in Chin et al. (2004), which found significantly higher rates just for lane closure. Data from NHTSA's FARS and GES databases indicate that roughly 75 percent of all crashes occur in the roadway, where they are almost certain to cause traffic disruption due to lane blockage. Another 20 percent occur in areas immediately adjacent to the roadway such as shoulders, medians, and roadsides, where lane closure is possible in cases where police or emergency equipment respond at the scene, and where rubbernecking is likely even if lanes remain open. These are also likely to cause some level of congestion. Finally, data taken directly from the from PA accident database indicate higher rates of lane closure for police-reported crashes than were found for rubbernecking in the Hampton Roads study. Under these circumstances, it seems likely that the lower rate found by Masinick and Teng (2004) may be a function of the specific traffic flow characteristics of the Hampton Roads freeway system they observed, or possibly their definition of a serious traffic impact. However, we should note that the method used by Chin et al. to decide which crash locations were likely to involve lane closure also involved a degree of subjective judgment.

Reduced Capacity due to Rubbernecking in the Opposite Lane (RCO)

As with the rate of rubbernecking in opposite lanes, there are only sparse data regarding the capacity reduction that is caused by such actions. The study cited above by Masinick and Teng (2004) also estimated average capacity reduction in their Hampton Roads study area. They found an average capacity reduction of 12.7 percent for all crashes. More recently, Knoop, Hoogendoorn, and van Zuylen (2008) conducted an aerial study (using a helicopter) of the impact of traffic crashes on opposite direction traffic flows. The authors found a 50-percent reduction in traffic flows in the opposite direction of one of the crashes. The differences between these two studies' results are difficult to interpret, but may be related to the specific locations. Hampton Roads, Virginia, is not a particularly congested area and may not be typical for urban roadways in most metropolitan areas. Likewise, although the Knoop et al. study appears to be quite accurate due to complete video documentation of the crash impacts, it represents only one crash on a divided rural interstate in Holland (a second crash was also observed but it occurred on a roadway where there was insufficient traffic to cause opposite direction rubbernecking.)

An alternative basis used in this study is to use the percent reduction that was derived for same direction rubbernecking (RCR). This produced estimates of reduced capacity that averaged 20 to24 percent, depending on roadway type. RCR was derived from data reflecting capacity reductions in crashes where lanes are not blocked, and thus represents a rubbernecking impact, albeit not specifically in opposite direction lanes. The rationale for this choice is that RCR is ultimately based on a sample of crash impacts (from Blumentritt et al., modified by Chin et al.) that represent a more diverse group of 28 roadways spread across 10 different States. This is a broader sample than the two narrower studies cited above. We also note that the RCR estimates are between the two extremes found in the two studies that specifically addressed opposite direction rubbernecking. Moreover, Knoop Hoogendoorn, and van Zuylen found similar impacts due to rubbernecking in both directions.

Table 3-17 summarizes the inputs and estimated reduced capacity from rubbernecking in crashes where same direction lanes were not closed and in opposite direction lanes for all crashes. As with lane closures, the results indicate relative impacts that reflect differences in both traffic density and crash severity. Fatal crashes have the greatest impact on congestion, followed by nonfatal injury crashes and PDO crashes. This reflects the added duration of the crash events to the more serious crashes involving death or injury. Within each severity category, urban interstates/expressways experience the highest per-crash congestion impact, followed by urban arterials and rural interstates/principal arterials. These impacts reflect the much higher traffic density found in urban roadways, and to a lesser extent on rural interstates.

The vehicle capacity reduction estimates in Tables 3-16 and 3-17 represent the number of vehicles that were prevented from passing the crash site due to congestion during the duration of the crash. However, the impacts of congestion are linked not to vehicles but to vehicle hours of delay. The value of travel delay to vehicle occupants as well as the impact on fuel consumption, greenhouse gases and criteria pollutants are all a function of the aggregate time that vehicles are delayed. To estimate delay hours we assume a constant rate of arrival by vehicles throughout the duration of the crash. Under this assumption, average delay hours are calculated as:

$$\sum_{n=1}^{v} D - \left(\frac{D}{v}\right) * n$$

Where D = crash duration and v = vehicle capacity reduction during D.

Table 3-17. Congestion Impacts of Rubbernecking in Open Lane Crashes and Opposite Directions (per Crash)

	Urban Interstates/ Expressways	Urban Arterials	Urban Other	Rural Interstate/ Principal Arterials	Rural Other
Fatal Crashes					
AAHT -1 direction	2,467	520	63	557	33
Crash duration (hours)	2.79	2.61	2.63	2.86	2.49
% Same direction rubbernecking	0.083	0.139	0.081	0.097	0.102
Capacity reduction %	0.23	0.33	0.45	0.35	0.52
S.D. Vehicle capacity reduction	108	40	2	41	3
% Opposite direction rubbernecking	50.7%	50.7%	50.7%	50.7%	50.7%
O.D. capacity reduction	787	217	36	279	21
Injury Crashes					
AAHT -1 direction	3,109	655	79	696	41
Crash duration (hours)	0.77	0.70	0.81	1.13	1.02
% Same direction rubbernecking	0.438	0.484	0.397	0.451	0.426
Capacity reduction %	0.23	0.33	0.45	0.35	0.52
S.D. vehicle capacity reduction	207	61	9	107	8
% Opposite direction rubbernecking	50.7%	50.7%	50.7%	50.7%	50.7%
O.D. capacity reduction	261	71	13	132	10
PDO Crashes					
AAHT -1 direction	3,150	664	80	703	41
Crash duration (hours)	0.68	0.62	0.71	1.04	0.93
% Same direction rubbernecking	0.635	0.577	0.527	0.706	0.606
Capacity reduction %	0.23	0.33	0.45	0.35	0.52
S.D. Vehicle capacity reduction	280	69	11	158	11
% Opposite direction rubbernecking	50.7%	50.7%	50.7%	50.7%	50.7%
O.D. capacity reduction	233	64	12	119	9

This produces an estimate that assumes vehicle arrivals beginning at *D/v*, or the full interval between vehicle arrivals after the crash occurs. An alternate assumption could be made reflecting vehicle arrival beginning at the time of the crash by modifying the equation to multiply by n-1 instead of n. However, it is likely that the arrival of vehicles will be spread randomly across this interval. To reflect this, vehicle hours are estimated as a simple average of the product of the number of vehicles affected and crash duration or *Dv*/2. This produces an estimate midway between the two more extreme assumptions. Table 3-18 summarizes the estimates of vehicle delay hours for congestion during crash duration, including lane blockage and rubbernecking, but excluding detours. Note that these calculations are based on the portion of vehicles delayed that did not detour to avoid congestion. Discussion of the methods used to calculate this portion is included in the next section.

Post-Crash Duration Queue Dispersal

The previous sections analyze the impacts of congestion caused by both lane blockage and rubbernecking during the period after the crash until the crashed vehicles have been removed and the emergency response vehicles have departed. Once the roadway is cleared of damaged vehicles, crash debris, and emergency vehicles, a residual queue of vehicles may still be present, which will then rapidly disperse past the accident scene. There are no studies which focus on this particular aspect of congestion activity. Therefore, an estimate of vehicle dispersal time will be made based on the results of the previous analysis of impacts during the accident duration phase.

Table 3-18. Vehicle Delay Hours from Congestion during Crash Duration

	Urban Interstates/ Expressways	Urban Arterials	Urban Other	Rural Interstate/ Principal Arterials	Rural Other
Fatal Crashes					
Total vehicle capacity reduction	4,386	986	161	1,308	85
Vehicles that didn't detour	3,223	926	157	1,188	84
Crash duration (hours)	2.79	2.61	2.63	2.86	2.49
Vehicle delay hours, non-detours	4,497	1,210	206	1,700	104
Injury Crashes					
Total vehicle capacity reduction	819	193	38	366	27
Vehicles that didn't detour	793	190	38	357	27
Crash duration (hours)	0.77	0.70	0.81	1.13	1.02
Vehicle delay hours, non-detours	305	67	15	201	14
PDO Crashes					
Total vehicle capacity reduction	587	160	29	280	22
Vehicles that didn't detour	573	157	29	275	22
Crash duration (hours)	0.68	0.62	0.71	1.04	0.93
Vehicle delay hours, non-detours	195	49	10	143	10

Dispersal time is a function of both time spent waiting in queue before proceeding and the added time spent accelerating to cruising speed. Time spent waiting in queue is a relatively straightforward calculation, but acceleration time requires consideration of acceleration, velocity, and the distance travelled while accelerating. The interaction between these factors is described in the following relationship.[19]

$$s - s_0 = v_0 t + \frac{1}{2} a t^2$$

where:

s = distance v = velocity
t = time (duration)
a = acceleration

Based on this relationship, the formula to estimate initial queue dispersal time is as follows.

$$\sum_{n=1}^{k} (n-1)i + \sqrt{2(n-1)\frac{d}{a}} - \frac{(n-1)d}{c} \leq t$$

$$\sum_{n=k+1}^{m} (n-1)i + \frac{(n-1)d - \frac{at^2}{2}}{c} + t - \frac{(n-1)d}{c} > t$$

where:

k = number of vehicles in queue at end of crash duration
i = interval (seconds) between initiated acceleration of each subsequent vehicle in queue
d = average total distance (feet) between the front bumper of each subsequent vehicle in queue
a = average vehicle acceleration rate (ft/sec sq)
c = average cruising speed (ft/sec) over distance travelled
t = average acceleration time to cruising speed (s/a) (seconds)

The sum of the results from the above formulas was divided by the average number of lanes on the roadway in the direction of the congestion impact to determine total vehicle hours of delay for each roadway crash scenario.

This formula defines queue dispersal time as the total time required to dissipate all traffic existing in the queue at the end of the crash duration period, plus the time required to dissipate any additional traffic that accumulates while the initial queue is dispersing. Dispersal time for each vehicle is defined as the time spent waiting for vehicles ahead in the queue to move so that it can proceed, plus the difference between the time it takes for the vehicle to accelerate up to the average cruising speed it would have been traveling if the crash had not occurred, and the time it would have taken to travel the distance covered while accelerating to cruising speed, at that cruising speed. For example, a vehicle that is 20 cars back in the queue may have to wait 40 seconds before the traffic in front of it has cleared enough for it to begin to accelerate through the now-open roadway. In addition, the vehicle loses time because it must accelerate back up to normal cruising speed over a distance of, for example, 350 feet, whereas had the crash not occurred, it would have cruised through this same distance in a shorter time period at cruising speed.

The first formula measures the difference in acceleration versus cruising time only up to the point where the crash occurred. Once vehicles cross this point they may still experience minor delay if they have not yet achieved cruising speed, but they would no longer influence the accumulation of secondary queues. This formula is thus applied to each vehicle in the queue that has not yet reached cruising speed at the time it crosses the crash site when determining the delay under which secondary queues can occur. Once vehicles in queue reach cruising speed prior to crossing the accident site, the second more general formula that measures time lost over the full distance required to achieve cruising speed is applied. This formula is also applied when measuring the full time loss for all vehicles in queue.

In the formula, k is assumed to equal the number of vehicles that were unable to pass through the crash site during the crash interval, which was calculated in the previous section, adjusted to reflect vehicles that would have diverted to secondary routes to avoid the delays caused by the crash. A number of studies have examined the tendency of drivers to divert to alternate routes when they are faced with traffic congestion. Most such studies are based on reactions to changeable roadway messaging systems (CMS). Harder, Bloomfield, and Chihak (2004) performed a study with 120 drivers using a driving simulator in which 56 percent of participants diverted to an alternate route when a CMS indicated that a crash had occurred in the roadway ahead and that drivers should use an upcoming exit. Another simulator study by Srinivasan and Krishnamurthy (2003) found that that CMSs can reduce congestion levels by 40 percent. Chatterjee and McDonald (2004) examined studies of the effectiveness of CMSs in six European countries found that an average of 8 percent of drivers diverted from their intended routes based on the information they received from CMS. Horowitz, Weisser, and Notbohm (2003) evaluated diversion from a rural work zone and found diversions of 7 percent to 10 percent, depending on the day of the week.

For this study, diversion was based on a 1996 study by Yim and Ygnace (1996) of drivers' responses to increasing congestion on Paris roadways. Yim and Ygnace derived probabilities that drivers would divert to an alternate roadway based on volume to capacity ratios. Although capacity information is not available for the roadways included in this study, Yim and Ygnace did include diversion rate results for a variety of queue length backups. These results were used to create a regression model which approximates the relationship found in their study.[20] This data, together with the regression model, are shown in Figure 3-F. The model predicts roughly 20 percent of vehicles would divert when queues reach 2 miles in length.

The interval "i" is an assumed value of 1.75 seconds, based on personal observations of vehicles progressing through a busy intersection during rush hour in Northern Virginia. Commencing with each green light at this intersection, the vehicles that advanced through lengthy queue dispersions were counted and timed to produce an estimate of about 1.75 seconds delay between each vehicle.[21] The distance between vehicles' front bumpers, d, is an assumed value of 24 feet, based on an examination of vehicle lengths in Ward's Automotive Reports Yearbook indicating roughly 200 inches across all light vehicle types, or 16.7 feet, plus an allowance for spacing. The acceleration rate, a, is assumed to be 10 ft/ sec^2, or roughly 0.3g.[22] Distance travelled in ft/sec at cruising speed is a function of the assumed roadway cruising speeds, which vary by roadway as follows.

Urban Interstates/Expressways – 55 mph (81 ft/sec)
Urban Arterials – 45 mph (66 ft/sec)
Urban Other – 35 mph (51 ft/sec)
Rural Interstates/Arterials – 55 mph (81 ft/sec)
Rural Other – 35 mph (51 ft/sec)

Secondary queues are estimated by measuring the number of vehicles that would join the initial queue before it disperses, which is a function of the arrival rate (AAHT) of additional vehicles during the time it takes for the queue to disperse. These vehicles then disperse while additional queues arrive. The process is repeated until the number of additional vehicles arriving before the previous queue disperses queue drops below 1, at which time the congestion impacts of the crash will have been completed. Note that AAHT for queue dispersal is different from AAHT for the initial crash since it occurs after the crash duration has ended. Queue dispersal AAHTs were calculated by repeating the process previously described for calculating AAHT, but with offsets of 1 hour for injury and PDO crashes, and 3 hours for fatal crashes. These offsets are based on the closest full hour increment to the crash durations for each severity category. This resulted in a small shift (1-2% decrease on weekdays and 1% decrease on weekends) in average AAHT levels one hour after the time of the crash, with more significant but still modest increases (2-5% on weekdays and 1-4% on weekends) 3 hours after the crash, reflecting the dynamics of daily driving cycles.

Table 3-19 summarizes the estimated vehicle hours calculated for queue dispersal for fatal crashes. Note that, unlike the previous aspects of delay, queue dispersion was directly calculated in terms of vehicle hours of travel delay rather than vehicles delayed.

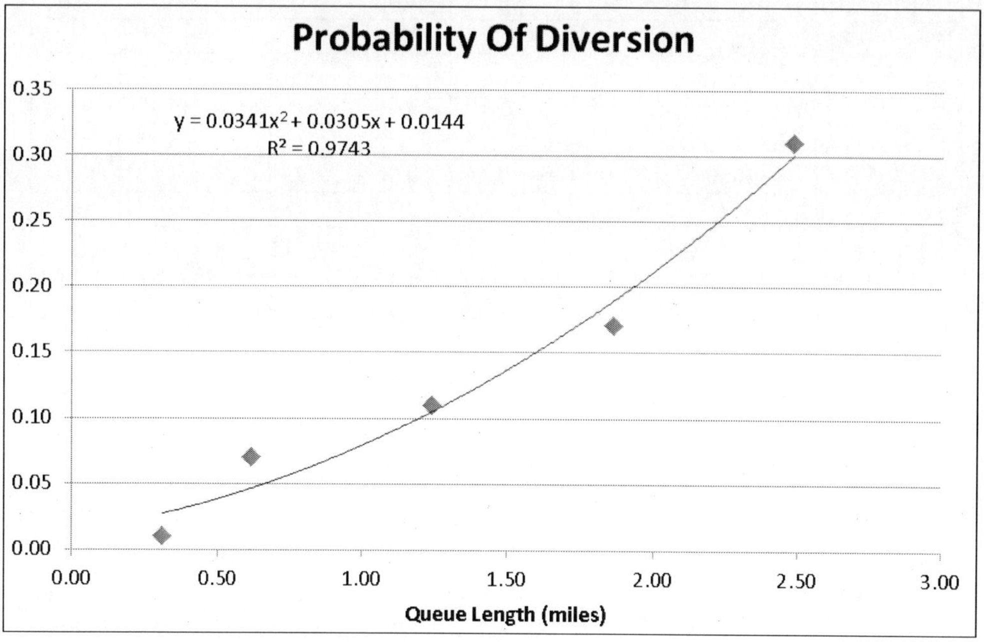

Figure 3-F. Probability of Diversion.

Table 3-19. Queue Dispersal Travel Delay Summary, Impact per Fatal Crash by Functional Roadway Type

	Urban Interstates/ Expressways	Urban Arterials	Urban Other	Rural Interstate/ Principal Arterials	Rural Other
Inputs					
AAHT (one way)	2,169	457	55	490	29
Average Speed	55	45	35	55	35
Average # Lanes (one way)	2.85	1.99	1.45	1.89	1.11
Results -direction of crash					
Initial Vehicles in Queue	2,475	714	122	917	63
E.T. -Initial Queue Dispersal (min)	25.30	10.49	2.45	14.17	1.66
Vehicle Hours, Initial Queue	522.29	43.55	1.28	71.86	0.35
Vehicle Hours, Secondary Queues	73.14	0.81	0.02	1.76	0.02
Total Vehicle Hours	595.43	44.36	1.30	73.62	0.37
Results -opposite direction of crash					
Initial Vehicles in Queue	748	212	35	271	21
E.T. -Initial Queue Dispersal (min)	7.65	3.12	0.71	4.20	0.56
Vehicle Hours, Initial Queue	47.85	3.87	0.11	6.33	0.04
Vehicle Hours, Secondary Queues	7.31	0.09	0.02	0.17	0.02
Total Vehicle Hours	55.16	3.96	0.12	6.50	0.06
Results -Total					
Initial Vehicles in Queue	3,223	926	157	1,188	84
E.T. -Initial Queue Dispersal (min)	25.30	10.49	2.45	14.17	1.66
Vehicle Hours, Initial Queue	570.14	47.42	1.39	78.18	0.39
Vehicle Hours, Secondary Queues	80.45	0.90	0.03	1.93	0.04
Total Vehicle Hours, Queue Disp.	650.59	48.32	1.42	80.11	0.43

Table 3-20 summarizes total vehicle hours of delay per crash during both crash duration and queue dispersal by crash severity and roadway functional category.

Detours and Non-linear Congestion Impacts

The methods described in the above sections account for most congestion impacts that result from traffic crashes. However, they do not account for the added impacts of vehicles that detour to alternate routes to avoid waiting in traffic congestion. In addition, they are based on average traffic densities for each roadway type. Delay impacts can occur disproportionately due to excess roadway capacity under low density traffic conditions or inadequate capacity under highly congested conditions. Moreover, average roadway capacity data is not collected under the HPMS, and thus could not be used in association with AADT statistics to predict average delay impacts. The current model thus assumes that changes in roadway capacity produce proportional changes in vehicle travel. This assumption is likely to produce an overestimate of travel delay in cases where roadway capacity is sufficient to handle existing traffic density even under diminished capacity. Thus, an estimate based solely on an average traffic density may not adequately capture the full impacts of congestion under all density conditions. This type of dynamic can only be measured through direct observation or estimated through traffic simulation measurements such as TSIS-CORSIM.

An advantage of using simulations, as was done with the recent FMCSA study, is that it can capture these conditions. FMCSA did attempt to measure the impacts of detouring, albeit with limited assumptions regarding the scope of potential detour routes. In addition, the FMCSA study examined a variety of traffic density scenarios under each roadway type, which enabled the capture of disproportionate impacts of congestion under high traffic density conditions. Moreover, the TSIS- CORSIM model examines a variety of scenarios that recognize the interaction of reduced roadway capacity with traffic density. However, the FMCSA study specifically examined commercial vehicle crashes, which primarily involve heavy trucks. Commercial truck crashes differ from other crashes in a number of ways. These include:

1) Truck crashes are more likely to result in lane closings. This is a function of the size and function of commercial vehicles. An overturned or jackknifed truck trailer can block many more lanes than a normal passenger vehicle. Moreover, if the trucks cargo spills, this can also spread over a larger area, and close more lanes.
2) Truck crashes close off roadways for a longer duration than non-truck crashes. This is a function of both the added lane closings, the need to clean up spilled cargo (and occasionally hazardous waste), and the added difficulty of clearing a larger vehicle and often its trailer from the roadway.
3) The diurnal profile of truck crashes is different from non-truck crashes. Commercial deliveries are less likely to occur during evening and weekend hours. The incidence of truck crashes is thus relatively underrepresented in late night and weekend hours compared to passenger vehicles. Because driving patterns and crash incidence have different frequency profiles during these times, the traffic density during a truck crash will differ, on average, from that in passenger vehicle crashes. This is most noticeable for fatal crashes, where late night weekend crashes involving passenger vehicles make up a much larger portion of fatal crashes involving passenger vehicles than heavy trucks. Since roadway densities are relatively light during these hours, the average density for a fatal truck crash is noticeably higher than that of a passenger vehicle crash.

Because of these factors, the results obtained in the FMCSA report are not representative of the broader universe of crashes that are addressed in this current study.[23] However, as noted above, the FMCSA report has the clear advantage of being able to address detour behavior, the interaction of traffic density with roadway capacity, and the impact of non–linear congestion effects. To adjust for these effects, results derived from the methods described previously for this report were re-computed using data that is specific to commercial truck crashes for crash duration, probability of lane closings, and diurnal crash profile. These results were then compared to the results already computed for all crash types, and the resulting factor was applied to the FMCSA results to derive an estimate of the overall impacts of congestion on all crash types. This approach essentially assumes that the FMCSA estimate for commercial trucks is a more complete estimate since it captures aspects not yet measured in the current effort (detours, capacity, and non-linear congestion impacts). In essence, the FMCSA estimate for commercial trucks is normalized down to reflect the impacts that would be expected for the full universe of crashes, which generally have less serious consequences for congestion.

An additional adjustment was made to reflect the fact that the FMCSA estimates were computed for roads with a specific number of lanes. This was necessary because the simulation required specific road design parameters. However, although the lane profile was selected to represent a typical roadway within each roadway category, the average number of lanes in most of these categories is somewhat different from the specifications used for the FMCSA simulations. AADT for two categories, Urban Other and Rural Other, was also adjusted to reflect the inclusion of several categories of roadways for which average AADT information could not be weighted by VMT. These include urban local, rural local, and rural minor collectors. This data was provided in summary formats which exclude the possibility of weighting by VMT. Therefore, they are not included in the FMCSA estimates. An alternate weighting method based on segment length is used for FHWA's published averages. We obtained estimates for both Urban Other and Rural Other roadway types from FHWA both including and excluding these 3 minor roadways types. From these, we computed a ratio to adjust the VMT weighted AADT values to include these minor roadways.

Table 3-20. Summary of Vehicle Delay Hours during Crash Duration and Queue Dispersal

	Urban Interstates/ Expressways	Urban Arterials	Urban Other	Rural Interstate/ Principal Arterials	Rural Other
Fatal Crashes					
Crash Duration (Hours)	2.79	2.61	2.63	2.86	2.49
Vehicle Hours, Crash Duration	4,497.11	1,209.94	206.46	1,700.19	104.40
Vehicle Hours, Queue Dispersal	650.59	48.32	1.42	80.11	0.43
Total Vehicle Hours/Crash	5,147.70	1,258.26	207.88	1,780.31	104.82
Injury Crashes					
Crash Duration (Hours)	0.77	0.70	0.81	1.13	1.02
Vehicle Hours, Crash Duration	304.76	66.79	15.30	201.43	13.79
Vehicle Hours, Queue Dispersal	40.53	1.77	0.10	6.25	0.08
Total Vehicle Hours/Crash	345.29	68.56	15.40	207.68	13.86
PDO Crashes					
Crash Duration (Hours)	0.68	0.62	0.71	1.04	0.93
Vehicle Hours, Crash Duration	195.13	48.75	10.24	142.68	10.26
Vehicle Hours, Queue Dispersal	19.86	1.19	0.07	3.57	0.07
Total Vehicle Hours/Crash	215.00	49.94	10.32	146.25	10.33

The adjusted all lane AADT was thus used for the all crash calculations, while the FMCSA AADT value was used for the commercial truck calculations. The resulting ratios thus represent both the different characteristics of commercial versus all crashes as well as the correction for all lane and all roadway AADT counts in the two "Other" categories.

Table 3-21 summarizes the normalization process and the final vehicle delay hours estimates for all crashes across the three crash severity and five roadway types. As would be expected, the results indicate lower average delay hours per crash for all crashes compared to the heavy vehicle crashes examined in the FMCSA report. The difference is least pronounced for urban interstates, and most pronounced for minor roadways. Note however, that a significant portion of the difference for the two "other" categories is due to inclusion of local roads in this study, which were not included in the truck study. We note that Lan and Hu (1999) in their study of urban interstate crashes in Minneapolis-St Paul, found crashes that did not involve heavy trucks had roughly 47 percent the delay hours that occurred in heavy-truck crashes.[24] This study finds ratios for urban interstates ranging from 34 percent to 60 percent, depending on the crash severity. The ratios for other roadway categories are smaller, in part due to the previously mentioned inclusion of local roadways, and in part due to the disproportionate impact that heavy-truck crashes can have on roadways with less capacity.

Table 3-21. Vehicle Delay Hours by Crash Severity and Roadway Type, Average for All Crashes

	Urban Interstates/ Expressways	Urban Arterials	Urban Other	Rural Interstate/ Principal Arterials	Rural Other
Fatal Crashes					
Total Vehicle Hours, All Crashes	5,147.70	1,258.26	207.88	1,780.31	104.82
Total Vehicle Hours, Truck Crashes	8,590.02	2,094.75	1,400.59	3,294.58	617.33
Ratio, All/Truck	0.60	0.60	0.15	0.54	0.17
FMCSA Truck Vehicle Hours	6,729.00	483.00	291.00	464.00	99.00
All Crashes Vehicle Hours	4,032.45	290.13	43.19	250.73	16.81
Injury Crashes					
Total Vehicle Hours, All Crashes	345.29	68.56	15.40	207.68	13.86
Total Vehicle Hours, Truck Crashes	1,022.25	145.66	87.83	711.47	108.99
Ratio, All/Truck	0.34	0.47	0.18	0.29	0.13
FMCSA Truck Vehicle Hours	2,522.00	137.00	108.00	159.00	34.00
All Crashes Vehicle Hours	851.85	64.48	18.94	46.41	4.32
PDO Crashes					
Total Vehicle Hours, All Crashes	215.00	49.94	10.32	146.25	10.33
Total Vehicle Hours, Truck Crashes	636.06	139.40	82.49	415.83	81.13
Ratio, All/Truck	0.34	0.36	0.13	0.35	0.13
FMCSA Truck Vehicle Hours	2,144.00	109.00	91.00	134.00	28.00
All Crashes Vehicle Hours	724.71	39.05	11.38	47.13	3.57

Environmental and Resource Impacts

Motor vehicle crashes result in significant time delays to other motorists who are inconvenienced by road blockage due to lane closures, police, fire, or emergency services activity, and general traffic slowdowns resulting from rubbernecking and chain reaction braking. This results in a significant time penalty for those affected, which can be valued based on wage rates and the value people place on their free time. However, it also results in wasted fuel, increased greenhouse gas production, and increased criteria pollutant emissions as engines idle while drivers are caught in traffic jams and slowdowns. These impacts are also created when drivers are forced to detour around a crash. Such detours can be a matter of blocks or miles, but in either case, more fuel is burned by other motorists as a direct result of the initial crash.

Unlike lost time, which is a function of the number of *people* affected by the crash, these resource and environmental impacts are a function of the number of *vehicles* affected in the crash. The reduced capacity impacts previously derived are presented in vehicle hours. Based on the previous analysis, individual crashes can cause an average of from 4 to 4,000 vehicle hours of delay depending on the circumstances and severity of the crash.

Fuel Burned by Vehicles Delayed in Traffic

Traffic jams usually involve a combination of idle time and very slow forward progress through the distance affected by the crash, followed by either a return to normal speeds or possibly to higher than normal speeds while drivers attempt to make up for lost time once they are past the crash site. Much of the time wasted in traffic jams caused by crashes is essentially spent at idle speeds. Technically, idle conditions occur when the motor is running but the car is not moving forward, however, the vast majority of cars in the US are equipped with automatic transmissions and inching forward in a traffic jam often involves nothing more than releasing the brake, which allows low speed acceleration through the automatic transmission drive trains. The portion of time drivers spend idling in traffic crashes, and the portion spent moving forward at low speeds in stop-start conditions is not known. However, an upper bound of the impact of idling can be derived assuming that all of the vehicle hours lost in traffic crashes is spent in an idle condition – either sitting still or inching forward without applying the gas pedal.

There is very little research available on idling fuel consumption. Researchers at Argonne National Laboratory (Gaines, Levinson, and McConnell, 2010) measured the fuel consumption rate for a variety of engines while at idle. Argonne researchers compiled these performance data into engine size groupings to produce plots of fuel consumption as a function of idling time. Fuel consumption is linear with time and increases with engine size. The resulting relationships are shown in Figure 3-G.

The relationships depicted in Figure 3-G indicate that idling consumes 0.165 gallons/hour for I-liter engines, 0.493 gallons/hour for 3-liter engines, and 0.822 gallons/hour in 5-liter engines. The average engine size for passenger vehicles in the US has changed over time as light trucks and SUVs, which carry larger displacement engines, became more popular. At the same time, engine size for specific vehicles tended to decrease as more advanced and powerful small engines allowed for better fuel economy. In estimating the impact of idling on a fleet of vehicles, the relevant metric is exposure rather than production. Newer vehicles are driven more miles than older vehicles, and are thus more likely to be caught up in congestion resulting from traffic crashes. To estimate the exposure adjusted engine size of the on-road passenger vehicle fleet, average engine size for six different passenger car body types and nine different light truck types were obtained from EPA's light duty fuel economy trends database (EPA, 2010) and combined with survival probability and VMT data by vehicle age derived by NHTSA (Lu, 2006). The results indicate an average on-road exposure adjusted displacement of 2.70 liters for passenger cars and 3.94 liters for light trucks, vans, and SUVs, with an average displacement of 3.31 liters for all passenger vehicles. Similar data was not available for heavy trucks and buses, which are much less prevalent than passenger vehicles. From the values in Figure 3-G, an imputed value of 0.542 gallons of fuel wasted for each hour of idling was derived for the average 3.3-liter engine.[25] This added fuel has societal cost implications through both out of pocket expenses and added health risks due to increased criteria pollutant emissions and greenhouse gases.

Resource Cost Impacts – Greenhouse Gases and Criteria Pollutants:

As previously discussed, driver response to crashes is a complex mix of slowing down, idling, accelerating, and often, seeking other alternate routes to detour around traffic congestion. The impact of these interactions on fuel consumption and emissions is difficult to quantify, even for properly observed crash samples. In their report on heavy truck crashes, FMCSA simulated traffic responses using the TSIS-CORSIM model developed by the University of Florida McTrans Center in 2010.26 They then linked these results to EPA's Motor Vehicle Emission Simulator (MOVES) (EPA, 2009) in order to estimate the full range of fuel consumption and emission impacts from traffic congestion resulting from the heavy truck crashes in their model. Truck crashes typically involve more lane closures and take more time to clear than other crashes. They thus reflect both longer crash durations and a higher rate of lane closure than would be estimated in this study. However, these factors can be largely muted by normalizing the results to a common per-vehicle-hour basis. This can be done using data in the FMCSA report. For this study, the fuel consumption and emissions impacts per vehicle hour from the FMCSA study will be applied to the vehicle hour estimates derived for the general crash population.

The environmental impacts estimated both here and in the FMCSA report reflect emissions of both greenhouse gases and criteria pollutants. These emissions are described below. The descriptions are quoted directly from the FMCSA report, but were originally taken from a variety of EPA sources.

Carbon dioxide is a greenhouse gas emitted naturally through the carbon cycle and through human activities like fossil fuel combustion. Since the Industrial Revolution in the 1700s, human activities, such as burning oil, coal, and gas, have increased CO_2 concentrations in the atmosphere. The release of greenhouse gases and aerosols resulting from human activities are changing the amount of radiation coming into and leaving the atmosphere, likely contributing to changes in climate.

Carbon monoxide is a colorless, odorless gas emitted from combustion processes. Nationally, the majority of CO emissions to ambient air come from mobile sources. CO can cause harmful health effects by reducing oxygen delivery to the body's organs (like the heart and brain) and tissues. At high levels, CO can cause death.

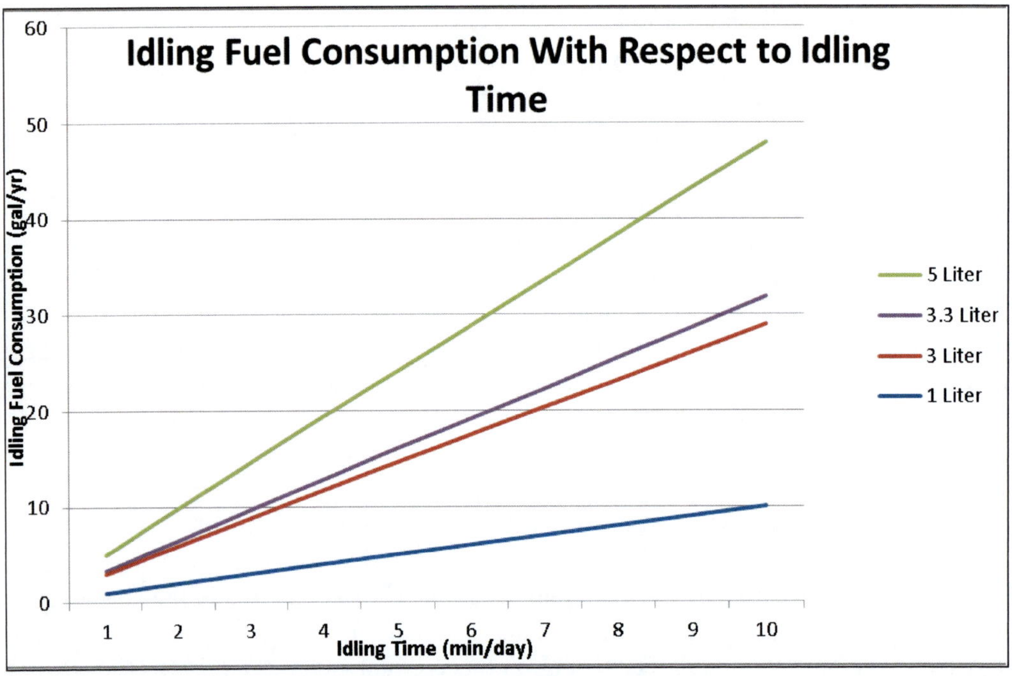

Figure 3-G. Idling Fuel Consumption With Respect to Idling Time.

Hydrocarbon and volatile organic compounds are a group of chemical compounds composed of carbon and hydrogen. When in gaseous form, hydrocarbons (HC) are called volatile organic compounds (VOCs). They are generated via incomplete gasoline combustion or are petrochemical industry by-products. HC/VOCs include methane, gasoline and diesel vapors, benzene, formaldehyde, butadiene, and acetaldehyde. All HC/VOCs are carcinogenic to some extent, fatal at high concentrations, harmful to crops, and bio-accumulate within the food chain. All HC/VOCs contribute to smog, ground level ozone, and acid rain formation.

Nitrous oxides (NO_x) are a group of highly reactive gases that include nitrogen dioxide (NO_2), nitrous acid, and nitric acid. NO_2 forms quickly from emissions from cars, trucks, and buses. In addition to contributing to the formation of ground-level ozone, and fine particle pollution, NO2 is linked to a number of adverse effects to the respiratory system.

Particulate matter (PM) is a complex mixture of extremely small particles and liquid droplets. Particle pollution is made up of a number of components, including acids (such as nitrates and sulfates), organic chemicals, metals, and soil or dust particles. The size of particles is directly linked to their potential to cause health problems; particles that are 10 micrometers in diameter or smaller generally pass through the throat and nose and enter the lungs. Once inhaled, these particles can affect the heart and lungs and cause serious health effects. EPA groups particulate pollution in two categories:

- Particulate matter smaller than 10 micrometers (PM10), and larger than 2.5 micrometers in diameter; a size referred to as "inhalable coarse particles."
- Particulate matter smaller than 2.5 micrometers in diameter (PM2.5), also known as "fine particle" emissions. These particles can be directly emitted from automobiles and react in the air.

Sulfur dioxide is one of a group of highly reactive gases known as "oxides of sulfur." The largest source of SO_2 emissions occur from fossil fuel combustion and the pollutant is linked to a number of adverse effects on the respiratory system.

The FMCSA report examined previous research to assign a societal cost to the health and environmental effects of these emissions. The values used in the FMCSA report are shown in Table 3-22.

Table 3-23 illustrates the net emissions/fatal crash calculated in the FMCSA report stratified by facility type. Both Tables 3-22 and 3-23 were taken directly from the FMCSA report.

Table 3-24 shows the estimated value/fatal crash of the emissions found in the FMCSA report. It represents the product of the values in Table 3-22 and the emissions in Table 23.

Table 3-25 shows the net emissions/vehicle hour in fatal crashes found in the FMCSA report. It was derived by dividing the net emissions/ fatal crash by the total vehicle hours fatal/crash listed in Table 3-35 of the FMCSA report. Those hours were 6,729 for Urban Interstate/Expressways, 483 for Urban Arterials, 291 for Urban Other, 464 for Rural Interstate/Principal Arterials, 99 for Rural Other, and 1,626 for all facility types.

Table 3-22. Emissions Costs Used in FMCSA Report

Emission	Cost per Short Ton (2010 Dollars)	Source
CO_2	$21	Interagency Working Group on Social Cost of Carbon
CO	$145	McCubbin and Delucchi (1999)
NO_x	$12,000	Fann et al. (2009)
PM10	$46,094	McCubbin and Delucchi (1999)
PM2.5	$270,000	Pope et al. (2003)
SO_2	$67,000	Fann et al. (2009)
VOC	$2,800	Fann et al. (2009)

Table 3-23. Estimated Net Tailpipe Emissions/Crash by Facility Type, FMCSA Fatal Commercial Vehicle Crashes (Short Tons)

Facility Type	CO_2	CO	NO_x	PM10	PM2.5	SO_2	Total HC	VOC
Urban Interstate/ Expressway	32.01443	0.22563	0.05816	0.00492	0.00475	0.00058	0.02456	0.02421
Urban Arterial	8.24277	0.05208	0.01474	0.00069	0.00065	0.00014	0.00355	0.00344
Urban Other	2.55629	0.01568	0.0047	0.00018	0.00017	0.00005	0.0009	0.00088
Rural Interstate/ Principal Arterials	5.34325	0.03512	0.02044	0.00112	0.00108	0.00008	0.00242	0.00238
Rural Other	2.11074	0.0128	0.00678	0.00035	0.00033	0.00003	0.00083	0.00081
Average for All Facility Types	10.20434	0.06922	0.02102	0.00145	0.0014	0.00018	0.00652	0.00641

Table 3-24. Estimated Value of Net Tailpipe Emissions/Crash, FMCSA Fatal Commercial Vehicle Crashes (2010 Dollars)

Facility Type	CO_2	CO	NO_x	PM10	PM2.5	SO_2	Total HC	VOC	All
Urban Interstate/ Expressway	$672	$33	$698	$227	$1,283	$39	$0	$68	$3,019
Urban Arterial	$173	$8	$177	$32	$176	$9	$0	$10	$584
Urban Other	$54	$2	$56	$8	$46	$3	$0	$2	$172
Rural Interstate/Principal Arterials	$112	$5	$245	$52	$292	$5	$0	$7	$718
Rural Other	$44	$2	$81	$16	$89	$2	$0	$2	$237
Average for All Facility Types	$214	$10	$252	$67	$378	$12	$0	$18	$951

Table 3-25. Estimated Net Tailpipe Emissions/Vehicle Hour by Facility Type, All Fatal Crashes (Short Tons)

Facility Type	CO_2	CO	NO_x	PM10	PM2.5	SO_2	Total HC	VOC
Urban Interstate/ Expressway	4.76E-03	3.35E-05	8.64E-06	7.31E-07	7.06E-07	8.62E-08	3.65E-06	3.60E-06
Urban Arterial	1.71E-02	1.08E-04	3.05E-05	1.43E-06	1.35E-06	2.90E-07	7.35E-06	7.12E-06
Urban Other	8.78E-03	5.39E-05	1.62E-05	6.19E-07	5.84E-07	1.72E-07	3.09E-06	3.02E-06
Rural Interstate/ Principal Arterials	1.15E-02	7.57E-05	4.41E-05	2.41E-06	2.33E-06	1.72E-07	5.22E-06	5.13E-06
Rural Other	2.13E-02	1.29E-04	6.85E-05	3.54E-06	3.33E-06	3.03E-07	8.38E-06	8.18E-06
Average for All Facility Types	6.27E-03	4.25E-05	1.29E-05	8.91E-07	8.60E-07	1.11E-07	4.01E-06	3.94E-06

Table 3-26 lists the estimated net emissions per crash by facility type for all fatal vehicle crashes. These values were derived by multiplying the net emissions per vehicle hour from Table 3-25 by the estimated "all crashes vehicle hours" previously derived in Table 3-21.

Table 3-26. Estimated Net Tailpipe Emissions/Crash by Facility Type, All Fatal Crashes (Short Tons)

Facility Type	CO_2	CO	NO_x	PM10	PM2.5	SO_2	Total HC	VOC
Urban Interstate/ Expressway	19.185129	0.135212	0.034853	0.002948	0.002847	0.000348	0.014718	0.014508
Urban Arterial	4.951222	0.031283	0.008854	0.000414	0.000390	0.000084	0.002132	0.002066
Urban Other	0.379413	0.002327	0.000698	0.000027	0.000025	0.000007	0.000134	0.000131
Rural Interstate/ Principal Arterials	2.887361	0.018978	0.011045	0.000605	0.000584	0.000043	0.001308	0.001286
Rural Other	0.358411	0.002173	0.001151	0.000059	0.000056	0.000005	0.000141	0.000138
Average for All Road Types	3.695575	0.024857	0.007931	0.000522	0.000501	0.000064	0.002248	0.002207

The emissions derived above were based on the emissions resulting from linkage of the TSIS-CORSIM model results to EPA's MOVES model. MOVES models emissions from mobile sources including exhaust emissions (tailpipe), evaporative emissions (both running and parked, including leaks and diurnal emissions), and refueling emissions (vapor displacement and spillage). However, in order to produce and distribute the added fuel burned due to congestion from traffic crashes, additional "upstream" emissions are produced. Upstream emissions from fuel extraction, production, and distribution are not currently modeled by MOVES. These emissions must thus be estimated separately. To estimate these emissions, we adopt upstream emissions/gallon values used by NHTSA and EPA in their analysis of 2017-2025 Corporate Average Fuel Economy Standards (CAFE). In that study, the agencies analyzed the upstream emissions associated with production of each gallon of fuel. These values were derived separately for gasoline and diesel fuel for each pollutant, and then combined based on the 2010 relative highway consumption of these fuels.[27] In the CAFE analysis, values for all pollutants except CO_2 were measured in short tons. CO_2 was measured in metric tons. However, for this analysis, CO_2 was converted to short tons for consistency with the approach taken in the FMCSA report. Table 3-27 summarizes the upstream emissions per gallon of gasoline and diesel, as well as for the combined fuels. Note that there are no values for PM10 or Total HC. Total HC was not estimated in the CAFE rulemaking nor valued in the FMCSA study. Most damage from particulate matter is caused by the finer particles in PM2.5, and damage caused by the larger particles is uncertain. For consistency with the recent CAFE studies, this study does not value PM10.

Table 3-27. Emissions per Gallon of Fuel (Short tons)

Pollutant	Gasoline	Diesel	Combined
CO_2	0.002274964	0.002290422	0.002278293
CO	0.000000541	0.000000556	0.000000544
NO_x	0.000001772	0.000001849	0.000001789
PM2.5	0.000000199	0.000000193	0.000000198
SO_2	0.000001014	0.000001025	0.000001016
VOC	0.000002758	0.000000425	0.000002255

Table 3-28. Estimated Upstream Emissions/Crash by Facility Type, Short Tons, All Fatal Crashes

Facility Type	CO_2	CO	NO_x	PM10	PM2.5	SO_2	Total HC	VOC
Urban Interstate/ Expressway	4.445342	0.001061	0.003490	0.000000	0.000386	0.001982	0.000000	0.004401
Urban Arterial	1.147236	0.000274	0.000901	0.000000	0.000100	0.000512	0.000000	0.001136
Urban Other	0.087913	0.000021	0.000069	0.000000	0.000008	0.000039	0.000000	0.000087
Rural Interstate/ Principal Arterials	0.669024	0.000160	0.000525	0.000000	0.000058	0.000298	0.000000	0.000662
Rural Other	0.083047	0.000020	0.000065	0.000000	0.000007	0.000037	0.000000	0.000082
Average for All Road Types	0.856293	0.000204	0.000672	0.000000	0.000074	0.000382	0.000000	0.000848

To estimate upstream emissions, the combined upstream emissions/gallon from Table 3-27 were applied to the net added gallons of fuel used due to congestion. Added fuel usage is discussed in a following section of this study (see Table 33). The results for fatal crashes are shown in Table 3-28.

Table 3-22 above listed the emissions values originally used in the FMCSA study. For all criteria pollutants in this analysis, we have adopted values consistent with the values used by NHTSA and EPA in their most recent analysis of CAFE for model years 2017-2025 (NCSA, 2012). However, those values reflect projected real growth in the value of reducing pollutants over the course of that rulemaking, which covers production from 2017 forward. To derive estimates of these values for 2010, the base year of this report, we interpolated back based on the rate of growth for each pollutant implied by base EPA estimates for 2015, 2020, and 2030. The implied growth rate over these years was roughly 1.6 to 1.9 percent, depending on the pollutant. These values are listed in Table 3-29, together with those applied in the FMCSA report and the published values from the 2017-2025 CAFE rulemaking that reflect conditions in roughly 2030. The source of the original criteria pollutant 2015, 2020, and 2030 values used for these imputations was Pope, Arden, Burnett, and Thun (2002). All values assume a 3-percent discount rate.

The original source of the CO_2 value used in the MY 2017-2025 CAFE analysis was the Interagency Working Group on Social Cost of Carbon (IWG, 2010). Discussion of the 2010 CO_2 value is included in the previously cited MY2017-2025 CAFE FRIA. However, in May 2013 the IWG published revised guidance on the social cost of carbon based on improved models of the impacts of climate change (IWG, 2013). These revised models now address explicit assessments of damages from sea level changes as well as other updates and improvements. This study adopts the revised SCC value from that paper for 2010 ($33 in 2007 dollars), adjusted to 2010 economics.

Note that we do not apply values for PM10 or CO. NHTSA and EPA did not include values for these two pollutants in their 2017-2025 CAFE analyses. PM10 was not included because virtually all the adverse health impacts from PM arise from fine particulates, defined as the fraction that is less than 2.5 microns in diameter (hence the notation PM2.5). These account for most of the particulate matter when measured by number of particles, although a much smaller fraction of it when measured by total mass of particulate emissions. So while particulate matter between 2.5 and 10 microns accounts for most of the mass, it does little of the health damage. Likewise, CO was not valued by NHTSA and EPA because at current exposure levels, there is no evidence that CO causes any adverse health impacts.

Table 3-29. Tailpipe Emissions Costs (2010 Dollars)

Emission Type	FMCSA Report	CAFE 2017-2025$_{x2}$ 2030 values	This Study 2010 Values[28]
CO_2	$21	$34	$35
CO	$145	$0	$0
NO_x	$12,000	$6,700	$4,646
PM10	$46,094	$0	$0
PM2.5	$270,000	$306,500	$254,015
SO_2	$67,000	$39,600	$27,300
VOC	$2,800	$1,700	$1,122

Table 3-29 summarizes the values used for tailpipe emissions in this study, as well as those used in the FMCSA and 2017-2025 CAFE studies. These same values were applied to upstream emissions except for NOx and PM2.5. EPA has determined that the different populations exposed to upstream emissions face slightly different health impacts than those exposed through tailpipe emissions. Upstream emissions for NOx are valued at $4,481 and those for PM2.5 are valued at $211,602.

Table 3-30. Estimated Value of Net Emissions/Crash by Facility Type, All Fatal Crashes (2010 Dollars)

Facility Type	CO_2	CO	NO_x	PM10	PM2.5	SO_2	VOC
Urban Interstate/ Expressway	$743.98	$0.00	$177.56	$0.00	$804.75	$63.61	$21.22
Urban Arterial	$192.00	$0.00	$45.17	$0.00	$120.26	$16.26	$3.59
Urban Other	$14.71	$0.00	$3.55	$0.00	$8.02	$1.27	$0.24
Rural Interstate/ Principal Arterials	$111.97	$0.00	$53.67	$0.00	$160.54	$9.33	$2.19
Rural Other	$13.90	$0.00	$5.64	$0.00	$15.76	$1.15	$0.25
Average for All Road Types	$143.31	$0.00	$39.86	$0.00	$143.04	$12.16	$3.43

Table 3-31. Estimated Value of Net Emissions/Crash by Facility Type, All Injury Crashes (2010 Dollars)

Facility Type	CO_2	CO	NO_x	PM10	PM2.5	SO_2	VOC
Urban Interstate/ Expressway	$157.18	$0.00	$37.51	$0.00	$169.98	$13.46	$4.48
Urban Arterial	$42.58	$0.00	$10.01	$0.00	$26.20	$3.61	$0.79
Urban Other	$6.45	$0.00	$1.55	$0.00	$3.38	$0.56	$0.11
Rural Interstate/Principal Arterials	$20.65	$0.00	$9.90	$0.00	$29.70	$1.74	$0.40
Rural Other	$3.53	$0.00	$1.43	$0.00	$3.94	$0.29	$0.06
Average for All Road Types	$30.81	$0.00	$8.45	$0.00	$30.06	$2.62	$0.73

Table 3-32. Estimated Value of Net Emissions/Crash by Facility Type, All PDO Crashes (2010 Dollars)

Facility Type	CO_2	CO	NO_x	PM10	PM2.5	SO_2	VOC
Urban Interstate/Expressway	$133.68	$0.00	$31.91	$0.00	$144.33	$11.39	$3.81
Urban Arterial	$25.80	$0.00	$6.07	$0.00	$16.48	$2.17	$0.48
Urban Other	$3.89	$0.00	$0.94	$0.00	$2.02	$0.32	$0.07
Rural Interstate/ Principal Arterials	$21.01	$0.00	$10.07	$0.00	$30.00	$1.72	$0.41
Rural Other	$2.95	$0.00	$1.20	$0.00	$3.23	$0.25	$0.05
Average for All Road Types	$24.41	$0.00	$6.92	$0.00	$25.03	$2.06	$0.59

Table 3-30 summarizes the total cost/fatal crash for the various emissions categories. These values represent the product of the values in Table 3-29 and the corresponding emissions categories in Table 3-26. These same values are listed for Injury Crashes and PDO crashes, respectively, in Tables 3-31 and 3-32.

Excess Fuel Consumption

Fuel consumption and CO_2 production are directly related. EPA's Clean Energy Web site (www.epa.gov/cleanenergy/) states that "To obtain the number of grams of CO_2 emitted per gallon of gasoline combusted, the heat content of the fuel per gallon is multiplied by the kg CO_2 per heat content of the fuel. The average heat content per gallon of gasoline is 0.125 mmbtu/gallon and the average emissions per heat content of gasoline is 71.35 kg CO_2/mmbtu."

This produces the following relationship

0.125 mmbtu/gallon * 71.35 kg CO_2/mmbtu * 1 metric ton/1,000 kg = 8.92*10-3 metric tons CO_2/gallon of gasoline.

So, there are .00892 metric tons of CO_2 in every gallon of gasoline.

Since the FMCSA emissions values are expressed in short tons, we must convert this relationship to a short ton basis. 1 short ton = 0.90718474 metric tons. Therefore, there are .00892/.90718474 = .00098326 short tons of CO_2 in every gallon of gasoline. Therefore 101.70233 gallons of gasoline contains 1 short ton of CO_2.[29]

Table 3-33. Net Increase in and Cost of Fuel Consumption, Fatal Crashes

Facility Type	CO_2 (short tons)	Gallons/ fuel	Value (2010 Dollars)
Urban Interstate/Expressway	19.185129	1951	$4,800
Urban Arterial	4.951222	504	$1,239
Urban Other	0.379413	39	$95
Rural Interstate/Principal Arterials	2.887361	294	$722
Rural Other	0.358411	36	$90
Average All Roadway Types	3.695575	376	$925

Table 3-34. Net Increase in and Cost of Fuel Consumption, Injury Crashes

Facility Type	CO_2 (short tons)	Gallons/ fuel	Value (2010 Dollars)
Urban Interstate/Expressway	4.053262	412	$1,014
Urban Arterial	1.098054	112	$275
Urban Other	0.166285	17	$42
Rural Interstate/Principal Arterials	0.532603	54	$133
Rural Other	0.091012	9	$23
Average All Roadway Types	0.794605	81	$199

Table 3-35. Net Increase in and Cost of Fuel Consumption, PDO Crashes

Facility Type	CO_2	Gallons/ fuel	Value (2010 Dollars)
Urban Interstate/Expressway	3.447093	351	$862
Urban Arterial	0.665271	68	$166
Urban Other	0.100257	10	$25
Rural Interstate/Principal Arterials	0.541695	55	$136
Rural Other	0.075970	8	$19
Average All Roadway Types	0.629492	64	$157

The average fuel price in 2010 was roughly $2.46/gallon.[30] In Tables 3-33, 3-34, and 3-35, the CO_2 emissions previously estimated are converted to gallon equivalents and valued using this average cost of $2.46/gallon.

Value of Travel Time

The added time spent by vehicle occupants stuck in or detouring around traffic at a crash site is an opportunity cost that represents a real cost to society. While the ability to travel is a valued asset that improves quality-of-life, consumers generally seek to minimize the time spent travelling because it reduces their opportunities to engage in more lucrative or enjoyable pursuits. Time spent travelling could instead be dedicated to production, which would yield monetary benefits to the travelers, their employers, or both. Alternately, it could be spent in recreation or other activities which the traveler would preferably choose to engage in. Finally, the conditions associated with traffic congestion and delay can cause frustration and tension which in themselves have a negative impact on vehicle occupants.

The USDOT has issued general guidance regarding valuing travel time.[31] This guidance lays out guidelines for valuing travel time under various surface modes, and for both business and personal travel. Generally, business travel is valued using wage rates while personal travel is valued using a variable percentage of wage rates, depending on mode and on whether travel is local or intercity. Based on this guidance and updated wage data from the U.S. Bureau of Labor Statistics, FMCSA derived average values of travel time by roadway type for their commercial vehicle study. These values, which are shown in Table 36 below, were weighted according to the prevalence of vehicle types on the roadway as well as average occupancy and are thus applicable for this study as well.

In Table 3-37 these values are combined with the total vehicle hours/crash from Table 3-21 to derive the travel time cost/crash by crash severity and roadway type. Average costs across all roadway types were derived based on FARS incidence for each roadway type for fatal crashes. These roadway definitions are not available in NHTSA's injury databases. Therefore, for nonfatal crashes, relative incidence by roadway type from the previously described PA database was normalized to the fatal crash weights from FARS to establish relative weights for injury and PDO crashes. These same weights, shown in Table 3-37a, are applied to all subsequent average cost calculations across the 5 roadway types.

Table 3-36. Average Value of Travel per Hour by Road Type (2010 Dollars)

Road Type	Average VOT
Urban Interstate/Expressway	$24.28
Urban Arterial	$23.91
Urban Other	$23.88
Rural Interstate/Principal Arterial	$26.05
Rural Other	$24.78
Total Rural and Urban	$24.34

Table 3-37. Value of Travel Time/Crash, by Crash Severity and Roadway Type, All Crashes

	Urban Interstates/ Expressways	Urban Arterials	Urban Other	Rural Interstate/ Principal Arterials	Rural Other	Average All Roadway Types
VOT/Vehicle Hour	$24.28	$23.91	$23.88	$26.05	$24.78	$24.34
Fatal Crashes						
Vehicle Hours/Crash	4,032.45	290.13	43.19	250.73	16.81	527.01
Total Cost/Crash	$97,908	$6,937	$1,031	$6,532	$417	$12,855
Injury Crashes						
Vehicle Hours/Crash	851.85	64.48	18.94	46.41	4.32	140.45
Total Cost/Crash	$20,683	$1,542	$452	$1,209	$107	$3,409
PDO Crashes						
Vehicle Hours/Crash	724.71	39.05	11.38	47.13	3.57	138.77
Total Cost/Crash	$17,596	$934	$272	$1,228	$88	$3,376

Table 3-37a. Relative Incidence Weights among Roadway Types

	Fatal	Injury	PDO
Urban Interstates/Expressways	0.10	0.13	0.16
Urban Arterials	0.21	0.37	0.30
Urban Other	0.13	0.17	0.18
Rural Interstates/Principal Arterials	0.18	0.11	0.14
Rural Other	0.37	0.23	0.22
Total	1.00	1.00	1.00

Congestion Cost Summary

Table 3-38 summarizes the various costs that are estimated to result from congestion caused by police- reported motor vehicle crashes. Total costs range from $14,121for fatal crashes to $3,673 for PDO crashes. The largest loss results from the opportunity cost of delay for vehicle occupants, but there are also significant impacts due to wasted fuel. Greenhouse gases and criteria pollutants are the least costly impact, but are still important given the large number of crashes that occur annually.

Table 3-38. Summary of Congestion Costs/Crash Due to Time Delay, Excess Fuel Burned, and Pollution Police-Reported Crashes

	Urban Interstates/ Expressways	Urban Arterials	Urban Other	Rural Interstate/ Principal Arterials	Rural Other	Average All Roadway Types
Fatal Crashes						
CO_2	$744	$192	$15	$112	$14	$143
CO	$0	$0	$0	$0	$0	$0
NO_x	$178	$45	$4	$54	$6	$40
PM10	$0	$0	$0	$0	$0	$0
PM2.5	$805	$120	$8	$161	$16	$143
SO_2	$64	$16	$1	$9	$1	$12
VOC	$21	$4	$0	$2	$0	$3
Total Emissions	$1,811	$377	$28	$338	$37	$342
Excess Fuel Burned	$4,800	$1,239	$95	$722	$90	$925
Value of Time	$97,908	$6,937	$1,031	$6,532	$417	$12,855
Total Congestion Costs	$104,519	$8,553	$1,154	$7,592	$543	$14,121
Injury Crashes						
CO_2	$157	$43	$6	$21	$4	$40
CO	$0	$0	$0	$0	$0	$0
NOx	$38	$10	$2	$10	$1	$10
PM10	$0	$0	$0	$0	$0	$0
PM2.5	$170	$26	$3	$30	$4	$36
SO_2	$13	$4	$1	$2	$0	$3
VOC	$4	$1	$0	$0	$0	$1
Total Emissions	$383	$83	$12	$62	$9	$90
Excess Fuel Burned	$1,014	$275	$42	$133	$23	$255
Value of Time	$20,683	$1,542	$452	$1,209	$107	$3,409
Total Congestion Costs	$22,080	$1,900	$506	$1,405	$139	$3,755
PDO Crashes						
CO_2	$134	$26	$4	$21	$3	$34
CO	$0	$0	$0	$0	$0	$0
NO_x	$32	$6	$1	$10	$1	$9
PM10	$0	$0	$0	$0	$0	$0
PM2.5	$144	$16	$2	$30	$3	$34
SO_2	$11	$2	$0	$2	$0	$3
VOC	$4	$0	$0	$0	$0	$1
Total Emissions	$325	$51	$7	$63	$8	$80
Excess Fuel Burned	$862	$166	$25	$136	$19	$217
Value of Time	$17,596	$934	$272	$1,228	$88	$3,376
Total Congestion Costs	$18,783	$1,151	$304	$1,426	$115	$3,673

Congestion costs have been estimated separately for fatal, injury, and PDO crashes. However, this report is primarily stratified according to injury severity for all injury crashes. As discussed previously, within injury crashes there are 5 nonfatal categories. For any given crash, congestion costs are a function of crash circumstances rather than injury severity. This implies an equal distribution of congestion costs among all involved parties, regardless of whether they died, were injured or were uninjured. To distribute costs among crash involved people for fatal crashes, the average cost/crash for fatal crash was divided by the average number of involved people /fatal crash. This data was obtained by examining FARS data for 2009 to 2011. From this data, the KABCO injury profile was obtained and run through an MAIS translator to reveal the average MAIS profile among fatal crashes. By definition, all fatalities occur in fatal crashes, so the average congestion cost per fatality was taken directly from the analysis of FARS crashes. The same approach was also applied to injury crashes. However, nonfatal injuries occur in both fatal and nonfatal injury crashes. The two nonfatal injury profiles were therefore weighted together based on the relative incidence of each injury severity in fatal or injury crashes. Since fatal crashes are relatively rare, the nonfatal injury crash estimate was heavily weighted towards the costs from injury crashes. Table 3-39 lists the weights, injuries per crash, and resulting congestion costs per injury for each injury severity for both fatal and injury crashes.

PDO crashes are expressed on a per damaged vehicle basis. Therefore the unit cost for PDO crashes was divided by the average number of vehicles damaged in PDO crashes. Again, this data was derived from 2009-2011 GES records, which indicated an average of 1.75 vehicles/PDO crash.

The results are summarized in Table 40. The nonfatal injury MAIS levels (MAIS 0 to 5) are the weighted average of these costs from fatal and injury crashes noted in the previous table. Congestion costs for nonfatal injuries decline gradually as injury severity decreases because a larger portion of less severe injuries occur in injury crashes, resulting in more weight being given to the less costly injury crashes. Note that the PDO unit cost is higher than nonfatal injury costs because it is expressed on a per vehicle basis. If it were adjusted for vehicle occupants, it would decline to $911/person.

Table 3-39. Allocation of Congestion Costs Across Involved People in Fatal and Injury Crashes

MAIS	Fatal Crashes			Injury Crashes		
	% of All Injuries	Injuries/ Crash	Cost/ Person	% of All Injuries	Injuries/ Crash	Cost/ Person
0	0.0083	0.6052	$5,720	0.9917	1.4342	$1,380
1	0.0105	0.5874	$5,720	0.9895	1.1015	$1,380
2	0.0162	0.1093	$5,720	0.9838	0.1318	$1,380
3	0.0254	0.0539	$5,720	0.9746	0.0410	$1,380
4	0.0302	0.0136	$5,720	0.9698	0.0086	$1,380
5	0.0343	0.0055	$5,720	0.9657	0.0031	$1,380
Fatal	1.0000	1.0937	$5,720	0.0000	0.0000	$1,380
Crash	0.0177	2.4687	$14,121	0.9823	2.7202	$3,755

Table 3-40. Final Congestion Cost/Severity Unit ($2010), Police-Reported Crashes

Severity	Cost/Injured[32]
MAIS0	$1,416
MAIS1	$1,426
MAIS2	$1,450
MAIS3	$1,490
MAIS4	$1,511
MAIS5	$1,529
Fatal	$5,720
PDO	$2,104

Unreported Crashes

Most crashes that involve either serious injury or significant roadway blockage are reported to police, by either the involved parties or by passing motorists. Police reports are filed in those cases where police respond to the crash and the crash severity passes a certain threshold, usually a specific amount of property damage, which varies by state. However, because they typically do not involve police or emergency vehicle presence, unreported crashes, even of the same nominal severity category, are unlikely to cause the same congestion impacts as police-reported crashes. Unfortunately, we were unable to find any research that directly addresses the issue of congestion caused by unreported crashes. To estimate these impacts, we assume that unreported crashes would have only half the probability of a lane being blocked and would be present on the roadway (crash duration) for only half as long as a police-reported crash. In addition we assume that the proportion of roadway blockage and probability of opposite direction rubbernecking is only half that of police-reported crashes.[33] These assumptions are based on the likelihood that any formal lane closing would require police presence and any significant informal lane closing (due to vehicle obstruction) would draw police attention and thus could become a reported crash. Nonetheless, unreported crashes would likely involve at least some level of temporary lane blockage and would cause rubbernecking until the vehicles are removed or driven away. An example might be a low speed crash in which one vehicle rear-ends another at a stoplight. If the damage is minor, the two drivers may contact their insurance companies, exchange insurance information and then drive away, but during the period they were examining their vehicles for damage and exchanging information the vehicles would have blocked the lane they were in. Alternately, this same crash might draw police attention, but, if the damage is minor, police may not file a formal report, and it would thus be an unreported crash. We note that all fatal and serious injury crashes are reported to the police. Therefore, only the minor injury and PDO congestion estimates are relevant to this estimate.

The impact of these assumptions is noted in Table 3-41. These assumptions imply that on average, unreported injury crashes result in congestion impacts that are roughly 15 percent of the impacts that occur in police-reported injury crashes, and unreported PDO crashes produce congestion impacts that are roughly 20 percent of the impacts that occur in police-reported PDO crashes.

Table 3-41. Summary of Congestion Costs/Crash Due to Time Delay, Excess Fuel Burned, and Pollution Unreported Crashes

	Urban Interstates/ Expressways	Urban Arterials	Urban Other	Rural Interstate/ Principal Arterials	Rural Other	Average All Roadway Types
Fatal Crashes						
CO_2	$61.38	$13.81	$0.84	$8.44	$1.00	$11.14
CO	$0.00	$0.00	$0.00	$0.00	$0.00	$0.00
NOx	$14.65	$3.25	$0.20	$4.04	$0.40	$3.08
PM10	$0.00	$0.00	$0.00	$0.00	$0.00	$0.00
PM2.5	$66.39	$8.65	$0.46	$12.10	$1.13	$11.23
SO_2	$5.25	$1.17	$0.07	$0.70	$0.08	$0.95
VOC	$1.75	$0.26	$0.01	$0.16	$0.02	$0.27
Total Emissions	$149.42	$27.13	$1.59	$25.44	$2.63	$26.67
Excess Fuel Burned	$396.01	$89.07	$5.43	$54.43	$6.42	$71.89
Value of Time	$8,077.71	$498.80	$58.97	$492.11	$29.83	$1,031.44
Total Congestion Costs	$8,623.14	$615.00	$65.98	$571.97	$38.88	$1,130.00
Injury Crashes						
CO_2	$19.76	$6.24	$0.86	$2.98	$0.53	$5.37
CO	$0.00	$0.00	$0.00	$0.00	$0.00	$0.00
NOx	$4.72	$1.47	$0.21	$1.43	$0.22	$1.37
PM10	$0.00	$0.00	$0.00	$0.00	$0.00	$0.00
PM2.5	$21.37	$3.84	$0.45	$4.28	$0.59	$4.78
SO_2	$1.69	$0.53	$0.08	$0.25	$0.04	$0.46
VOC	$0.56	$0.12	$0.01	$0.06	$0.01	$0.12
Total Emissions	$48.10	$12.18	$1.61	$9.00	$1.39	$12.10
Excess Fuel Burned	$127.49	$40.23	$5.54	$19.22	$3.43	$34.65
Value of Time	$2,600.27	$225.75	$60.23	$174.42	$16.15	$443.93
Total Congestion Costs	$2,775.86	$278.16	$67.38	$202.64	$20.97	$490.68
PDO Crashes						
CO_2	$24.36	$4.51	$0.68	$4.28	$0.55	$6.13
CO	$0.00	$0.00	$0.00	$0.00	$0.00	$0.00
NOx	$5.82	$1.06	$0.16	$2.05	$0.22	$1.62
PM10	$0.00	$0.00	$0.00	$0.00	$0.00	$0.00
PM2.5	$26.30	$2.88	$0.35	$6.11	$0.60	$6.17
SO_2	$2.07	$0.38	$0.06	$0.35	$0.05	$0.52
VOC	$0.69	$0.08	$0.01	$0.08	$0.01	$0.15
Total Emissions	$59.25	$8.92	$1.26	$12.88	$1.42	$14.59
Excess Fuel Burned	$157.17	$29.10	$4.39	$27.62	$3.52	$39.54
Value of Time	$3,206.84	$163.25	$47.53	$250.23	$16.36	$616.56
Total Congestion Costs	$3,423.26	$201.27	$53.18	$290.73	$21.30	$670.69

Table 3-42. Final Congestion Cost/Severity Unit ($2010), Unreported Crashes

	Cost/Crash	Injured/Crash	Cost/Injured
MAIS0	$491	2.72	$180
MAIS1	$491	2.72	$180
MAIS2	$491	2.72	$180
MAIS3	$491	2.72	$180
MAIS4	$491	2.72	$180
MAIS5	$491	2.72	$180
Fatal	$1,130	2.47	$458
PDO	$671	1.75	$384

The final MAIS distribution for unreported crashes, which is summarized in Table 3-42, is based on the average person involvement rates from police-reported crashes. As previously discussed, it is possible that unreported crashes have lower person involvement rates than reported crashes, since the presence of more than one driver is likely to increase the chances of the crash being reported. We do not have data on involvement rates for unreported crashes, but it is likely that basing these unit costs on police- reported rates produces a conservative estimate of these costs for unreported crashes. Note that there is no need to average congestion costs from both fatal and nonfatal crashes when allocating nonfatal injury costs because all fatal crashes are reported to police. Although costs are shown for each injury category, virtually all unreported crashes involve either minor injury or property damage only.

Average and Total Congestion Costs, Reported and Unreported Crashes

The average cost/crash across both police-reported and unreported crashes was calculated by weighting each category's costs according to the relative incidence within each severity category. For all injury categories this was based on the incidence of injured people. For PDOs, it is based on the incidence of damaged vehicles. These definitions are consistent with the stratification used throughout this report. Incidence was derived from the incidence chapter of this report. Table 3-43 summarizes this process and its results. Table 3-44 summarizes the total costs of congestion. In 2010, motor vehicle crashes are estimated to have caused $28 billion in travel delay, excess fuel consumption, and health and other economic impacts from added criteria pollutants and greenhouse gases.

Discussion

While this analysis is designed to represent average nationwide experience, it is substantially influenced by data collected from Pennsylvania. The Pennsylvania traffic records database contains information on all police-reported crashes reported to the State. This data was used to estimate a number of critical factors in this analysis, including crash duration and the probability-of-lane closings by roadway type. The use of this database was largely a function of availability. Very few States maintain publically available traffic records databases at this level of detail. Nonetheless, using a single State's experience as a proxy for a nationwide projection raises the question of how representative the crash experience in that State is of average nationwide experience. To consider this issue, we examined data contained in FHWA's publication Highway Statistics 2009.[34] FHWA gathered data for the specific purpose of identifying "peer States" with similar characteristics that might influence various roadway measures. This data is presented in a table PS-1 titled "Selected Measures for Identifying Peer States." These measures include the urban/rural breakdown of net land area, population, annual VMT and lane miles, as well as personal income, gross State product per capita, annual VMT per capita, AADT per lane mile, and the proportion of VMT accounted for by trucks. In table 45 we summarize this data for both Pennsylvania and the country as a whole.

Table 3-43. Average Congestion Costs for All Crashes

	Incidence				Unit Costs (2010 Dollars)		
	Police-Reported	Unreported	Total	%PR	Police-Reported	Unreported	Combined
MAIS0	2,147,857	2,435,409	4,583,265	46.86%	$1,416	$180	$760
MAIS1	2,578,993	880,207	3,459,200	74.55%	$1,426	$180	$1,109
MAIS2	271,160	67,570	338,730	80.05%	$1,450	$180	$1,197
MAIS3	96,397	4,343	100,740	95.69%	$1,490	$180	$1,434
MAIS4	17,086	0	17,086	100.00%	$1,511	$180	$1,511
MAIS5	5,749	0	5,749	100.00%	$1,529	$180	$1,529
Fatal	32999	0	32,999	100.00%	$5,720	$458	$5,720
PDO	7,454,761	11,053,871	18,508,632	40.28%	$2,104	$384	$1,077

Table 3-44. Total Congestion Costs, 2010

	Police-Reported	Unreported	Combined
MAIS0	$3,042,349,414	$439,378,392	$3,481,727,806
MAIS1	$3,677,250,555	$158,800,405	$3,836,050,960
MAIS2	$393,360,852	$12,190,477	$405,551,329
MAIS3	$143,696,818	$783,532	$144,480,350
MAIS4	$25,826,478	$0	$25,826,478
MAIS5	$8,791,521	$0	$8,791,521
Fatal	$188,789,259	$0	$188,789,259
PDO	$15,687,384,220	$4,248,011,358	$19,935,395,578
Total	$23,167,449,117	$4,859,164,164	$28,026,613,281

Table 3-45. Demographic and Roadway Characteristics, Pennsylvania versus United States, 2008[35]

	Percent Urban U.S. Total	Percent Urban Pennsylvania	Value U.S. Total	Value Pennsylvania
Population Annual VMT	79%	77%		
Lane Miles	67%	64%		
Net Land Area	28%	38%		
% Truck VMT Rural	5%	28%		
% Truck VMT Urban			12%	15%
Personal Income			5%	7%
Gross State			$38,615	$38,793
Product/Capita			$46,593	$44,448
Annual VMT/Capita			9,728	8,664
AADT/Lane, Rural			1,456	1,327
AADT/Lane, Urban			6,744	5,097
AADT/Lane, All			2,771	2,499

Overall, Pennsylvania appears to be quite similar to the country as a whole in most of the categories that are examined. The only category where there is a significant difference is in the urban portion of land area, where Pennsylvania has a noticeably larger proportion of land designated as urban then the United States as whole. However, this is arguably the least relevant metric for purposes of roadway crash analysis, and it is largely offset by the close matches in more relevant measures such as population, VMT, truck travel, and AADT. Close matches are also found for the two wealth measures – personal income and gross State product per capita. These measures are less relevant than roadway measures, but are still important in that they influence the types of vehicles and transportation systems that residents choose to invest in. Overall, Pennsylvania's characteristics appear to be a reasonably close match to those of the country as a whole, and there do not appear to be any significant contra-indicators in this data for using statewide Pennsylvania experience for crash duration and lane closings as a proxy for the United States

Nonetheless, we note that this study, like all studies of this nature, is inherently dependent on the accuracy of the inputs and assumptions that are adopted. We have attempted to be as transparent as possible regarding the variation in various data sources and the basis we have used for our methodology. However, we do not claim that the end result is definitive. With 14 million traffic crashes every year, validation of this or any model of congestion impacts would require a massive data gathering effort that would not be practical or affordable. There is a reason why most studies of traffic crashes only focus on one aspect of the crash in one jurisdiction. There is clearly room for speculation regarding confidence intervals around our results, but since this data was not statistically derived, assignment of such intervals would be arbitrary. Further, this study is directly linked to the TSIS-CORSIM model, discussed in the body of this analysis. While we believe that this model is a step forward in estimating traffic dynamics, any faults it contains will spill over into this study as well.

Finally, we note that The Texas Transportation Institute (TTI) makes annual estimates of traffic congestion costs. However, the TTI study examines the cost of congestion from all sources, which is primarily a function of routine rush hour delays and normal traffic congestion. It is thus not directly comparable to this study, which only measures the costs of congestion caused by traffic crashes. We note that the TTI estimate of total congestion costs is $120 billion in 2010, compared to our estimate of $28 billion for crash-caused congestion. It might be tempting to ratio these costs and conclude that traffic crashes are responsible for 23 percent of total congestion costs, but we would discourage this because the two studies use different methodologies as well as different inputs. For example, this study uses values based on the current USDOT guidance on valuing travel time, weighted by vehicle types on the roadway as well as average occupancy, which sets the cost per vehicle hour of delay at $24. The TTI study uses a value of roughly $21/hour of vehicle delay. To draw conclusions regarding the relative impact of traffic crashes to congestion, common methods and inputs would be required. Thus, although the results of the two studies do not necessarily appear inconsistent, we do not draw conclusions based on a comparison of these two studies results.

4. Lost Quality-of-Life

The human capital costs documented in the first chapter represent the tangible losses that result from motor vehicle crashes. They define the value of resources that are used or that would be required to restore crash victims, to the extent possible, to their pre-crash physical and financial status. These are resources have been diverted from other more productive uses to merely maintain the status quo. These costs, which can be estimated in a fairly direct manner through empirical measurements, include medical care, lost productivity, legal and court costs, insurance administrative costs, legal costs, workplace costs, travel delay, and property damage.

However, in cases of serious injury or death, medical care cannot fully restore victims to their pre-crash status and human capital costs fail to capture the relatively intangible value of lost quality-of-life that results from these injuries. In the case of death, victims are deprived of their entire remaining lifespan. In the case of serious injury, the impact on the lives of crash victims can involve extended or even lifelong impairment or physical pain, which can interfere with or prevent even the most basic living functions. These more intangible effects can be valued using studies that examine the willingness of consumers to pay to avoid risk of death or injury. Assessing the value of these impacts provides a more complete basis for quantifying the harmful impacts of motor vehicle crashes on society.

Value of a Statistical Life

The value of a statistical life (VSL) is a measure of the implied value consumers place on their lives as revealed by the price they are willing to pay to avoid risk of death. A wide range of estimates of the value of VSL have been derived from numerous studies conducted over the past three decades. These "willingness to pay" studies (WTP) are most frequently based on wage rate differentials for risky jobs, or on studies of the prices consumers pay for products that reduce their risk of being fatally injured. The individual studies are too numerous to document here, but a number of authors have attempted to evaluate these studies as a group through systematic reviews or meta-analysis, which applies normalizing parameters and statistical weighting techniques to draw conclusions from related studies.

In 1990, Miller conducted a systematic review of 67 of these studies. In this study, Milller selected 47 studies that were the most methodologically sound, adjusted them to a common discount rate, and made adjustments for errors in perceived risk levels. The VSLs found in these 47 studies had both a mean and median value of $2.2 million in 1988 dollars with a standard deviation of $0.65 million. In 2000, Miller published another meta-analysis examining VSL estimates across differing countries. In this study he examined 68 studies, including the original 47 he had examined in 1990. Based on this study, Miller estimated the VSL in the United States to be $3.67 million in 1995 dollars.

Viscusi has also published a number of WTP reviews. In 1993 Viscusi found that most VSL estimates are clustered in the $3 million to $7 million range. In 2003, Viscusi and Aldy published a worldwide review of VSL studies and estimated a median value of $7 million in 2000 dollars. In 2004 Viscusi published his own estimate of WTP based on wage-risk premiums resulting in a $5 million VSL (using 2000 dollars).

Other reviews include those by Mrozek and Taylor (2002), who found VSL estimates ranging from $1.5 million to $2.5 million in 1998 dollars, and a 2003 meta-analysis by Kochi, Hubbell, and Kramer (2006), which produced a mean VSL estimate of $5.4 million in 2000 dollars.

It is apparent that there are a wide range of estimates regarding the implied VSL from WTP studies. This range is reflected in guidance supplied by the Office of Management and Budget in Circular A-4, issued on September 17, 2003, which recommends values between $1 million and $10 million be used by government agencies when evaluating the impacts of proposed regulations that affect fatality risk. In recent years, government agencies such as NHTSA, the FDA, EPA, the Consumer Product Safety Commission, the Department of Agriculture, and the Occupational Safety and Health Administration have used values ranging from $5 to $7 million in evaluating their regulations.

In February 2008, based on a review of the studies cited above, the Office of the Secretary of the Department of Transportation issued guidance setting a VSL of $5.8 million for use in Departmental regulatory programs (T. Duvall, Assistant Secretary for Transportation Policy, & D.J. Gribbon, General Council. *Treatment of the Economic Value of a Statistical Life in Departmental Analysis.* Memorandum to Secretarial Officers and Modal Administrators, Department of Transportation, February 5, 2008). This value was updated for inflation twice, most recently in July 2011 to a value of $6.2 million (P. Trottenberg, Assistant Secretary for Transportation Policy, & R. S. Rivkin, General Council. "Treatment of the Economic Value of a Statistical Life in Departmental Analysis – 2011 Interim Adjustment." Memorandum to Secretarial Officers and Modal Administrators, July 19, 2011). In February 2013, USDOT again updated their VSL guidance to a value of $9.1 million in 2012 dollars (P. Trottenberg, Assistant Secretary for Transportation Policy, & R. S. Rivkin, General Council. "Treatment of the Economic Value of a Statistical Life in Departmental Analysis – 2011 Interim Adjustment." Memorandum to Secretarial Officers and Modal Administrators, July 19, 2011). This latest update was based exclusively on studies that used the Census of Fatal Occupational Injuries, a complete census of occupational fatalities conducted by the Bureau of Labor Statistics. For a variety of reasons outlined in that guidance, USDOT considered studies based on this data to be superior to those that used other sources. This study adopts this current guidance for assessing the monetary value of fatalities caused by motor vehicle crashes. However, we also acknowledge the uncertainty evident in the wide range of results that are found in the literature. Since this study examines 2010 in detail, we obtained the 2010- based VSL, adjusted for both economics and real changes in wages, from OST. This value, $8.86 million, will be used in this report. However, the literature on VSL estimates indicates a wide range of measured estimates of VSLs – some as low as a few million dollars, some as high as over $30 million. The U. S. DOT guidance memorandum (U.S. Department of Transportation (1997), *Departmental Guidance for the Valuation of Travel Time in Economic Analysis*. Memorandum from the Office of the Secretary of Transportation, U.S. Department of Transportation. Available at http://ostpxweb.dot.gov/policy/- Data/VOT97guide.pdf) discusses a feasible range of VSLs for sensitivity analysis from $5.2 million to $12.9 million. Appendix A provides a sensitivity analysis consistent with this range.

Lost Quality-of-Life for Nonfatal Injuries

While WTP studies can be used to value loss of life, nonfatal injuries, which are a far more prevalent occurrence in motor vehicle crashes, require a more complex examination to reflect the diversity of possible outcomes. When a life is lost prematurely in a motor vehicle crash, the victim loses all of his remaining life, and this can be quantified in terms of life years by comparing the victim's age at death to expected remaining lifespan. However, when the victim is injured but survives, the loss to the victim is a direct function of the extent to which the victim is disabled or made to suffer through physical pain or emotional distress, as well as the duration through which these impacts occur.

The metric commonly used to value these nonfatal injury losses is the quality-adjusted life year (QALY). A QALY is a health outcome measure that assigns a value of 1 to a year of perfect health and a value of 0 to death (Gold, Stegel, Russel & Weinstein, 1996). QALY loss is determined by the duration and severity of the health problem, with a full year of QALY loss being equivalent to the loss of a full year of life in perfect health. QALYs are used in evaluating the outcomes of clinical trials of medical interventions, in approval of pharmaceuticals, and in studies of the return on investment in preventive health and safety measures (Miller, 2000). NHTSA routinely uses QALY based valuations to determine the relative value of nonfatal injuries when measuring the cost effectiveness of regulatory alternatives. The QALY valuations used by NHTSA were originally derived from work by Miller, Pindus, Douglass and Rossman) (1995). These values were adopted for the previous report on the cost of crashes issued by NHTSA (Blincoe et al.) in 2002, and incorporated in subsequent regulatory evaluations conducted by NHTSA. Miller, Pindus, Douglass and Rossman based their QALY valuations on the Injury Impairment Index (III). The III is based on physician estimates of impairment across six functional dimensions (cognitive, mobility, bending/grasping/lifting, sensory, pain, and cosmetic), originally developed for physician use by Hirsh et al. (1983), but subsequently enhanced to include permanent and partial work related disability by Miller, Pindus, Douglass and Rossman (1995).

It has been over 15 years since the 1995 study by Miller and his colleagues and NHTSA was concerned that subsequent advances in medical technology could have shifted the relative values of QALYs associated with motor vehicle injuries. In preparation for this current study, NHTSA contracted with the Pacific Institute for Research and Evaluation to re-examine the III injury preference weights based on the most recent literature and reflect any changes that may have occurred due to shifts in the injury case mix. The resulting study by Spicer and Miller (2010) provides the basis for the nonfatal injury QALY values used in this report. The report found slightly different QALY values for all injury levels, reflecting both the revised preference weights and the larger and more recent database examined in the new report. The results of this effort are summarized by MAIS injury severity level for a variety of discount rate assumptions in Table 5-1.

QALY values for the most serious injuries (MAIS5) are thus roughly 60 percent of a full remaining life, while minor injuries (MAIS1) are valued at less than 1 percent of a full remaining life. QALYs rise progressively with the severity of the injury. This reflects both the severity and longevity of injury consequences at each severity level. For example, serious brain injury, spinal cord injury, and other injuries likely to involve long term debilitation are typically classified in the higher MAIS categories, while less debilitating injuries with shorter recovery times tend to be classified in the lower MAIS categories. Note that the impact of discount rates on QALY values is relatively limited. Shifts in discount rates affect both the MAIS levels and the full life values, which minimizes the relative impact on QALYs.

Although the impact of discount rates on QALYs is minor, a single value must still be adopted for this analysis. Ideally, QALY values would reflect the discount rate implicit in consumer valuations used to measure the VSL, which these QALYs will be applied to. Estimates of this rate vary as widely as estimates for the VSL. Aldy and Viscusi (2007) cite a range of implicit discount rates of from 1 to 17 percent across five different studies that examined VSLs or VSLYs.[36] Hartwick (2008) derived implicit discount rates of between 3 percent and 4 percent for people who die from ages 30 to 40 with a VSL of $6.3 million. Based on 2007-2009 FARS data, the median age of a person killed in a motor vehicle crash is 38. On this basis, either a 3-percent or a 4-percent discount rate appears to be appropriate, and the difference in QALYs between these two rates is extremely small. The USDOT has adopted QALYs based on a 4-percent discount rate as an intermediate value between the 3-percent and 7-percent rates recommended by OMB (Trottenberg & Rivkin, 2013). Since this report is based on a 3-percent discount rate, we use the 3 percent values to retain consistency with the rest of the report. A separate table consistent with the OMB recommendation (i.e., with nonfatal injury QALYs based on a 4-percent discount rate), can be derived using Table 1 above.

Table 4-1. QALY Values for Injured Survivors by Discount Rate and MAIS

Discount Rate	***0%***	***3%***	***4%***	***7%***	***10%***
Injury Severity					
MAIS1	0.3%	0.3%	0.3%	0.4%	0.4%
MAIS2	3.5%	4.4%	4.7%	5.6%	6.5%
MAIS3	10.1%	10.4%	10.5%	10.8%	11.1%
MAIS4	25.5%	26.3%	26.6%	27.5%	28.4%
MAIS5	58.3%	59.1%	59.3%	60.1%	60.9%

Comprehensive Costs

The VSL and QALY measures discussed in the previous section represent an average valuation of the lost quality-of-life that would be lost to crash victims. However, it does not include the economic costs that result from an unexpected event such as death or injury resulting from a motor vehicle crash. Those costs, which include medical care, legal costs, emergency services, insurance administrative costs, workplace costs, congestion impacts, and property damage, were previously estimated in Chapters 2 and 3 of this study. The full societal impact of crashes includes both the intangible impacts represented by VSL and QALY estimates, and the economic impacts that result directly from the crash. Combining these impacts – the direct economic costs that result from the crash and the value of lost quality-of-life experienced by injured crash victims, results in a measure of the comprehensive cost to society from death or injury.

The economic cost estimates developed previously include lost market and household productivity. WTP based valuations of life, which encompass the entire expected life experience of consumers, theoretically encompass after-tax wages (the portion of wages actually received by the employee) and household productivity.[37] Since these measures are hypothetically already included under WTP valuations, combining measures of economic costs and lost quality-of-life requires an adjustment to avoid double counting these components. In Table 4-2 below, the components that make up comprehensive costs are listed in the left column. These consist of the various economic cost components with an additional line for QALYs. Because lost after- tax market and household productivity are separate line items that are implicitly included in QALYs, the QALY values are reduced by these values so that the separate components can be added to produce the total comprehensive cost for each injury severity level.

Comprehensive costs have been used by NHTSA and other agencies to evaluate regulatory programs for over a decade. They provide a convenient basis for measuring the full societal benefits of regulations against their costs, and they are the appropriate basis for valuing benefits in a cost-benefit context where societal impacts are the overriding concern. However, in some circumstances, users may wish to measure only the tangible economic value of goods and services lost and out of pocket expenses incurred that result of motor vehicle crashes. Economic impacts are commonly considered by policymakers and public interest groups when safety issues are being debated. These more tangible economic costs are both more easily understood and more reliably measured than lost quality-of-life, which, as noted previously, is subject to a wide range of estimates and the uncertainty implicit in this range. This report provides estimates of impacts under both bases to facilitate either approach.

Table 4-2 summarizes the total unit cost of crashes stratified by injury level and cost category, and Figure A illustrates the relative contribution of economic costs and quality-of-life to the total comprehensive cost for each injury severity level. The total Comprehensive cost for a fatality is $9.1 million, with roughly 97 percent of this due to components that influence the VSL (QALY and lost productivity), but roughly 85 percent representing lost quality-of-life in excess of these factors. The portion of total comprehensive costs represented by economic costs decreases as the severity of the injury increases. Economic costs represent 15 percent of fatal comprehensive costs, 19-25 percent of the more serious nonfatal injury costs, 46 percent of minor injury costs, and 100 percent of MAIS0 and PDO costs. This reflects the relatively small values for lost quality-of-life found for less severe injuries.

Table 4-2. Comprehensive Unit Costs, Reported and Unreported Crashes (2010 Dollars)

	PDO	MAIS0	MAIS1	MAIS2	MAIS3	MAIS4	MAIS5	Fatal
Medical	$0	$0	$3,801	$24,481	$70,902	$196,911	$441,618	$11,317
EMS	$28	$21	$89	$194	$416	$838	$855	$902
Market Prod.	$0	$0	$2,858	$39,637	$110,088	$174,728	$352,178	$933,262
Household Prod.	$60	$45	$941	$11,921	$33,505	$44,593	$107,070	$289,910
Insurance Adm.	$191	$143	$3,935	$9,370	$24,348	$37,372	$79,967	$28,322
Workplace	$62	$46	$341	$2,644	$5,776	$6,361	$11,091	$11,783
Legal Costs	$0	$0	$1,410	$6,739	$19,645	$35,307	$91,197	$106,488
Subtotal	$341	$255	$13,375	$94,986	$264,680	$496,110	$1,083,976	$1,381,984
Congestion	$1,077	$760	$1,109	$1,197	$1,434	$1,511	$1,529	$5,720
Prop. Damage	$2,444	$1,828	$5,404	$5,778	$10,882	$16,328	$15,092	$11,212
Subtotal	$3,521	$2,588	$6,513	$6,975	$12,316	$17,839	$16,621	$16,932
Total Economic	$3,862	$2,843	$19,888	$101,961	$276,996	$513,949	$1,100,597	$1,398,916
QALYs	$0	$0	$23,241	$340,872	$805,697	$2,037,483	$4,578,525	$7,747,082
Comp. Total	$3,862	$2,843	$43,129	$442,833	$1,082,693	$2,551,432	$5,679,122	$9,145,998

Note: Unit costs are on a per-person basis for all injury levels. PDO costs are on a per-damaged-vehicle basis.

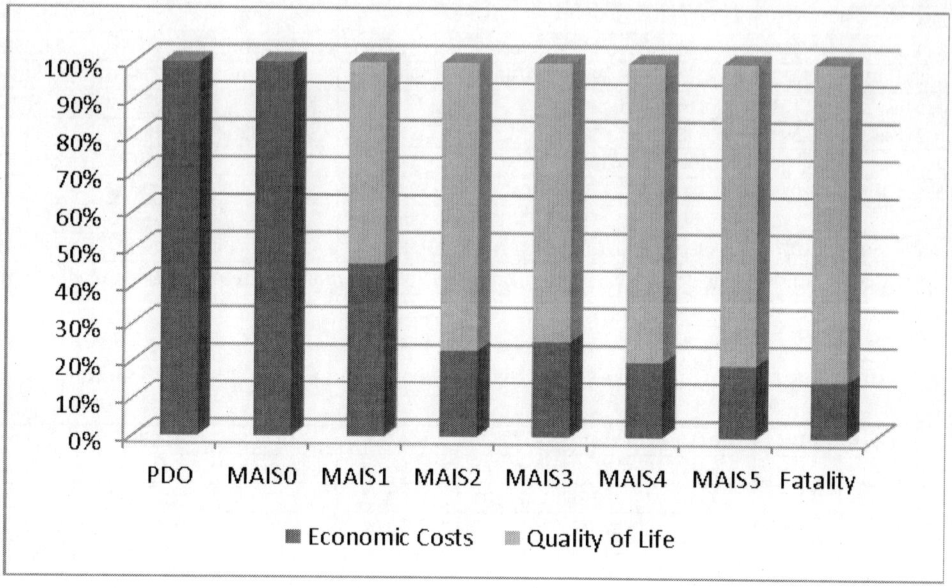

Figure 4-A. Relative Distribution of Comprehensive Costs.

The "Subtotal Injury" line represents components associated with injuries. Costs on this line are thus useful in analyzing the economic benefits of safety countermeasures that prevent injury in the event of a crash. The "Economic Total" line is useful for estimating the economic benefits from countermeasures that prevent crashes from occurring. To examine the total societal harm prevented by either countermeasure type, the value on the QALY line should be added to the appropriate economic values.

Over half of all PDO crashes and about a quarter of all non-fatal injury crashes are not reported to police. However, analyses of safety countermeasures frequently rely only on police-reported incidence data. Crashes that get reported to police are likely to be more severe than unreported crashes because vehicles are more likely to require towing and occupants are more likely to require hospitalization or emergency services. These crashes are typically also likely to require more time to investigate and clear from roadways than unreported crashes. Analysis based solely on police-reported crashes should thus be based on unit costs that are specific to police-reported crashes. For injury related costs, this is more or less automatically accounted for by the shift in the injury severity profile. Unreported crashes have a lower average severity profile than do reported crashes. However, for non-injury related cost components – property damage and congestion costs – there is no profile to shift. In addition, emergency services have higher involvement rates for police-reported crashes.

Table 4-3. Comprehensive Unit Costs, Police-Reported Crashes (2010 Dollars)

	PDO	MAIS0	MAIS1	MAIS2	MAIS3	MAIS4	MAIS5	Fatal
Medical	$0	$0	$3,801	$24,481	$70,902	$196,911	$441,618	$11,317
EMS	$59	$38	$109	$221	$416	$838	$855	$902
Market	$0	$0	$2,858	$39,637	$110,088	$174,728	$352,178	$933,262
Household	$60	$45	$941	$11,921	$33,505	$44,593	$107,070	$289,910
Insurance	$191	$143	$3,935	$9,370	$24,348	$37,372	$79,967	$28,322
Workplace	$62	$46	$341	$2,644	$5,776	$6,361	$11,091	$11,783
Legal Costs	$0	$0	$1,410	$6,739	$19,645	$35,307	$91,197	$106,488
Subtotal	$372	$272	$13,395	$95,013	$264,680	$496,110	$1,083,976	$1,381,984
Congestion	$2,104	$1,416	$1,426	$1,450	$1,490	$1,511	$1,529	$5,720
Prop. Damage	$3,599	$2,692	$7,959	$8,510	$16,027	$16,328	$15,092	$11,212
Subtotal	$5,704	$4,108	$9,385	$9,960	$17,517	$17,839	$16,621	$16,932
Total	$6,076	$4,380	$22,779	$104,974	$282,197	$513,949	$1,100,597	$1,398,916
QALYs	$0	$0	$23,241	$340,872	$805,697	$2,037,483	$4,578,525	$7,747,082
Total	$6,076	$4,380	$46,020	$445,846	$1,087,894	$2,551,432	$5,679,122	$9,145,998

Note: Unit costs are on a per-person basis for all injury levels. PDP costs are on a per-damaged-vehicle basis.

Table 4-4. Comprehensive Unit Costs, Unreported Crashes (2010 Dollars)

	PDO	MAIS0	MAIS1	MAIS2	MAIS3	MAIS4	MAIS5	Fatal
Medical	$0	$0	$3,801	$24,481	$70,902	$196,911	$441,618	$11,317
EMS	$7	$6	$32	$84	$416	$838	$855	$902
Market	$0	$0	$2,858	$39,637	$110,088	$174,728	$352,178	$933,262
Household	$60	$45	$941	$11,921	$33,505	$44,593	$107,070	$289,910
Insurance	$191	$143	$3,935	$9,370	$24,348	$37,372	$79,967	$28,322
Workplace	$62	$46	$341	$2,644	$5,776	$6,361	$11,091	$11,783
Legal Costs	$0	$0	$1,410	$6,739	$19,645	$35,307	$91,197	$106,488
Subtotal	$320	$240	$13,318	$94,876	$264,680	$496,110	$1,083,976	$1,381,984
Congestion	$384	$180	$180	$180	$180	$180	$180	$458
Prop. Damage	$1,224	$916	$2,707	$2,894	$5,451	$16,328	$15,092	$11,212
Subtotal	$1,609	$1,096	$2,888	$3,075	$5,632	$16,508	$15,272	$11,670
Total	$1,928	$1,337	$16,205	$97,951	$270,312	$512,618	$1,099,248	$1,393,654
QALYs	$0	$0	$23,241	$340,872	$805,697	$2,037,483	$4,578,525	$7,747,082
Comp. Total	$1,928	$1,337	$39,446	$438,823	$1,076,009	$2,550,101	$5,677,773	$9,140,736

Note: Unit costs are on a per-person basis for all injury levels. PDP costs are on a per-damaged-vehicle basis.

For this report, costs specific to police-reported and unreported crashes have been developed. The changes in unit costs are all due to economic cost factors and these are discussed in detail in the Human Capital chapter. The results of this analysis on comprehensive costs are presented in Tables 4-3 and 4-4. The differences seem negligible at the more severe injury levels due to the overwhelming costs of factors such as lost productivity and medical care which do not vary by reporting status, except through the shift in injury profiles. However, at lower severity levels the unit costs are significant. For PDO vehicles and MAIS 0s, police-reported crashes have costs that are three times those of unreported crashes. For minor (MAIS1) injuries, reported crashes cost 16 percent more than unreported crashes. These ratios decline as injury severity increases. Note that for MAIS 4s, MAIS5s, and Fatalities, property damage costs are identical under both reported and unreported cases. All injuries at these levels are believed to be reported to police, and the original property damage cost estimate is thus assumed to represent police-reported cases. These same costs are thus listed under both scenarios.

5. Incidence

Crash costs are driven by the incidence of fatalities, injuries, and damaged vehicles that result from motor vehicle crashes. Most serious crashes are reported in police records within individual States and jurisdictions, but many less serious crashes are either not reported to police, or are reported but not recorded because their severity falls below a local reporting threshold. In this section we estimate the incidence of both the police-reported and unreported crashes that occur annually on our roadways.

Fatalities

The incidence of fatalities that result from motor vehicle crashes is derived from the Fatality Analysis Reporting System (FARS). FARS is an annual census of all fatal roadway crashes. FARS collects data on all fatal traffic crashes within the 50 States, the District of Columbia, and Puerto Rico. To be included in FARS, a crash must involve a motor vehicle travelling on a roadway customarily open to the public and result in the death of a person (occupant of a vehicle or a nonoccupant) within 30 days of the crash. FARS collects information on over 100 different coded data elements that characterize the crash, the vehicle, and the people involved.

Over the past decade, fatal crashes have declined due to a variety of factors including safer vehicles, safer roadways, improved driver behavior such as increased seat belt use and decreased impaired driving, increased congestion, which reduces travel speeds, more use of mass transit, and, since 2007, reduced economic activity. In 2000, when NHTSA last examined this issue, there were roughly 42,000 fatalities from motor vehicle crashes. By 2010, that total has dropped to roughly 33,000 – a 21-percent decline. This is an encouraging trend, and is even more impressive in light of the increased population and generally rising rates of travel over time. Some aspects of this decline are likely to remain and even accelerate as the older on-road fleet is replaced by more modern vehicles with advanced safety features such as electronic stability control and advanced air bag systems. However, as the economy rebounds from the recession, increased economic activity is likely to offset some of this progress.

Table 5-1. Fatalities and Fatality Rates, 1949-2011

Year	Fatalities	Fatality Rate per 100M VMT	Year	Fatalities	Fatality Rate per 100M VMT
1949	30,246	7.13	**1981**	49,301	3.17
1950	33,186	7.24	**1982**	43,945	2.76
1951	35,309	7.19	**1983**	42,589	2.58
1952	36,088	7.03	**1984**	44,257	2.57
1953	36,190	6.65	**1985**	43,825	2.47
1954	33,890	6.03	**1986**	46,087	2.51
1955	36,688	6.06	**1987**	46,390	2.41
1956	37,965	6.05	**1988**	47,087	2.32
1957	36,932	5.71	**1989**	45,582	2.17
1958	35,331	5.32	**1990**	44,599	2.08
1959	36,223	5.17	**1991**	41,508	1.91
1960	36,399	5.06	**1992**	39,250	1.75
1961	36,285	4.92	**1993**	40,150	1.75
1962	38,980	5.08	**1994**	40,716	1.73
1963	41,723	5.18	**1995**	41,817	1.73
1964	45,645	5.39	**1996**	42,065	1.69
1965	47,089	5.3	**1997**	42,013	1.64
1966	50,894	5.5	**1998**	41,501	1.58
1967	50,724	5.26	**1999**	41,717	1.55
1968	52,725	5.19	**2000**	41,945	1.53
1969	53,543	5.04	**2001**	42,196	1.51
1970	52,627	4.74	**2002**	43,005	1.51
1971	52,542	4.46	**2003**	42,884	1.48
1972	54,589	4.33	**2004**	42,836	1.44
1973	54,052	4.12	**2005**	43,510	1.46
1974	45,196	3.53	**2006**	42,708	1.42
1975	44,525	3.35	**2007**	41,259	1.36
1976	45,523	3.25	**2008**	37,423	1.26
1977	47,878	3.26	**2009**	33,883	1.15
1978	50,331	3.26	**2010**	32,999	1.11
1979	51,093	3.34	**2011**	32,367	1.1
1980	51,091	3.35			

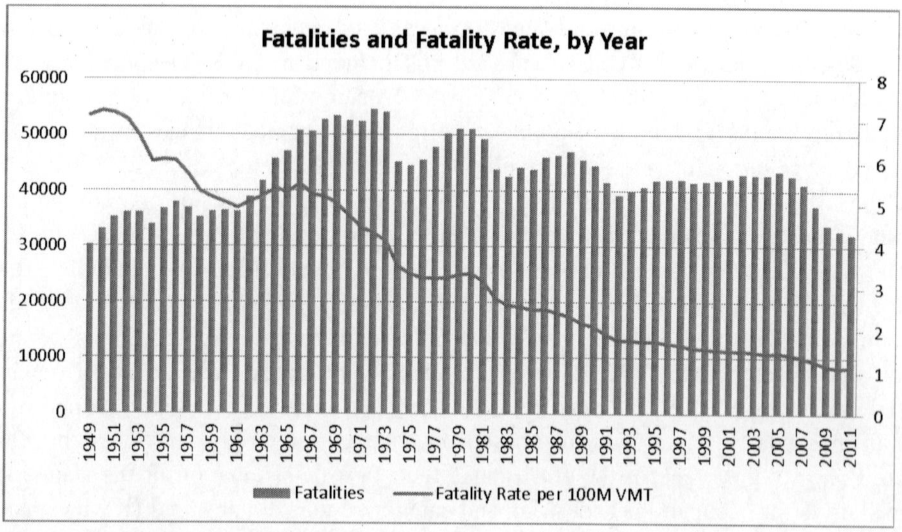

Figure 5-A. Fatalities and Fatality Rate, by Year.

Table 5-1 and Figure 5-A illustrate the trend in fatalities over the past 60 years, along with the trend in the fatality rates. Over this period, fatality rates have exhibited steady decline. Fatality counts rose during the 1960s in response to rapid increase in the driving population associated at least in part by the demographic shift of the baby boom generation into the driving cohort. For the first half of the past decade, fatality counts were relatively steady while fatality rates continued their decline, but a noticeable decline in fatalities occurred beginning in 2006-2007, which accelerated in the following years as the country weathered the recession. Although an economic recovery has been in place for several years, 2011 fatalities continued to decline to their lowest level since 1949. However, preliminary data for 2012 indicate that fatalities may now begin to increase from these historic low points.

Nonfatal Police-Reported Injuries

While FARS provides a dependable census of all fatal crashes, there is no equivalent data source for nonfatal injuries. Nonfatal injuries are estimated in several NHTSA databases. The Crashworthiness Data System (CDS) is a nationally representative sample of roughly 5,000 crashes annually. CDS contains detailed information on police-reported injuries incurred by passengers of towed passenger vehicles. CDS employs trained crash investigators to obtain data from police-reported crash sites, studying evidence such as skid marks, fluid spills, broken glass, and bent guard rails. They locate the vehicles involved, photograph them, measure the crash damage, and identify interior locations that were struck by the occupants. These researchers follow up on their on-site investigations by interviewing crash victims and reviewing medical records to determine the nature and severity of injuries. This enables researchers to properly categorize injury severity based on the Abbreviated Injury Scale (AIS), the basis for stratifying injury severity in this report. Crashes covered by the CDS represent about 62 percent of all police-reported injuries and typically involve the most serious injuries to vehicle occupants.

Injuries that occur in non-tow-away crashes, to occupants of large trucks, buses, motorcycles, bicyclists, or to pedestrians, are not included in CDS. The incidence of these injuries can be derived from the General Estimates System (GES). Data for GES come from a nationally representative sample of police- reported motor vehicle crashes of all severity and vehicle types. In order for a crash to be eligible for the GES sample a police accident report (PAR) must be completed, it must involve at least one motor vehicle traveling on a traffic way, and it must have resulted in property damage, injury, or death.

These accident reports are chosen from 60 areas that reflect the geography, roadway mileage, population, and traffic density of the GES data collectors make weekly visits to approximately 400 police jurisdictions in the 60 areas across the United States, where they randomly sample about 50,000 PARs each year. No other data are collected beyond the selected PARs. As a result, the only severity stratification in GES is that obtained from the PAR. In most States this is typically based on what is commonly known as the KABCO system. Police at the scene of the crash make their best determination of the status of each involved driver or occupant or pedestrian and categorize it as either killed (K), incapacitating injury (A), non-incapacitating injury (B), possible injury (C), or uninjured (O). Unlike the AIS severity stratification that can be obtained from CDS which is derived from medical records, these designations reflect only the initial opinion of responders who are not medical specialists. The KABCO results from GES thus provide only vague and sometimes inaccurate information regarding injury severity.

To address this problem, translators have been developed to convert KABCO ratings into specific MAIS ratings. These translators are developed from 1982-1986 data from the National Accident Sampling System (NASS). NASS was the primary injury data system used by NHTSA through 1986. It was replaced in 1989 by the current GES and CDS systems. Both NASS and CDS contain severity designations on MAIS and KABCO bases, which allows for an examination of the actual injury severity levels that are contained in each KABCO category. An example of these translators is shown in Table 5-2 for Non-CDS cases involving nonoccupants and motorcyclists. The results indicate the importance of expressing injuries on an MAIS basis rather than relying on the KABCO ratings from PARs. About 36 percent of the cases that police coded as uninjured were actually injured. 4.3 percent of the cases coded as possible injury were actually uninjured, as were 2.2 percent of cases coded Non-incapacitating, 0.5 percent of those coded as incapacitating, and 3.2 percent of those coded injured, but severity unknown. In addition, 21.6 percent of those coded Unknown (if injured) were uninjured while 78.4 percent were injured. There are also significant differences in the distribution of severities among those who are injured. 30.1 percent of cases coded as incapacitating, the most severe injury category under the KABCO system, actually experienced only a minor injury (MAIS1) and another 27.8 percent only experienced a moderate injury (MAIS2).

Although CDS contains the more accurate MAIS injury severity estimates, its smaller sample size makes it a less reliable indicator of aggregate incidence. To derive a National non-fatal injury profile for 2010, CDS was used to establish an initial incidence and distribution for cases fitting the CDS profile – crashes involving at least one towed passenger vehicle. These cases were then increased by the ratio of CDS equivalent injury cases from the GES to the CDS total. This normalization process acknowledges the smaller standard error that results from the more robust sample that GES uses.

Table 5-2. Translator for Non-CDS Cases, Nonoccupants and Motorcyclists

Translator for Non-CDS Cases, Nonoccupants and Motorcyclists						
MAIS	**O**	**C**	**B**	**A**	**ISU**	**Unknown**
MAIS0	0.640	0.043	0.022	0.005	0.032	0.216
MAIS1	0.308	0.572	0.618	0.301	0.433	0.469
MAIS2	0.044	0.164	0.158	0.278	0.085	0.117
MAIS3	0.004	0.040	0.059	0.270	0.042	0.064
MAIS4	0.001	0.001	0.002	0.025	0.005	0.008
MAIS5	0.000	0.001	0.000	0.026	0.000	0.004
ISU	0.004	0.179	0.140	0.089	0.403	0.098
Fatality	0.000	0.000	0.000	0.006	0.000	0.024
Total	1.000	1.000	1.000	1.000	1.000	1.000

A different approach was used for cases not covered by CDS. Non-CDS cases were isolated from the 1982-86 NASS files and split according to their seat belt status. Belt status was examined separately because belts have a significant impact on injury profiles and belt use has increased significantly since the 1982-86 period. Four separate categories were examined: belted occupants, unbelted occupants, unknown belt status, and nonoccupants including motorcyclists. A separate translator was developed for each of these categories. These translators were applied to their corresponding non-CDS equivalent cases from the 2010 GES file to estimate total non-CDS equivalent injuries by MAIS level.

The combined CDS and non-CDS cases represent police-reported injuries as estimated in these systems. While the data systems noted previously estimate National level totals of injuries based on samples, individual States collect police-reported injury totals from the various jurisdictions within that State. State data systems thus provide a potential census of all crashes for which a police report was filed. At one time this data was gathered together and published by FHWA, however, FHWA no longer compiles this data so they must be obtained from other sources.

Since the early 1980s, NHTSA has been obtaining from various States computer data files coded from data recorded on police accident reports (PARs). A PAR is completed by a police officer at a motor vehicle traffic crash scene and contains information describing characteristics of the crash, the vehicles, and the people involved. The data recorded on these forms are computerized into a central crash data file at the State level. Information will vary from State to State because each State has different data collection and reporting standards. NHTSA refers to the collection of these computerized State crash data files as the State Data System (SDS).

The State crash data files are requested annually from the State agencies that maintain the files. In most instances, this agency is the State police, the State Highway Safety Department, or the State Department of Transportation. The data is received in various formats and converted to Statistical Analysis System (SAS) data files. The SAS files are placed on NHTSA's Local Area Network, where they are available for the analytical needs of the NHTSA staff. The State crash data files in SDS are not available for research outside NHTSA unless permission has been granted from the State to release the crash data. The State crash data files are obtained to support NHTSA's efforts to identify traffic safety problems, to help develop and implement vehicle and driver countermeasures, to evaluate motor vehicle standards, and to study crash avoidance issues, crashworthiness issues, and regulations.

Because only 34 States participate in this system, SDS data was supplemented by directly contacting or accessing the Web sites of non-participating States.

Table 5-3. Motor Vehicle Police-Reported Traffic Crashes by State

Motor Vehicle Police-Reported Traffic Crashes by State							
State	**# Crashes**	**Year**	**Source**	**State**	**# Crashes**	**Year**	**Source**
Alabama	123503	2009	SDS	**Montana**	21971	2008	SDS
Alaska	12890	2009	ST Web site	**Nebraska**	34664	2009	SDS
Arizona	106767	2009	ST Web site	**Nevada**	53151	2009	ST Website
Arkansas	62808	2009	SDS	**New Hampshire**	33265	2009	ST Web site
California	426228	2009	SDS	**New Jersey**	301233	2009	SDS
Colorado	105000	2009	SDS	**New Mexico**	46213	2009	SDS
Connecticut	103719	2009	SDS	**New York**	314974	2009	SDS
Delaware	16723	2009	SDS	**North Carolina**	209695	2009	SDS
DC	16841	2009	ST Web site	**North Dakota**	17686	2009	SDS
Florida	235803	2009	SDS	**Ohio**	299040	2009	SDS
Georgia	318531	2009	SDS	**Oklahoma**	71218	2009	ST Website
Hawaii	10000	2009	ST Web site	**Oregon**	41271	2009	ST Web site
Idaho	22992	2009	ST Web site	**Pennsylvania**	121298	2009	SDS
Illinois	292437	2009	SDS	**Rhode Island**	41788	2009	ST Web site
Indiana	189983	2009	SDS	**South Carolina**	106864	2009	SDS
Iowa	55488	2009	SDS	**South Dakota**	16994	2009	ST Web site
Kansas	61119	2009	SDS	**Tennessee**	155099	2009	ST Website
Kentucky	126237	2009	SDS	**Texas**	428667	2009	SDS
Louisiana	155857	2009	SDS	**Utah**	51367	2009	ST Website
Maine	33118	2009	ST Web site	**Vermont**	12640	2009	ST Website
Maryland	96391	2009	SDS	**Virginia**	116742	2009	SDS
Massachusetts	136384	2009	ST Web site	**Washington**	110070	2009	SDS
Michigan	293403	2009	SDS	**West Virginia**	39906	2009	ST Website
Minnesota	73498	2009	SDS	**Wisconsin**	121736	2009	SDS
Mississippi	74122	2009	ST Web site	**Wyoming**	15507	2009	SDS
Missouri	153015	2009	SDS	**TOTAL**	6085916		

Previous analysis comparing State police reports to GES counts have found that actual police-reported injuries exceed those accounted for in the GES by 10 to 15 percent (Blincoe & Faigin, 1992, Blincoe et al., 2002). These previous analyses have focused on the difference in State injury counts and GES estimates. A similar attempt was made to examine these counts for the current analysis, however, it was found that State injury reporting practices have now become too dissimilar and fragmented to produce a reliable injury count for this comparison. For example, definitions of specific injury levels often overlap between States, hospital follow-up requirements vary by jurisdiction, and use of the "Unknown" severity appears to vary by jurisdiction as well. Instead, a comparison was made between total police-reported crashes in all States to those derived from GES.[38] Due to significant widespread delays in finalizing records within many State reporting systems, the latest year for which complete crash counts were available from most States was 2009.[39] Based on this data, there were 6,085,916 police-reported crashes in State records for 2009. These counts are shown in Table 5-3. By contrast, the 2009 GES estimated a total of 5,497,506 crashes in 2009. The ratio of State to GES crashes is thus 1.107, implying that GES understated total crashes by 10.7 percent. This is consistent with past estimates based on injury counts, which were in the 10- to 15-percent range. Our final estimate of police-reported injuries is derived by inflating the non-fatal injury profile derived above by this 1.107 factor. The results are shown in Table 5-4.

An estimated 2.97 million injuries were documented in police reports in 2010. This is a decrease from the 4.1 million estimated for 2000 in the previous version of this report. This decline is illustrated in Figure 5-B. Note however that Figure B only reflects the raw injury counts from the GES system. It does not reflect adjustments for MAIS injury translation or GES undercounting. Nonetheless, it does illustrate the steady decrease in injuries and injury rates over the past decade. This decrease reflects a variety of factors in including safer vehicles, safer roads, increased belt use, increased enforcement of alcohol countermeasures, and, after 2007, an economic slowdown that reduced driving and exposure.

Table 5-4. Estimated Police-Reported Non-Fatal Injury Profile, 2010

Severity	CDS	Non-CDS Unbelted	Non-CDS Belted	Non-CDS Unk Belt	GES Non-occupant	Total	Adjusted for States
MAIS0	892899	54730	838576	103563	50425	1940194	2147857
MAIS1	1364841	50678	684262	73701	156165	2329646	2578993
MAIS2	179341	4360	17559	3338	40345	244943	271160
MAIS3	56419	566	9291	214	20587	87077	96397
MAIS4	13129	76	233		1996	15434	17086
MAIS5	3932	34			1227	5193	5749
Total	2510562	110443	1549921	180816	270745	4622488	5117242
Total Injuries	1617663	55713	711345	77253	220320	2682294	2969386
% Total	60.31%	2.08%	26.52%	2.88%	8.21%	100.00%	

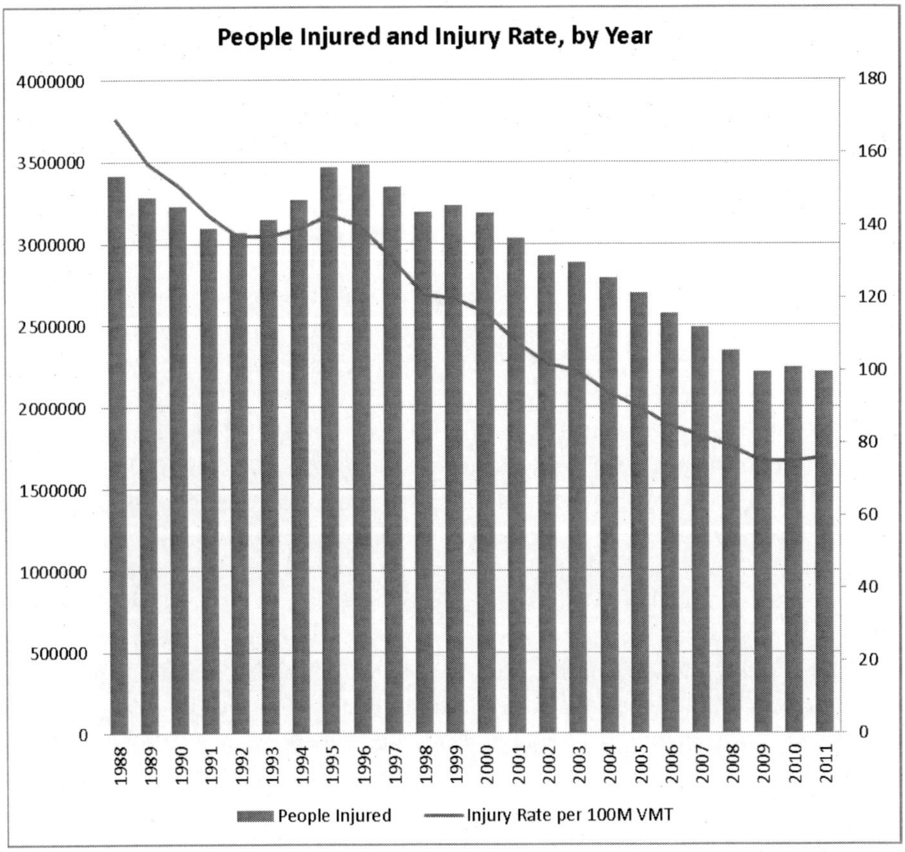

Figure 5-B. People Injured and Injury Rate, by Year.

Unreported Crashes and Injuries

The primary basis for incidence estimates used in this report are databases maintained by NHTSA that examine police-reported crashes. As discussed above, FARS is a census of all fatal crashes, while GES and CDS sample a broader set of police-reported crashes, including nonfatal crashes as well. These sources provide a basis for estimating the incidence of all police-reported crashes nationwide. However, a significant number of crashes are not reported to police.

In previous NHTSA analysis of this issue (Blincoe & Faigin, 1992), unreported injury crashes were estimated from data derived from Rice (1989), and Miller (1991) from the National Health Interview Survey (NHIS), while unreported property damage crashes were estimated based on a comparison of insurance claims data to National estimates of police-reported PDOs (Blincoe & Faigin, 1992). In subsequent NHTSA analysis, (Blincoe 1996, Blincoe et al., 2002), the unreported PDO estimate was retained but the unreported injury basis was derived from a study by Greenblatt, Merrin, and Morgenstein (1981). The switch in the unreported injury basis occurred because the Department of Health and Human Services announced that it had discovered a programming error that affected all motor vehicle injury estimates in the NHIS from 1982 through 1994. The most recent estimates of unreported crashes were thus based on injury survey data that is currently over 30 years old, and PDO insurance data that is over 20 years old.

NHTSA was concerned that changes in police reporting practices, insurance coverage, vehicle costs, litigation practices, real incomes, and the proliferation of cell phones may have shifted the unreported crash proportion over the past two to three decades. To address this concern, NHTSA contracted with M. Davis and Company (MDAC) to conduct a comprehensive nationally representative survey of households to determine the relative incidence of reported and unreported crashes. In late 2009 and the first half of 2010, MDAC conducted interviews with roughly 2,300 households where the respondent had experienced a motor vehicle crash during the previous 12 months. The interviews addressed the rate of reporting to police, the rate of reporting to insurance agencies, the severity of the crash, the location of vehicle damage, the types of injuries experienced in the crash, the cost of medical care, vehicle repair costs, the reasons why the crash was not reported, the crash location, and the number of vehicles involved in the crash. Most data elements were stratified separately for injury crashes and PDO crashes.

The results of the survey are contained in a report (NHTSA, 2011c) that details both methods and survey results. Some of the findings include:

- 29.3 percent of all crashes are not reported to police;
- 15.4 percent of injury crashes are not reported to police;
- 35.6 percent of property-damage-only crashes are not reported to police;
- 18.5 percent of all crashes are not reported to insurance companies;
- 12.3 percent of injury crashes are not reported to insurance companies;
- 20.8 percent of property-damage-only crashes are not reported to insurance companies;
- Medical care costs for police-reported crashes are roughly 9 times higher than for unreported crashes; and
- Vehicle repair costs for police-reported crashes are roughly 5 times higher than for unreported crashes.

Respondents gave a variety of reasons why they did not report the crashes they were in. The most common was that the extent of vehicle damage or injury was not serious/severe enough to warrant reporting the crash, which together made up 56.7 percent of the responses for unreported injury crashes and 71.7 percent of the responses for PDO crashes. None of the other 10 possible responses was offered by more than 8 percent of the respondents. However, the survey data also indicate that involvement of other vehicles was a large determinant for reporting. Single-vehicle crashes represent only 15.6 percent of reported injury crashes but 27.0 percent of unreported injury crashes. This discrepancy is even more pronounced for property damage crashes. For PDOs, single-vehicle crashes represent 14.7 percent of reported crashes, but 39 percent of unreported crashes. Thus, the presence of a second driver, which automatically raises the issue of fault, increases the likelihood that crashes will be reported. In single-vehicle crashes, especially in cases where there is no injury, the driver has no incentive to report a crash in which they were clearly at fault.

Table 5-5. State PDO Reporting Thresholds

State	PDO Reporting Thresholds	State	PDO Reporting Thresholds
Alabama	$250	**Missouri**	$500
Alaska	$2,000	**Montana**	$1,000
Arizona	Not required	**Nebraska**	$1,000
Arkansas	$1,000	**Nevada**	$750
California	$750	**New Hampshire**	$1,000
Colorado	Not required	**New Jersey**	$500
Connecticut	Not required	**New Mexico**	$500
D.C.	Not required	**New York**	$1,000
Delaware	$500	**North Carolina**	$1,000
Florida	$500	**North Dakota**	$1,000
Georgia	$500	**Ohio**	$400
Hawaii	$3,000	**Oklahoma**	$300
Idaho	$1,500	**Oregon**	$1,500
Illinois	$1,500	**Pennsylvania**	towed vehicles
Indiana	$1,000	**Rhode Island**	$1,000
Iowa	$1,000	**South Carolina**	$1,000
Kansas	$1,000	**South Dakota**	$1,000
Kentucky	$500	**Tennessee**	$400
Louisiana	All Crashes	**Texas**	$1,000
Maine	$1,000	**Utah**	$1,000
Maryland	Not required	**Vermont**	$1,000
Massachusetts	$1,000	**Virginia**	$1,000
Michigan	$1,000	**Washington**	$700
Minnesota	Not required	**West Virginia**	$500
Mississippi	Not required	**Wisconsin**	$1,000
		Wyoming	$1,000

Source: http://dmvanswers.com/questions/356/When-do-I-have-to-file-an-accident-report accessed on April 6, 2012.

The unreported rates found in the current survey differ from estimates used in previous studies. For example, in both the 2002 and 1996 reports, it was estimated that 21.4 percent of injury crashes and 48 percent of PDO crashes are not reported to the police. These estimates reflect data from two to three decades ago. The current survey indicates that 15.4 percent of injury crashes and 35.6 percent of PDO crashes are not reported to police.

It is not clear to what extent the change in non-reporting rates represents an actual shift in reporting habits as opposed to a more accurate estimate. We note that the MDAC survey, in addition to being more current, is also much larger than the 1981 survey that was the basis for previous injury crash estimates, which surveyed only 279 households compared to 2,299 in the MDAC survey. However, it seems likely that this difference represents, at least in part, an actual change. The proliferation of cell phones makes reporting easy, not just for the crash involved driver, but also for the other drivers who witness the crash or see a disabled vehicle. It is likely that in many cases where a driver might previously have chosen not to report a crash to police, other passing drivers have already notified police and foreclosed their option not to report.

A limitation of this survey is that it only reflects the knowledge of the crash involved drivers. Reporting a crash to police does not assure that a police accident report will actually be filed. Individual police jurisdictions typically have reporting thresholds, especially for crashes that only involve property damage. Consumers may report crashes, but if police determine that the crashes do not meet the damage threshold, they may not file an accident report. Reporting thresholds vary by State and sometimes by jurisdiction. Table 5 lists damage reporting thresholds by State.

Damage thresholds vary from $250 to $3,000, but seven States have no requirement for reporting crashes unless there is bodily injury, while one State requires reports only for tow-aways and one State requires all crashes to be reported. The most common threshold is $1,000. These thresholds are established for individual drivers, but police are likely to take these thresholds into account when deciding whether to file a crash report for PDO crashes, although special circumstances such as adverse weather conditions or natural disasters can influence police decisions as well.

In an effort to further understand the relationship between crash incidence and police reports, we queried a number of police jurisdictions that participate as primary sampling units for the National Accident Sampling System (NASS) regarding the relationship between police accident responses and accident reports. Only six jurisdictions responded with data, most of which was already publically available. The results are summarized in Table 5-6.

The reporting rates across this small sample vary widely. Often crashes occur in locales where both State and local police have jurisdiction, and it is likely that in those jurisdictions where the reporting rate is extremely low, many of the missing crashes were documented in reports by State police. A simple average of these six jurisdictions indicates that nearly half of all crashes that police investigate do not get into police reports that would be sampled by NHTSA to estimate nationwide police-reported crash totals, but this proportion is likely over stated because of the aforementioned jurisdictional overlap.[40] This sample is too small to use for National projections, but this does give an indication that the number of crashes in police records can differ significantly from the number that police respond to.

There are thus two bases for non-police reported crashes: crashes that are not reported to police, and crashes that *are* reported to police, but that are not documented in police records. The MDAC survey is useful for estimating crashes not reported to police, but not for those that police don't document.

Table 5-6. Sample Data of Various Jurisdictions Accident Reporting Rates

	Accident Responses	Written Reports	Written Report Rate	Unreported Rate
Wake County, NC	1,086	8	0.7%	99.3%
Knox County, TN	4,546	1,719	37.8%	62.2%
Golden, CO	866	633	73.1%	26.9%
Muskegon, MI	1,695	1,139	67.2%	32.8%
Cary, NC	4,269	4,064	95.2%	4.8%
Henrico County, VA	12,522	6,474	51.7%	48.3%
Simple Average			54.3%	45.7%

Table 5-7 summarizes National estimates of injured people in injury crashes and drivers of PDO vehicles from the MDAC study.

These estimates exceed those that would be derived from GES by significant margins. MDAC estimates nearly three times as many injured people as does GES, and 80 percent more PDO involved vehicles than does GES. Within the MDAC study itself, the survey estimates indicate 20 percent more injuries, and 59 percent more PDO vehicles than are reported to police. Since the MDAC survey includes both reported and unreported crashes, this might be expected. However, the police reported subset from the MDAC survey also exceeds the GES estimate by a significant margin. For injury crashes, the MDAC police reported total is more than double the GES police-reported total. Since police are universally required to report all injury crashes that they investigate, it might be expected that National estimates from independent surveys using valid methodologies would yield somewhat similar results, but this is not the case here. There are several possible explanations for these differences.

- Recall perception bias – the survey was designed to collect injuries that had occurred within the past 12 months. If survey respondents underestimate the time that has elapsed since an injury, they would overstate the frequency of injuries within this time frame.
- Delayed injury recognition – drivers may report a crash to police, but at the time when queried as to whether they were injured, they may have said no, only to find later that day or some days later that they had actually sustained a minor non-debilitating injury (such as chest bruising from the seat belt or minor whiplash). At the time of the survey, however, respondents remember that they were, in fact, injured. Under these circumstances, many of the crashes that survey respondents now characterize as injury crashes may have been considered by police to be PDO crashes. This would cause an underestimation in police records of injury crashes, and an overestimation in police records of PDO crashes.
- Motorist initiated reports – injury crashes that are investigated by police are typically included in any count of police-reported injury crashes. However, drivers can report crashes to police after the crash as well. This is often done for insurance purposes when police were not present at the crash scene. While many States require motorists to fill out reports if they are injured in a motor vehicle crash, the treatment of these reports in compiling injury data varies by State, and even by local jurisdiction. Since the GES system is confined to police reports, these types of injuries may not be included in the GES total and to some extent may be missing from State compilations as well. In a previous examination of this issue (Blincoe & Faigin, 1992), motorist reports were reflected in published injury counts in 8 of 22 States that were queried, and in the 8 States that did count motorist reports, they accounted for a highly variable portion of the State injury counts (ranging from less than 1 percent to 39 percent of total State injury counts.) In the remaining 14 States, motorist reports were not included in injury counts at all. These cases would show up as police-reported injury cases in the MDAC survey, but would be missing entirely from GES and to a lesser extent from State data.

- Unknown survey design weaknesses or data collection inaccuracies. Telephone surveys are only as accurate as their participant's responses, and GES is known to underestimate fatal crashes, although it is unclear whether this affects estimates of more common injury or PDO crashes.

Within the MDAC survey roughly 33 percent of police-reported injuries and 68 percent of unreported injuries did not seek medical care for their injuries. Another 16 percent of police-reported injuries and 5 percent of unreported injuries had zero medical cost. These could be considered to be cases where injuries were so minor that no treatment was sought,[41] and which may not even be noticed at the crash scene. What is notable here is that this proportion is much higher for cases that were reported to the police, the opposite of what would be expected given that logically, less serious crashes are more likely to go unreported. Again, assuming that no treatment was sought in these cases, they might be considered a proxy for the minimum rate of delayed injury recognition. Other cases that did experience medical costs might also reflect this as well, since latent discovery of injuries such as whiplash could result in subsequent visits to doctors or hospitals.

It is also worth noting that the ratio of total injuries to police-reported injuries internal to the MDAC study is similar to the results previously obtained from the 1981 study (1.20 in MDAC versus 1.21 in Greenblatt et al.). In previous studies, this markup was applied directly to the police-reported total obtained from NHTSA databases. In this case, with the separate MDAC police-reported total indicating a higher level of these injuries as well, the option exists to recognize some level of misapportionment of injury versus PDO cases within police files due to delayed injury recognition. The other two possible explanations – recall perception bias and/or unknown survey weaknesses or data inaccuracies, are not measurable.

We also examined National estimates of injuries from several other sources. The National Health Interview Survey (NHIS) is a cross-sectional household interview survey designed to monitor the health of the U.S. population. It captures information on a variety of health issues, including injuries from motor vehicle crashes. Motor vehicle injuries captured in the NHIS were those for which medical treatment was sought, including both police-reported and unreported crashes. In 2010, the NHIS estimate for motor vehicle injuries for which medical treatment was sought totaled 2.5 million. Of these, roughly 1.7 million cases were treated in emergency rooms (ER).

Table 5-7. Estimated Injured People and Drivers of PDOs from MDAC Study, by Police Reporting Status

Type of Crash	Injuries, or Drivers for PDOs				
	Number	Reported to Police	Unreported to Police	Unknown Status	Reported + Unreported
all crashes	20535814	14212974	5893978	428862	20106952
vehicle damage only	14178900	8911047	4932537	335316	13843584
injury crashes	6356914	5301927	961441	93546	6263368
injured as driver	4073484	3613854	409382	50249	4023236
injured as passenger	1475318	1280366	159474	35478	1439840
injured as pedestrian	808112	407707	392586	7819	800293

Another source of injury data is the National Electronic Injury Surveillance System – All Injury Program (NEISS). NEISS surveys 66 U.S. hospital emergency departments to produce National estimates of nonfatal injuries, including motor vehicle injuries, treated in ERs. In 2010, NEISS estimated that there were 3.3 million motor vehicle injuries treated in an ER.

Table 5-8 summarizes the various injury estimates from these 5 sources.

There are thus 5 separate categories of injuries estimated across these 5 data sources. The MDAC study is based on driver responses, and thus hypothetically covers the universe of possible injury reporting circumstances. Subsets of the MDAC study are thus comparable to their corresponding subsets from other sources.

GES, MDAC, and State Data provide estimates for police-reported injuries, but MDAC includes police reporting by drivers, which may or may not have resulted in a police report being filed, while GES reflects actual police reports only and State data counts reflect actual police reports as well as some motorist reports. The large difference between MDAC and the other sources likely reflects the factors discussed above, including recall perception bias, delayed injury recognition, and motorist reports.

The difference between the GES estimate and State counts continues the historic pattern that has been recognized since the 1990s. This pattern was discussed in detail in a previous NHTSA report (Blincoe & Faigin, 1992, pp. III-28-III-65) where differences ranging from 6 to 18 percent were documented between 1983 and 1990. These differences persisted through the 1996-1999 period when the average difference over the 4 year span was found to be 13 percent (Blincoe et al., 2002). The current (as of 2010) difference is 10.7 percent. As noted previously, due to inconsistencies among State data bases, the current difference is based on crash ratios rather than injury ratios. Blincoe and Faigin found a variety of causes for the difference between GES and State data including the inclusion of motorist reports in some State totals, misapportioned Injury/PDO cases (which may be caused by delayed injury recognition), undercounted police reports, and undercounted injuries/crash.

Estimates of medically treated injuries are available from the MDAC and NHIS surveys. There is considerable disagreement between the two surveys with the MDAC survey estimating 57 percent more medically treated injuries than NHIS. There does not appear to be an obvious theoretical reason for this difference.[42] Both surveys result from personal interviews and injuries severe enough to require medical treatment should be memorable enough to assume a reasonable level of accuracy on the part of respondents. There may be unknown design weaknesses or weighting issues that affect one or both studies. However, when the ER Treatment totals from these two studies are compared along with the NEISS estimate of ER treated injuries, two of the sources, MDAC and NEISS, are in fundamental agreement. The MDAC estimates 3.0 million ER injuries while NEISS estimates 3.3 million. By contrast, NHIS estimates only 1.7 million ER injuries or roughly half those found in MDAC and NEISS. In fact, both of these studies estimate more ER treated injuries than NHIS estimates for all treatment sources combined. Even the upper confidence bound for the NHIS is outside the confidence interval for either the MDAC or the NEISS surveys. There is thus reasonable agreement between two of the three surveys regarding ER treated injuries, which lends some credence to their findings and suggests the conflicting estimate from NHIS may be understated.

The MDAC estimate of injuries for which there was no medical treatment can be divided by police- reported status. Respondents indicated that roughly two-thirds of all injuries that were not reported to the police and about one-third of injuries that were reported to the police were not serious enough to warrant treatment. Overall, 38 percent of all injuries were so minor that drivers did not seek medical treatment.

A feasible characterization for the conflicting injury estimates among the various data sources is presented in Table 5-9. This table adopts the basic MDAC estimate, which encompasses the universe of injury categories, and interprets the differences between this estimate and the State data counts. The 2.97 million injuries estimated to be counted in State data files represent injuries included in police reports, supplemented in some States by motorist reports. All injuries in this category represent cases where the motorist indicated that the crash was reported to police and an injury was involved. The 1.8 million MDAC police-reported injuries for which no medical treatment was sought are likely to be minor injuries. Logically, they are consistent with a subset of "delayed injury recognition" cases, where drivers reported the crash to police but did not initially recognize that they were injured. The crash was thus initially reported to police as a PDO crash, and therefore did not get into the State injury counts. These two categories account for 4.7 million of the 5.4 million police-reported injuries in the MDAC survey. The remaining 635,000 injuries may represent a combination of motorist reports that are not included in State data totals and delayed recognition cases that eventually did seek medical treatment. An additional 975,000 injuries are not reported to police. As noted above, about two-thirds of these injuries were minor enough that the drivers did not seek medical treatment. In all, a total of 6.4 million people are injured to some extent, but 2.4 million, or 38 percent of these injuries, required no medical treatment, while the remaining 3.9 million injuries did require medical treatment.

It should be acknowledged that the above characterization is uncertain. There may be alternate explanations for portions of each category. In addition, since the MDAC data is derived from a survey, these estimates cannot be regarded with the certainty that is available in a true census such as FARS.

The ratio of total injuries to driver reported injuries from MDAC is 1.18 slightly below the 1.27 ratio found in the previous study. However, the ratio of total injuries to police-reported injuries from State data is 2.20. The difference primarily represents factors discussed above including delayed injury recognition, motorist reports, etc. The largest portion of the difference represents minor injuries for which no medical treatment was sought. Thus, although there is a significant difference between the estimates from the MDAC survey and police reports, most of this difference involves cases where injury and thus costs, were minimal.

Table 5-8. 2010 Injury Estimates

2010 Injury Estimates					
Source	**All Police Reported**	**Not Police Reported**	**Any Medical Treatment**	**ER Medical Treatment**	**No Medical Treatment**
GES/CDS	2,682,294				
State PARS/GES PARS	2,969,386				
NHIS			2,499,016	1,708,741	
NEISS				3,258,889	
MDAC Total	5,381,113	975,801	3,922,216	2,973,040	2,435,409
MDAC PR	5,381,113		3,605,346	2,750,879	1,775,767
MDAC NOT PR		975,801	316,159	223,841	659,641

Table 5-9. Summary of Motor Vehicle Injuries Compiled From Various Sources

Summary of Motor Vehicle Injuries Compiled From Various Sources	
2,969,386	State Data police-reported Injuries (PARs partially supplemented by motorist reports in some States)
1,775,767	Minor injuries possibly initially reported to police as PDOs, no treatment sought
635,960	Motorist reports not included in State PAR counts and minor injuries reported to police as PDO that later sought treatment
5,381,113	Total injuries reported by drivers
975,801	Injuries not reported to police or on motorist reports
6,356,914	Total Injuries based on MDAC
1.18	Ratio of total injuries to all driver reported injuries
2.14	Ratio of total injuries to police reports
1.34	Ratio of total injuries to driver reported injuries on police reports
1.81	Ratio of all driver reported injuries to police-reported injuries

As noted above, the survey found that unreported injuries were significantly less expensive to treat and that repair damage from unreported crashes was significantly less costly than for reported crashes. The study also documented the most serious injury and body region of these injuries. In the MDAC study, survey participants that were injured were asked a series of questions about their injury, the symptoms and level of treatment they received. These questions were designed so that a lay person would easily understand and be able to respond yes or no, yet the responses could be used to determine the AIS level of the injury. Using a probability algorithm based on injury characteristics in the Abbreviated Injury Scale (AIS), NHTSA examined the records for each injury case to assign the injury an MAIS designation consistent with the coding structure used for nonfatal injuries throughout this analysis. Based on this, a separate unreported rate was derived for each nonfatal MAIS severity category. From Table 8 above, the 2,435,409 minor injuries for which no medical treatment was sought will be treated as MAIS 0s.[43]659,641 of these were not reported to the police, while the remainder were arguably not in police records due to misreporting or misclassification as PDOs. The 316,159 injuries that were not reported to police but that did receive medical treatment were distributed according to the injury profile of non- hospitalized non-police-reported cases in the MDAC study, while the 635,960 cases that were reported by drivers but not reflected in PAR counts were distributed according to the injury profile of non- hospitalized police-reported cases in the MDAC study. Table 5-10 summarizes the unreported crash rates adopted for nonfatal injuries in this study.

Property-Damage Only Crashes

While crashes that involve death or injury produce the most serious consequences, they are relatively rare events. The vast majority of crashes are low-speed crashes that damage vehicles but leave vehicle occupants unharmed. Although these crashes impose a lower unit cost on society, their frequency makes this the most costly single type of crash overall.

Although police records include a large number of PDO crashes, they tend to be significantly undercounted in police records due to a variety of factors including relatively high reporting thresholds in various States, as well as the failure of drivers to report them to police. A full analysis of PDOs must therefore address not only police records but other sources as well.

The starting point for our estimate of PDOs is police reports. Because injury is not involved, the primary cost from PDO crashes is damage to the vehicle. Therefore, PDOs are analyzed on a per-damaged- vehicle basis. Data from the GES for 2010 indicate that there were 7,819,632 vehicles damaged without injury caused to either the vehicle's occupants or to pedestrians. Of these, 6,734,006 occurred in crashes where nobody was injured, while 1,085,626 occurred in crashes where an injury occurred, but not to the vehicle's occupant. These later cases are classified in this current report as MAIS 0 injuries so they will not be addressed as PDOs. The PDO category thus will ultimately represent only vehicles damaged in PDO crashes. However, a variety of other sources combine undamaged vehicles in both PDOs and injury crashes. Therefore, the initial analysis will treat these cases as a group.

Table 5-11 lists PDO estimates from three sources. The first is the GES police-report-based estimate previously discussed. This estimate is modified using the same 10.7 percent markup factor previously discussed for non-fatal injuries to reflect the undercounting of State police reports inherent in GES files. These sources imply a total of 8.7 million vehicles were damaged in crashes where the occupants were not injured, with 86 percent of these occurring in PDO crashes.

The second source is the MDAC survey, which gathered data on police and insurance reporting for both injuries and for damaged vehicles. MDAC found a total of 9.1 million vehicles damaged in crashes where the driver or occupant of that vehicle wasn't injured that were reported to police, and an additional 5.1 million that were not reported to police. MDAC also reported that of these 14.2 million cases, 11.2 million were reported to insurance companies and 3.0 million were not. MDAC published a table illustrating the interaction of these cases. Table 12 reproduces this table, which indicates that 64.2 percent of all cases were reported to police while 78.8 percent were reported to insurance and 58.4 percent were reported to both police and insurance.

Table 5-10. Nonfatal Injury Summary by Police Report Status

Severity	Police-Reported	Not Police-Reported	Total	Percent Unreported
MAIS0	2147857	2435409	4583265	53.1%
MAIS1	2578993	880207	3459200	25.4%
MAIS2	271160	67570	338730	19.9%
MAIS3	96397	4343	100740	4.3%
MAIS4	17086	0	17086	0.0%
MAIS5	5749	0	5749	0.0%
Total	5117242	3387528	8504771	39.8%
Total Injuries	2969386	952120	3921505	24.3%
% Total	75.7%	24.3%	100.0%	

The third source is insurance data gathered for this report and described previously in Chapter 2 (see Table 2-10 in Chapter 2). This data indicate that there were 16.9 million vehicle claims in crashes where the vehicle owners were not injured. These are shown in the "Insurance Reported" column of Table 5-11. Because this data was gleaned from actual insurance records, we consider them to be the more accurate estimate of insurance reported PDOs. Crashes not reported to insurance companies must be estimated from other sources. Using the proportions specified in the MDAC study, we estimate that, based on these insurance data, a total of 21.5 million vehicles were actually damaged in crashes where the occupants were not injured, with 4.6 million of them not reported to insurance. From these same data we can also estimate that 13.8 million of these vehicles were reported to police by their drivers, while another 7.7 million were not.

As noted previously in the discussion of injuries, for a variety of reasons actual police report counts do not match driver's responses. Specifically, even though a driver may report a crash to police, the police may not fill out an accident report if the crash damage is below a certain threshold. Report writing rates in a small sample of jurisdictions averaged only 45 percent. For this analysis, we assume the National police-reported PDO counts produced by GES are reasonably representative of the cases for which a report is actually made. Further, we assume that the best estimate of total PDO vehicles is derived from insurance data. In Table 5-11, the row marked "Insurance Adjusted for PR" re-distributes the difference between the GES police-reported counts and the implied police-reported counts derived from insurance data to non-police-reported status. Essentially, these 5.1 million cases are considered to be cases where drivers reported crashes to police, but where police did not fill out a report. The results indicate that 40 percent of PDOs are reflected in police reports while 60 percent are not, and that 37 percent of cases

Table 5-11. PDO Vehicle Summary

PDO Vehicles					
Source	Police Reported	Not Police Reported	Insurance Reported	Not Insurance Reported	Total
GES	7819632				7819632
State/GES Adjustment	8656584				8656584
MDAC	9126888	5052012	11224825	2954075	14178900
Insurance (implied from MDAC)	13798192	7694319	16936098	4556412	21492510
Insurance adjusted for PR	8656584	12835926	16936098	4556412	21492510
Percent of Total	40.3%	59.7%	78.8%	21.2%	100.0%

Table 5-12. MDAC PDO Reporting Status

PDO Reported to Police?				
		Yes	No	Total
Reported to Insurance?	Yes	58.4%	20.4%	78.8%
	No	5.8%	15.3%	21.2%
	Total	64.2%	35.8%	100.0%

A final adjustment to the PDO estimate was made to accomplish the exclusion of MAIS 0 cases referred to earlier in this section. GES data from 2010 indicates that 86.12 percent of PDO vehicles occurred in PDO crashes, but the remaining 13.8 percent represent vehicles with uninjured occupants in injury crashes. Since these cases are already counted as MAIS 0 injuries, they were removed from the PDO total. No separate accounting for the nature of the crash was possible from either the MDAC study or the insurance data, so the GES proportion was applied to all PDO categories. The results are summarized in Table 5-13. This data indicates that roughly 7.5 million vehicles were damaged in PDO crashes that were documented in police reports, but another 11.0 million occurred which were not reflected in police reports, for a total of 18.5 million vehicles damaged in all property damage-only crashes. Roughly 40 percent of all PDO crashes are thus reflected in police reports. A combination of factors previously discussed, including crashes unreported by drivers and high damage reporting thresholds within police jurisdictions are responsible for this low reporting rate.

Table 5-13. PDO Vehicle Summary Adjusted to Remove MAIS 0 Cases

PDO Vehicles					
Source	**Police Reported**	**Not Police Reported**	**Insurance Reported**	**Not Insurance Reported**	**Total**
GES	6734006				6734006
State/GES Adjustment	7454761				7454761
MDAC	7859771	4350624	9666445	2543950	12210395
Insurance (implied from MDAC)	11882542	6626090	14584802	3923830	18508632
Insurance adjusted for PR	7454761	11053871	14584802	3923830	18508632
Percent of Total	40.3%	59.7%	78.8%	21.2%	100.0%

Table 5-14. 2010 Incidence Summary

Severity	Police- Reported	Not Police- Reported	Total	Percent Unreported
MAIS0	2,147,857	2,435,409	4,583,265	53.1%
MAIS1	2,578,993	880,207	3,459,200	25.4%
MAIS2	271,160	67,570	338,730	19.9%
MAIS3	96,397	4,343	100,740	4.3%
MAIS4	17,086	0	17,086	0.0%
MAIS5	5,749	0	5,749	0.0%
Fatalities	32,999	0	32,999	0.0%
Total	5,150,241	3,387,528	8,537,770	39.7%
Total Injuries	3,002,385	952,120	3,954,504	24.1%
PDO	7,454,761	11,053,871	18,508,632	59.7%

Note: All injury categories reflect injured people. PDO reflects damaged vehicles.

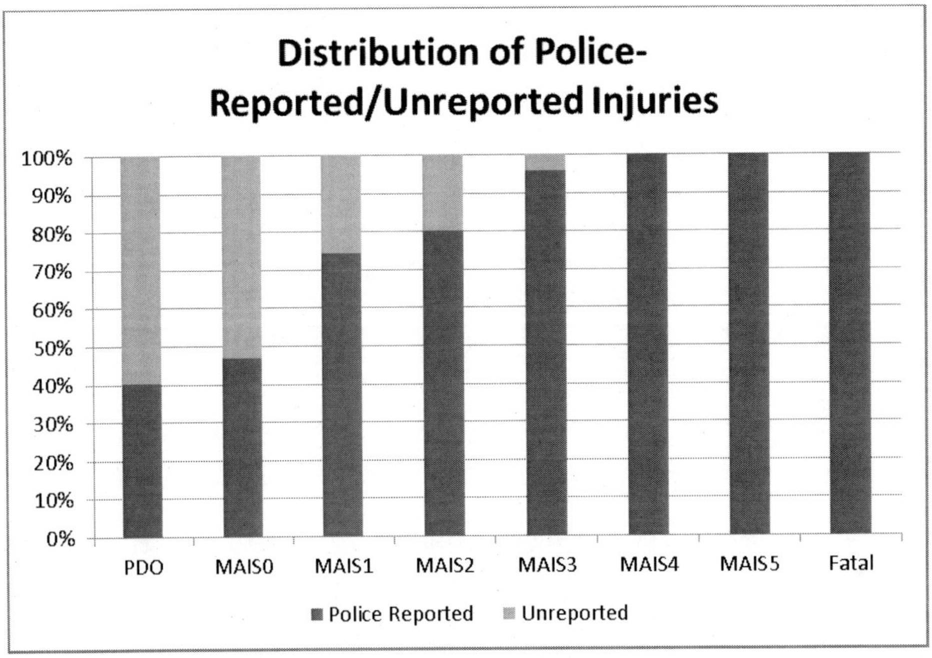

Figure 5-C. Distribution of Police-Reported/Unreported Injuries.

Overall, the 40 percent reported rate for PDOs is the lowest among all injury severity categories. This is expected given the relatively minor nature of these crashes. Figure 5-C illustrates the reporting rates of each severity level. MAIS 0 injuries (uninjured people in injury crashes) have a reporting rate of only 47 percent, while 73 percent of MAIS1s, 79 percent of MAIS2s, and 95 percent of MAIS3s are reflected in police records. All of the more serious MAIS4 and MAIS5 injuries, as well as all fatalities, are estimated to be accounted for in police records. Table 5-14 summarizes reported and unreported incidence for all severity categories.

6. State Costs

In recent years, States have continued to increase their involvement in establishing and enforcing laws related to motor vehicle safety. This is due, in part, to Federal legislation enacted to promote highway safety such as The Safe, Accountable, Flexible, Efficient Transportation Equity Act: A Legacy for Users (SAFETEA-LU) which was enacted in 2005 and which provided one time grants to States that enacted and are enforcing a conforming primary seat belt law for all passenger motor vehicles. SAFETEA-LU authorized a total of $770 million in grant money over a six-year period to address roadway and driver behavioral safety activities, especially those designed to increase belt use.

State legislators are often interested in the societal and economic cost of motor vehicle injury as they consider new traffic safety laws, changes to existing laws and funding for enforcement of the laws. This information can assist them in making the case to their constituencies as to the relevance of the laws designed to make the population safer.

A State-specific distribution of total economic costs has been prepared as follows:

- The year 2010 fatalities were obtained by State from FARS. The portion of total National fatalities in each State was then applied directly to the total fatality cost ($46.2 billion).
- State crash incidence data was obtained from individual States for 2008-2010. In cases where data was not available, a factor based on the trend in fatalities within the State was used to estimate crashes from the last years for which complete data was available. The portion of total National crashes in each State was applied to the total cost of all nonfatal injuries, PDOs, and uninjured occupants ($230.9 billion).
- The total costs for each State were then adjusted to reflect locality cost differences based on the ratio of costs in each State to the National total. Medical costs were adjusted based on data obtained from the ACCRA Cost of Living Index and cited by Miller and Galbraith (1995). Lost productivity, travel delay and workplace costs were adjusted based on 2010 per-capita income. Insurance administration and legal costs were adjusted using a combination of these two inflators weighted according to the relative weight of medical and lost productivity administrative costs. All other cost categories were adjusted using a composite index developed by ACCRA (also provided by Miller).

These four adjustment factors were applied separately to the fatal and nonfatal costs for each State. Weights to combine each factor were derived separately from the relative importance of each cost category to nationwide fatal and nonfatal total costs. The sum of fatal and nonfatal costs for each State was then adjusted to force the sum of all States' costs to equal the National total.

The results of this analysis are depicted in Table 6-1. There is considerable variation in costs among the States with New York, for example, having costs that are 17 times higher than those for Idaho. This is primarily due to the higher incidence of death and injury in New York (a function of population), but also to the higher cost levels in that State. However, as noted by Miller and Galbraith (1995), cost comparisons between States that are based on State injury totals can be misleading because injury totals do not capture differences in nonfatal injury severity between States. This would tend to under State costs in rural States relative to urban States, which typically have lower average speeds and consequently less severe injuries. Ideally, State costs would be based on individual State injury profiles, but these are not available for many States.

Differences between States may also result from different reporting practices that result in more or less complete recording of injuries from State to State. Differences in roadway characteristics and State of repair may account for some of this discrepancy, though it seems likely that variation in injury reporting is also a contributing factor. Finally, the impact of crash costs must be viewed in the context of each State's economy. Smaller, less populated States may have lower absolute costs, but they may also have fewer resources available to address these costs. A significant portion of these costs is borne by the general public through State and local revenue, or through private insurance plans. The per capita costs for each State vary from roughly $500 to $1,700 compared to the nationwide average of almost $900. This represents 1.1 to 4.1 percent of the per capita income for each State, with an overall average of 2.2 percent.

Table 6-1. Estimated 2010 Economic Costs Due to Motor Vehicle Crashes by State

State	(Millions 2010 Dollars)	% Total	Cost per Capita	% per Capita Personal Income
ALABAMA	$5,076	1.8%	$1,062	3.1%
ALASKA	$682	0.2%	$961	2.2%
ARIZONA	$4,753	1.7%	$744	2.1%
ARKANSAS	$2,692	1.0%	$923	2.8%
CALIFORNIA	$22,653	8.2%	$608	1.4%
COLORADO	$4,804	1.7%	$955	2.2%
CONNECTICUT	$5,635	2.0%	$1,577	2.8%
DELAWARE	$782	0.3%	$871	2.2%
DIST. OF COL.	$999	0.4%	$1,659	2.3%
FLORIDA	$12,079	4.4%	$642	1.6%
GEORGIA	$12,485	4.5%	$1,289	3.6%
HAWAII	$640	0.2%	$470	1.1%
IDAHO	$1,001	0.4%	$638	2.0%
ILLINOIS	$12,636	4.6%	$985	2.3%
INDIANA	$7,362	2.7%	$1,135	3.2%
IOWA	$2,489	0.9%	$817	2.1%
KANSAS	$2,783	1.0%	$975	2.5%
KENTUCKY	$4,988	1.8%	$1,150	3.4%
LOUISIANA	$6,536	2.4%	$1,442	3.8%
MAINE	$1,495	0.5%	$1,126	3.0%
MARYLAND	$5,097	1.8%	$883	1.8%
MASSACHUSETTS	$6,784	2.4%	$1,036	2.0%
MICHIGAN	$11,115	4.0%	$1,125	3.2%
MINNESOTA	$3,502	1.3%	$660	1.5%
MISSISSIPPI	$3,077	1.1%	$1,037	3.3%
MISSOURI	$6,381	2.3%	$1,065	2.9%
MONTANA	$1,018	0.4%	$1,030	2.9%
NEBRASKA	$1,483	0.5%	$812	2.1%
NEVADA	$2,277	0.8%	$843	2.3%
NEW HAMPSHIRE	$1,585	0.6%	$1,204	2.7%
NEW JERSEY	$14,848	5.4%	$1,689	3.3%
NEW MEXICO	$2,010	0.7%	$976	2.9%
NEW YORK	$17,447	6.3%	$900	1.8%
NORTH CAROLINA	$9,049	3.3%	$949	2.7%
NORTH DAKOTA	$807	0.3%	$1,199	3.0%
OHIO	$11,702	4.2%	$1,014	2.8%
OKLAHOMA	$3,287	1.2%	$876	2.4%
OREGON	$2,009	0.7%	$524	1.4%
PENNSYLVANIA	$6,542	2.4%	$515	1.3%
RHODE ISLAND	$1,858	0.7%	$1,764	4.1%
SOUTH CAROLINA	$4,594	1.7%	$993	3.0%

State	(Millions 2010 Dollars)	% Total	Cost per Capita	% per Capita Personal Income
SOUTH DAKOTA	$816	0.3%	$1,003	2.6%
TENNESSEE	$6,461	2.3%	$1,018	2.9%
TEXAS	$19,424	7.0%	$772	2.0%
UTAH	$1,979	0.7%	$716	2.2%
VERMONT	$613	0.2%	$979	2.4%
VIRGINIA	$5,707	2.1%	$713	1.6%
WASHINGTON	$5,174	1.9%	$769	1.8%
WEST VIRGINIA	$1,680	0.6%	$907	2.8%
WISCONSIN	$5,239	1.9%	$921	2.4%
WYOMING	$885	0.3%	$1,570	3.3%
Total	$277,020	100.0%	$897	2.2%

7. ALCOHOL

Alcohol consumption is a major cause of motor vehicle crashes and injury. Over the past two decades, about 40 percent of all motor vehicle fatalities occur in crashes in which a driver or nonoccupant has consumed a measurable level of alcohol prior to the crash, and of these cases, 86 percent involved a level of consumption which met the current typical legal definition for intoxication or impairment, a blood alcohol concentration of .08 grams per deciliter or higher. Over the past two decades, there has been an increased awareness of the problems caused by impaired driving. Many groups from NHTSA to Mothers Against Drunk Driving (MADD), Students Against Destructive Decisions (SADD), and State and local agencies, have promoted the enactment of laws and implemented public awareness campaigns to assist in combating this problem. Legal measures such as administrative license revocation/suspension have been enacted in numerous States. As a result, there has been a marked decrease in the number of fatalities resulting from alcohol-involved crashes. Table 7-1 displays the share of fatalities associated with alcohol involvement (BAC>.01 g/dL) and the current definition of legal intoxication (illegal per se, .08 g/dL) since 1982. Alcohol involvement in fatal crashes has declined from 60 percent of all fatalities in 1982 to roughly 40 percent in 2010, while legal intoxication (defined as a BAC of .08 g/dL or greater) has declined from 53 percent to 35 percent over the same period. While these declines are encouraging, alcohol still remains a significant causative factor in motor vehicle crashes.

All 50 States, the District of Columbia, and Puerto Rico define legal intoxication, the level at which DWI convictions can be made, as having a BAC of .08 or higher. FARS data indicates that fatalities involving legally intoxicated drivers or nonoccupants account for 86 percent of the fatalities arising from all levels of alcohol involvement.

Fatalities

FARS provides detailed information about all traffic fatalities that occur within 30 days of a crash on a public road. Each case is investigated and documentation regarding alcohol involvement is included. Alcohol involvement can be indicated either by the judgment of the investigating police officers or by the results of administered BAC tests. Cases where either of these factors is positive are taken as alcohol-involved and any fatalities that result from these crashes are considered to be alcohol-involved fatalities. In addition, there are a large number of cases where alcohol involvement is unknown. In 1986, NHTSA's National Center for Statistics and Analysis (NCSA) developed an algorithm based on discriminant analysis of crash characteristics that estimates the BAC level for these cases (Klein, 1986). In 1998, NHTSA developed a more sophisticated technique to accomplish these estimates using multiple imputation (Rubin, Schafer, & Subramanian, 1998), and substituted this method beginning with the 2001 FARS file. NHTSA has recomputed previous FARS files using this method and alcohol involvement rates based on the new method are routinely published by NHTSA and used in this report. The total number of alcohol-involved fatalities by BAC level is shown in Table 7-1 from 1982 through 2011. In 2010, about 86 percent of all fatalities that occurred in alcohol-involved crashes were in cases where a driver or pedestrian had a BAC of .08 or higher.

Table 7-1. Alcohol-Involved and Intoxicated Traffic Fatalities, Highest BAC in CRASH

Year	Total		BAC=.00		BAC=.01 to .07		BAC=.08+		BAC=.01+	
	Number	Percent	Number	Percent	Number	Percent	Number	Percent	Number	Percent
1982	43,945	100%	17,773	40%	2,927	7%	23,246	53%	26,173	60%
1983	42,589	100%	17,955	42%	2,594	6%	22,041	52%	24,635	58%
1984	44,257	100%	19,496	44%	3,046	7%	21,715	49%	24,762	56%
1985	43,825	100%	20,659	47%	3,081	7%	20,086	46%	23,167	53%
1986	46,087	100%	21,070	46%	3,546	8%	21,471	47%	25,017	54%
1987	46,390	100%	22,297	48%	3,398	7%	20,696	45%	24,094	52%
1988	47,087	100%	23,254	49%	3,234	7%	20,599	44%	23,833	51%
1989	45,582	100%	23,159	51%	2,893	6%	19,531	43%	22,424	49%
1990	44,599	100%	22,012	49%	2,980	7%	19,607	44%	22,587	51%
1991	41,508	100%	21,349	51%	2,560	6%	17,599	42%	20,159	49%
1992	39,250	100%	20,960	53%	2,443	6%	15,847	40%	18,290	47%
1993	40,150	100%	22,242	55%	2,361	6%	15,547	39%	17,908	45%
1994	40,716	100%	23,409	57%	2,322	6%	14,985	37%	17,308	43%
1995	41,817	100%	24,085	58%	2,490	6%	15,242	36%	17,732	42%
1996	42,065	100%	24,316	58%	2,486	6%	15,263	36%	17,749	42%
1997	42,013	100%	25,302	60%	2,290	5%	14,421	34%	16,711	40%
1998	41,501	100%	24,828	60%	2,465	6%	14,207	34%	16,673	40%
1999	41,717	100%	25,145	60%	2,321	6%	14,250	34%	16,572	40%
2000	41,945	100%	24,565	59%	2,511	6%	14,870	35%	17,380	41%
2001	42,196	100%	24,796	59%	2,542	6%	14,858	35%	17,400	41%
2002	43,005	100%	25,481	59%	2,432	6%	15,093	35%	17,524	41%
2003	42,884	100%	25,779	60%	2,427	6%	14,678	34%	17,105	40%

Year	Total		BAC=.00		BAC=.01 to .07		BAC=.08+		BAC=.01+	
	Number	Percent	Number	Percent	Number	Percent	Number	Percent	Number	Percent
2004	42,836	100%	25,918	61%	2,325	5%	14,593	34%	16,919	39%
2005	43,510	100%	25,920	60%	2,489	6%	15,102	35%	17,590	40%
2006	42,708	100%	24,970	58%	2,594	6%	15,144	35%	17,738	42%
2007	41,259	100%	24,101	58%	2,554	6%	14,603	35%	17,158	42%
2008	37,423	100%	21,974	59%	2,191	6%	13,258	35%	15,449	41%
2009	33,883	100%	19,704	58%	2,031	6%	12,149	36%	14,179	42%
2010	32,999	100%	19,676	60%	1,861	6%	11,462	35%	13,323	40%
2011	32,367	100%	19,212	59%	1,758	5%	11,397	35%	13,155	41%

Alcohol use by drivers is the focus of most behavioral programs and State laws. Drivers are involved in the vast majority of alcohol-related traffic crashes, but a significant number of crashes occur where pedestrians or bicyclist alcohol use was indicated, while drivers were not drinking. Table 7-2 summarizes the incidence of alcohol-related crashes based on driver BAC, while Table 7-3 shows the incidence of fatalities where pedestrians or bicyclists were using alcohol, but not drivers. In 2010, 85 percent of all fatalities that occurred in alcohol-involved crashes were in cases where a driver had a BAC of .08 or higher. About 5 percent of all alcohol-related traffic fatalities involve alcohol use by pedestrians rather than motor vehicle drivers. Of these cases, over 90 percent involve alcohol impairment (BAC = .08 or higher) on the part of the pedestrian.

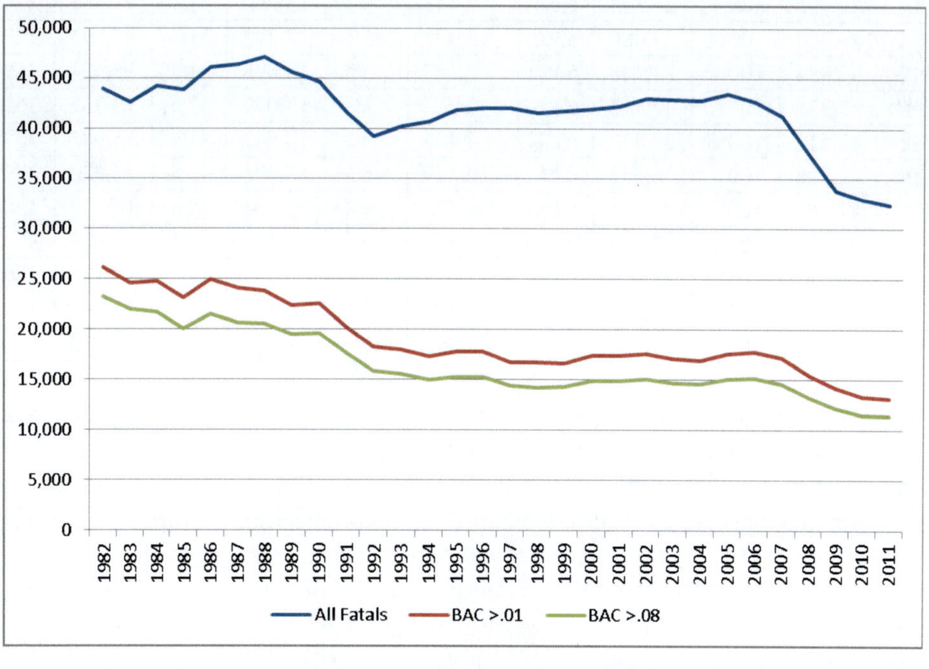

Figure 7-A. Historical Trend of Fatalities, Alcohol-Involved Fatalities, and Alcohol-Impaired Fatalities.

Table 7-2. Alcohol-Involved and Intoxicated Traffic Fatalities, Highest Driver BAC

Year	Total*		BAC=.00		BAC=.01 to .07		BAC=.08+		BAC=.01+	
	Number	Percent	Number	Percent	Number	Percent	Number	Percent	Number	Percent
1982	43,945	100%	19,771	45%	2,912	7%	21,113	48%	24,025	55%
1983	42,589	100%	19,787	46%	2,588	6%	20,051	47%	22,639	53%
1984	44,257	100%	21,429	48%	3,007	7%	19,638	44%	22,645	51%
1985	43,825	100%	22,589	52%	2,974	7%	18,125	41%	21,098	48%
1986	46,087	100%	22,896	50%	3,487	8%	19,554	42%	23,041	50%
1987	46,390	100%	24,186	52%	3,238	7%	18,813	41%	22,051	48%
1988	47,087	100%	25,164	53%	3,156	7%	18,611	40%	21,767	46%
1989	45,582	100%	25,152	55%	2,793	6%	17,521	38%	20,314	45%
1990	44,599	100%	23,823	53%	2,901	7%	17,705	40%	20,607	46%
1991	41,508	100%	23,025	55%	2,480	6%	15,827	38%	18,307	44%
1992	39,250	100%	22,726	58%	2,352	6%	14,049	36%	16,401	42%
1993	40,150	100%	23,979	60%	2,300	6%	13,739	34%	16,039	40%
1994	40,716	100%	24,948	61%	2,236	5%	13,390	33%	15,626	38%
1995	41,817	100%	25,768	62%	2,416	6%	13,478	32%	15,893	38%
1996	42,065	100%	26,052	62%	2,415	6%	13,451	32%	15,866	38%
1997	42,013	100%	26,902	64%	2,216	5%	12,757	30%	14,973	36%
1998	41,501	100%	26,477	64%	2,353	6%	12,546	30%	14,899	36%
1999	41,717	100%	26,798	64%	2,235	5%	12,555	30%	14,790	35%
2000	41,945	100%	26,082	62%	2,422	6%	13,324	32%	15,746	38%
2001	42,196	100%	26,334	62%	2,441	6%	13,290	31%	15,731	37%
2002	43,005	100%	27,080	63%	2,321	5%	13,472	31%	15,793	37%
2003	42,884	100%	27,328	64%	2,327	5%	13,096	31%	15,423	36%
2004	42,836	100%	27,413	64%	2,212	5%	13,099	31%	15,311	36%
2005	43,510	100%	27,423	63%	2,404	6%	13,582	31%	15,985	37%
2006	42,708	100%	26,633	62%	2,479	6%	13,491	32%	15,970	37%
2007	41,259	100%	25,611	62%	2,494	6%	13,041	32%	15,534	38%
2008	37,423	100%	23,499	63%	2,115	6%	11,711	31%	13,826	37%
2009	33,883	100%	21,051	62%	1,972	6%	10,759	32%	12,731	38%
2010	32,999	100%	21,005	64%	1,771	5%	10,136	31%	11,906	36%
2011	32,367	100%	20,752	64%	1,633	5%	9,878	31%	11,510	36%

Figure 7-A illustrates the historical trend of overall fatalities plotted against alcohol-related and alcohol impaired fatalities. Their general trends are similar, but there was a noticeable decline in alcohol-related fatalities as a proportion of total fatalities during the 1990s. Overall alcohol-related fatalities declined from 60 percent of total fatalities in 1982 to about 40 percent by 1997. Since that time, the proportion has remained roughly constant. A similar trend is evident for fatalities in crashes involving alcohol impairment. Alcohol impaired fatalities declined from 53 percent of all fatalities in 1982 to about 34 percent in 1997, and have remained at roughly 35 percent through 2010.

Table 7-3. Pedestrian and Bicyclist Alcohol Use Related Traffic Fatalities

Year	Total Including Occupants		BAC=.01 to .07		BAC=.08+		BAC=.01+	
	Number	Percent	Number	Percent	Number	Percent	Number	Percent
1982	43,945	100%	15	0.0%	2,133	5%	2,148	5%
1983	42,589	100%	6	0.0%	1,990	5%	1,996	5%
1984	44,257	100%	39	0.1%	2,077	5%	2,117	5%
1985	43,825	100%	107	0.2%	1,961	4%	2,069	5%
1986	46,087	100%	59	0.1%	1,917	4%	1,976	4%
1987	46,390	100%	160	0.3%	1,883	4%	2,043	4%
1988	47,087	100%	78	0.2%	1,988	4%	2,066	4%
1989	45,582	100%	100	0.2%	2,010	4%	2,110	5%
1990	44,599	100%	79	0.2%	1,902	4%	1,980	4%
1991	41,508	100%	80	0.2%	1,772	4%	1,852	4%
1992	39,250	100%	91	0.2%	1,798	5%	1,889	5%
1993	40,150	100%	61	0.2%	1,808	5%	1,869	5%
1994	40,716	100%	86	0.2%	1,595	4%	1,682	4%
1995	41,817	100%	74	0.2%	1,764	4%	1,839	4%
1996	42,065	100%	71	0.2%	1,812	4%	1,883	4%
1997	42,013	100%	74	0.2%	1,664	4%	1,738	4%
1998	41,501	100%	112	0.3%	1,661	4%	1,774	4%
1999	41,717	100%	86	0.2%	1,695	4%	1,782	4%
2000	41,945	100%	89	0.2%	1,546	4%	1,634	4%
2001	42,196	100%	101	0.2%	1,568	4%	1,669	4%
2002	43,005	100%	111	0.3%	1,621	4%	1,731	4%
2003	42,884	100%	100	0.2%	1,582	4%	1,682	4%
2004	42,836	100%	113	0.3%	1,494	3%	1,608	4%
2005	43,510	100%	85	0.2%	1,520	3%	1,605	4%
2006	42,708	100%	115	0.3%	1,653	4%	1,768	4%
2007	41,259	100%	60	0.1%	1,562	4%	1,624	4%
2008	37,423	100%	76	0.2%	1,547	4%	1,623	4%
2009	33,883	100%	59	0.2%	1,390	4%	1,448	4%
2010	32,999	100%	90	0.3%	1,326	4%	1,417	4%
2011	32,367	100%	125	0.4%	1,519	5%	1,645	5%

Nonfatal Injuries

NHTSA collects crash data though a two-tiered system, a system that was redesigned in 1988 to replace the former NASS; the NASS Crashworthiness Data System and the General Estimates System comprise this new method.

The CDS is a probability sample of a subset of police-reported crashes in the United States It offers detailed data on a representative, random sample of thousands of minor, serious, and fatal crashes. The crash in question must be police-reported and must involve property damage and/or personal injury resulting from the crash in order to qualify as a CDS case. It must also include a towed passenger car or light truck or van in transport on a public road or highway. Injuries in vehicles meeting these criteria are analyzed at a level of detail not found in the broader GES.

In contrast, the GES collects data on a sample of all police-reported crashes, without a specific set of vehicle and severity criteria. Although GES collects data on a broader array of crashes, it collects less information on each crash, limiting possible analysis of alcohol involvement. Cases are restricted to a simple "yes," "no," or "unknown" alcohol indication on the police crash report, as observed by the reporting police office. Actual BAC test results are not available through the GES sample.

The GES provides a sample of U.S. crashes by police-reported severity for all crash types. GES records injury severity by person on the KABCO scale (National Safety Council, 1990) from police crash reports as discussed at the beginning of Chapter 2.

KABCO ratings are coarse and inconsistently coded between States and over time. The codes are selected by police officers without medical training, typically without benefit of a hands-on examination. Some of the injured are transported from the scene before the police officer who completes the crash report even arrives. Miller, Viner, et al. (1991) and Blincoe and Faigin (1992) documented great diversity in KABCO coding across cases. O'Day (1993) more carefully quantified variability in use of the A-injury code between States. Viner and Conley (1994) probed how differing State definitions of A-injury contributed to this variability. Miller, Whiting, et al. (1987) found police-reported injury counts by KABCO severity systematically varied between States because of differing State crash reporting thresholds (rules governing which crashes should be reported to the police). Miller and Blincoe (1994) found that State reporting thresholds often changed over time.

Thus police reports inaccurately describe injuries medically and crash databases inaccurately describe motor vehicle crash severity. We adopted a widely used method to refine crash and injury severity. Developed by Miller and Blincoe (1994), numerous studies have used this method, notably in impaired- driving cost estimates in Blincoe (1996); Miller, Lestina, and Spicer (1998); Blincoe et al. (2002); and Zaloshnja and Miller (2009).

To minimize the effects of variability in severity definitions by State, reporting threshold, and police perception of injury severity, the method uses NHTSA data sets that include both police-reported KABCO and medical descriptions of injury in the Occupant Injury Coding system (OIC; AAAM, 1990, 1985). OIC codes include AIS severity score and body region, plus more detailed injury descriptors. We used both 2008–2010 CDS and 1984–1986 NASS data (NASS; NHTSA, 1987). CDS describes injuries to passenger vehicle occupants involved in tow-away crashes. The 1984–1986 NASS data provides the most recent medical description available of injuries to medium/heavy truck and bus occupants, nonoccupants, and others in non-CDS crashes. The NASS data was coded with the 1980 version of AIS, which differs slightly from the 1985 version; but NHTSA made most AIS 85 changes well before their formal adoption. CDS data was coded in AIS 90/98 with coding shifting to AIS 2005 Update 2008 in 2011. We differentiated our analysis of the two versions of AIS because AIS 90/98 scores and OIC codes differ greatly from codes and scores in AIS 85, especially for brain and severe lower limb injury. Garthe, Ferguson, and Early (1996) find that AIS scores shifted for roughly 25 percent of all OICs between AIS 85 and AIS 90/98.

We used 2008–2010 CDS and GES non-CDS weights to weight the CDS and NASS data, respectively, so that they represent estimated counts of people injured in motor vehicle crashes during 2008–2010. In applying the GES weights to old NASS, we controlled for police-reported injury severity, restraint use, alcohol involvement, and occupant type (CDS occupant, non-CDS occupant, and nonoccupant). Weighting NASS data to GES restraint use and alcohol involvement levels updates the NASS injury profile to reflect contemporary belt use and alcohol-involvement levels, although it is imperfect in terms of its representation of airbag use in non-tow-away crashes. At completion of the weighting process, we had a hybrid CDS/NASS casualty-level file—that is, we had an appropriately reweighted NASS record for each injured survivor in each non-CDS crash. Similarly, we reweighted the 2008–2010 CDS file to match GES counts in order to get appropriately weighted unit records for CDS sample strata. From this file we obtained counts of alcohol cases based on all indicators of alcohol use to obtain an initial count of alcohol involved crashes from police-reported crashes. The results are shown in the upper part of Table 4 below:

As noted in chapter 5, GES has historically undercounted police-reported crashes on the order of 10 to 13 percent. Our most recent analysis indicates an undercounting of roughly 10.7 percent for 2010. We therefore multiplied incidence by 1.107 to adjust for systematic undercounting in GES of police crash reports. Also as previously noted, a significant portion of crashes are not reported to police. We assume that these underreporting rates apply to alcohol-involved crashes as well as to overall crashes. We thus divided by estimated fractions reported to the police: 1.0 for people with critical to fatal injuries, 0.953 for people with MAIS3 injuries, 0.794 for MAIS2, 0.725 for MAIS1, 0.469 for uninjured people in injury crashes, and 0.406 for crashes without injuries.[44] The results of these adjustments are shown in the lower half of Table 7-4.

Table 7-4. Alcohol Involvement Identified in Police-Reported Crashes

Alcohol Involvement in Police-Reported Crashes Initially Derived from CDS/GES			
Injury severity	**Total Incidence**	**Alcohol Involved**	**Percentage Alcohol Involved**
PDO	6,187,743	410,414	6.63%
MAIS0	1,782,823	118,235	6.63%
MAIS1	2,204,294	104,230	4.73%
MAIS2	220,982	17,783	8.05%
MAIS3	74,235	8,455	11.39%
MAIS4	13,131	1,574	11.99%
MAIS5	3,861	574	14.86%
Fatal	32,999	13,323	40.37%
Adjusted for GES Undercount and Unreported			
Injury severity	**Total Incidence**	**Alcohol Involved**	**Percentage Alcohol Involved**
PDO	17,007,212	1,128,037	6.63%
MAIS0	4,211,513	279,302	6.63%
MAIS1	3,273,070	154,767	4.73%
MAIS2	305,594	24,592	8.05%
MAIS3	85,883	9,781	11.39%
MAIS4	14,537	1,742	11.99%
MAIS5	4,274	635	14.86%
Fatal	32,999	13,323	40.37%

Underreported Alcohol

Although police accident reports typically include an indication of whether alcohol was involved, the nature of accident investigations often precludes an accurate assessment of alcohol involvement at the crash site. Police underreporting of alcohol involvement has been well documented in numerous studies. Typically, studies on underreporting compare the results of BAC tests administered in medical care facilities to police reports of alcohol involvement. In a 1982 study of injured drivers, Terhune found that police correctly identified 42 percent of drivers who had been drinking. These rates of identification improved at higher BAC levels, ranging from only 18.5 percent of those with BACs of .01to .09, to 48.9 percent for those with BACs of .10 or greater. In a 1990 study, Soderstrom, Birschbach, and Dischinger found that police correctly identified alcohol use in 71 percent of legally intoxicated, injured drivers. Earlier studies by Maull, Kunning, and Hickman in 1984 and Dischinger and Cowley in 1989, found that police correctly identified 57.1 percent and 51.7 percent of intoxicated drivers, respectively. The Dischinger and Cowley study also found a lower identification rate for "involved but not intoxicated" drivers of 28.6 percent. In a 1991 study of injured motorcycle drivers, Soderstrom, Birschbach, and Dischinger found that police correctly identified only half the drivers with positive alcohol readings later identified by the hospital.

These early studies demonstrate that during the late 1980s and early 1990s, the police were identifying approximately half of all legally intoxicated drivers, and about one quarter of all drivers who were alcohol involved, but not legally intoxicated. It is clear from the studies that police are more accurate in identifying alcohol involvement as the BAC rate increases. This may reflect the more obvious nature of impaired behavior on the part of drivers who have higher BAC levels, as well as a tendency to investigate more thoroughly the more serious crashes that result from higher BACs.

In several previous versions of this report (Blincoe & Faigin, 1992, and Blincoe, 1996) the studies cited above were used to estimate the impact of police underreporting of alcohol involvement. In the most recent version (Blincoe et al., 2002), more updated information was used. However, those studies are over a decade old, and when applied to current data, they produced results that imply a higher rate of alcohol involvement in less severe injuries than in fatalities and more severe injuries. This is both counter-intuitive and at odds with historical alcohol involvement patterns. Moreover, over the last decade there has been a concerted effort on the part of Federal, State and local governments to reduce alcohol-related crashes, and this may have improved the rate of alcohol reporting during accident investigations. Data that was more recent was therefore needed to make this adjustment for 2010 data.

The Crash Outcome Data Evaluation System (CODES) is a system that links existing crash and injury data so that specific person, vehicle, and event characteristics can be matched to their medical and financial outcomes. At the time of the 2002 study there were 25 States participating in this program and 17 of these States are part of a data network supporting NHTSA highway safety programs. An effort was made to contact all States participating in NHTSA's CODES project to determine whether data was available that could be used to estimate current alcohol reporting rates. For a variety of reasons, only one State, Maryland, had data that was properly linked to allow a comparison between alcohol assessments in police reports and actual measured BACs. The Maryland data represented 2,070 cases admitted to the R Adams Cowley Shock Trauma Center between 1997 and 1999. The basis for this data was thus similar to most of the studies cited above from the late 80s and early 90s.

An analysis of this data indicated that police were correctly identifying 74 percent of all alcohol involved cases where BACs equaled or exceeded .10 g/dL, and 46 percent of all cases where BACs were positive, but less than .10. This represents a significant improvement from the corresponding rates of only 55 percent and 27 percent that were found in the earlier studies. This was consistent with the expectation that reporting rates have improved, and, when applied to police-reported rates in the NHTSA data bases, the more recent factors produce overall estimates that are consistent with FARS rates of involvement for fatal crashes. However, although this data produce logical results, they were gathered from only one State and there are no data to confirm whether the Maryland experience is typical of the Nation. These estimates were thus subject to the caveat that these results have not been verified by broader studies from more diverse regions. One of the previous studies (Soderstrom, Birschbach, & Dischinger, 1990) was conducted at this same facility and found a higher rate of alcohol recognition than the other studies previously discussed. A second caveat is that, because this data was collected at a trauma unit, they may reflect the more serious cases rather than a sample of all injury levels. There are two different, somewhat offsetting biases that could result from this. Trauma unit cases are more likely to involve emergency transport and treatment which may occur before police are able to gain access to drivers to determine alcohol involvement. This could result in police missing a larger portion of trauma unit cases. On the other hand, the severity of the crash may prompt a more thorough investigation by the police, resulting in a higher rate of correct alcohol identification. It is not clear what the net effect of these biases would be.

Given these caveats, this current paper is based on a more recent study that analyzed what portion of U.S. nonfatal crashes are alcohol-involved and how well police and hospitals detect involvement (Miller et al., 2012). In that study, a capture recapture model estimated alcohol involvement from levels detected by police and hospitals and the extent of detection overlap. The authors analyzed 550,933 Crash Outcome Data Evaluation System driver records from 2006-2008 police crash report censuses probabilistically linked to hospital inpatient and emergency department (ED) discharge censuses for Connecticut, Kentucky (admissions only), Maryland, Nebraska, New York, South Carolina, and Utah. They then computed National estimates from NHTSA's General Estimates System.

Nationally an estimated 7.5 percent of drivers in nonfatal crashes and 12.9 percent of nonfatal crashes were alcohol-involved. (Crashes often involve multiple drivers but rarely are two alcohol-involved.) Police correctly identified an estimated 32 percent of alcohol-involved drivers in non-fatal crashes including 48 percent in injury crashes. Excluding Kentucky, police in the six States reported 47 percent of alcohol involvement for cases treated in EDs and released and 39 percent for admitted cases. In contrast, hospitals reported 28 percent of involvement for ED cases and 51 percent for admitted cases. Underreporting varied widely between States. Police-reported alcohol involvement for 44 percent of those who hospitals reported were alcohol-involved, while hospitals reported alcohol involvement for 33 percent of those who police reported were alcohol-involved. Police alcohol reporting completeness rose with police-reported driver injury severity. At least one system reported 62 percent of alcohol involvement. Based on the combined results from the 6 States that had both admitted and ED data, police records account for 30 to45 percent of total actual alcohol involvement, depending on injury severity. These rates and the resulting estimates of alcohol involvement are summarized in Table 7-5. Note that although fatalities are listed in Table 7-5, they were not examined in the capture-recapture analysis. As noted previously fatal crashes are investigated much more thoroughly than nonfatal crashes and NHTSA's FARS, through both documentation of police and medical records and through modeling for unreported cases, is believed to account for all alcohol involvement in fatal crashes.

BAC Levels

BAC levels are difficult to determine from injury data. Although there are some indications of BAC included in CDS data, the GES has no such indicators. To determine BAC levels, an initial assessment was made that virtually all police-reported BACs for nonfatal crashes represent BACs that are at the .05 BAC level or higher. It is illegal *per se* in every State to drive a motor vehicle with a BAC of .08 or higher. Some State laws establish lesser included offenses at lower BAC levels (most typically at .05 BAC). Unless a crash involves a fatality, police generally do not test or use the alcohol checkbox unless they suspect the driver might be near these levels. In fact, except for fatal crashes, some States do not even allow testing unless a BAC over .08 is suspected. Low BAC levels (especially below .05) are thus unlikely to be registered in police records. An examination of available data from NHTSA's CDS and NASS data systems bears this out. For nonfatal crashes, less than half of 1 percent of nonfatal injuries were recorded as BACS being between .01 and.04 g/dL. However, this primarily represents a limitation in data gathering rather than an indication of near complete absence of crashes at these lower BAC levels. An estimate of crashes at these BAC levels was thus derived from crash probabilities.

Subcategories of BAC levels were calculated as a function of odds ratios for crashes at each specific BAC level compared to exposure at those levels. Odds ratios were derived from a study of relative crash risk conducted by Dunlap and Associates (Blomberg, Peck, Moskowitz, Burns and Fiorentino, 2005). In this study over 2,800 crashes and nearly 15,000 drivers in Long Beach, California and Fort Lauderdale, Florida were sampled to determine the relative risk of crashes at different BAC levels. Logistic regression techniques were used to create a relative risk model which indicated a notable dose-response relationship beginning at 0.04 percent BAC and increasing exponentially at >=.10 percent BAC. The results of this model are summarized in Figure 7-B below.

Table 7-5. Total Alcohol Involvement Adjusted for Unreported Cases

Injury severity	Total Incidence	Percent Identified	Alcohol Involved	Percent Involved
PDO	18,508,632	42.90%	2,629,458	14.21%
MAIS0	4,583,265	42.90%	651,054	14.21%
MAIS1	3,459,200	45.40%	340,897	9.85%
MAIS2	338,730	42.60%	57,728	17.04%
MAIS3	100,740	39.70%	24,638	24.46%
MAIS4	17,086	40.60%	4,292	25.12%
MAIS5	5,749	30.10%	2,110	36.70%
Fatal	32,999	100.00%	13,323	40.37%

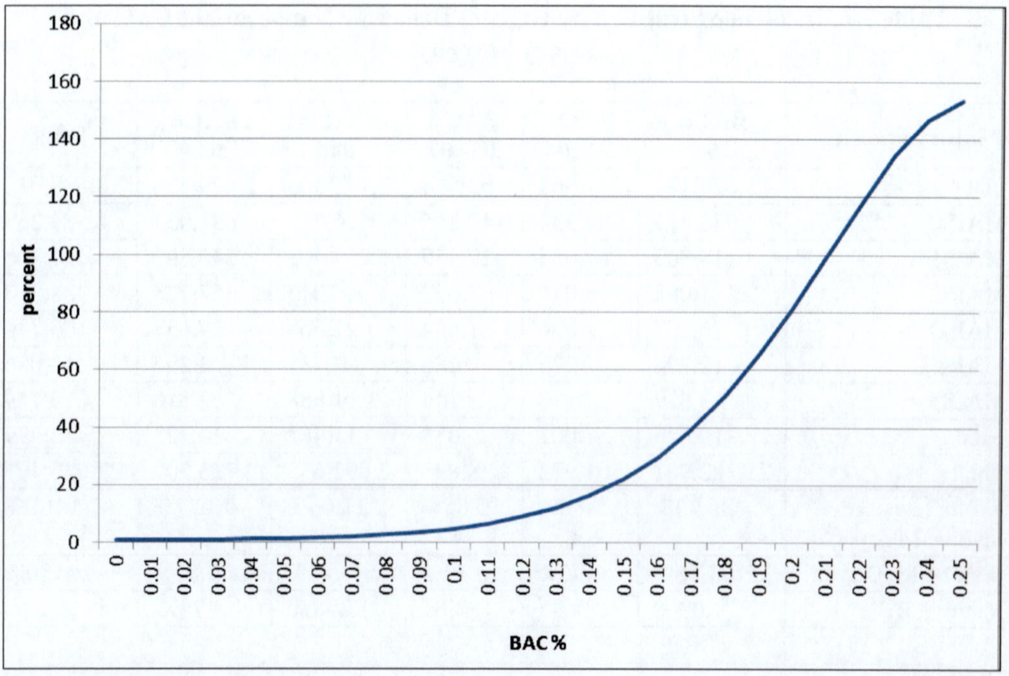

Figure 7-B. Relative Risk of Crash by Blood Alcohol Concentration (Source: Blomberg, Peck, Moskowitz, Burns & Fiorentino, 2005).

The authors found some level of added crash risk beginning at roughly .04 BAC, but this risk rises noticeably at .08 BAC and rises exponentially from .10 BAC and beyond. For example, at .04 BAC the risk of a crash is 18 percent higher than at zero BAC, but at .08 BAC the risk of a crash is 2.69 times as high and at .10 BAC it is 4.79 times as high. To determine BAC distributions, the relative risk ratios of each individual BAC category were combined with exposure data from the same study to estimate the relative risk factor for each grouped BAC category. These grouped relative risk factors were then combined with National exposure data from Lacey et al. to determine the distribution of each grouped BAC category as follows:.

$$rn*en/ry*ey$$

where: rn = relative risk ratio of specific BAC category
en = exposure of specific BAC category
ry = relative risk of broader BAC category
ey = exposure of broader BAC category

The broader categories are those derived above for nonfatal injuries, which were all assumed to be BAC>=.05, and the difference between these and the total incidence, which represent 0-.04 BAC. Essentially, this divides alcohol BAC cases into two broad categories at the .05 BAC level. The .08+BAC category was then derived using the above formula from the >=.05 BAC total and the .01-.04 BAC category was derived from the <.05 BAC category. The inputs used for each category and the resulting BAC distributions are shown in Table 7-6.

Table 7-6. Incidence Stratified by Highest Driver or Nonoccupant BAC and Injury Severity

Injury severity	BAC= 0	BAC= .01-.04	BAC= .05-.07	BAC>= .08	BAC= .01+	Total
PDO	15,879,174	341,369	162,584	2,125,505	2,629,458	18,508,632
MAIS0	3,932,211	84,534	40,255	526,265	651,054	4,583,265
MAIS1	3,118,303	67,037	19,459	254,401	340,897	3,459,200
MAIS2	281,002	6,041	3,672	48,015	57,728	338,730
MAIS3	76,102	1,636	1,634	21,368	24,638	100,740
MAIS4	12,794	275	286	3,731	4,292	17,086
MAIS5	3,639	78	144	1,888	2,110	5,749
Fatal	19,676	1,002	859	11,462	13,323	32,999
Total	23,322,901	501,972	228,893	2,992,635	3,723,500	27,046,401
% of Crash-Involved People	86.23%	1.86%	0.85%	11.06%	13.77%	100%
% of Miles Driven	97.18%	1.96%	0.39%	0.47%	2.82%	100%
Relative Risk	1.0000	1.0645	1.6581	17.9870	4.7477	

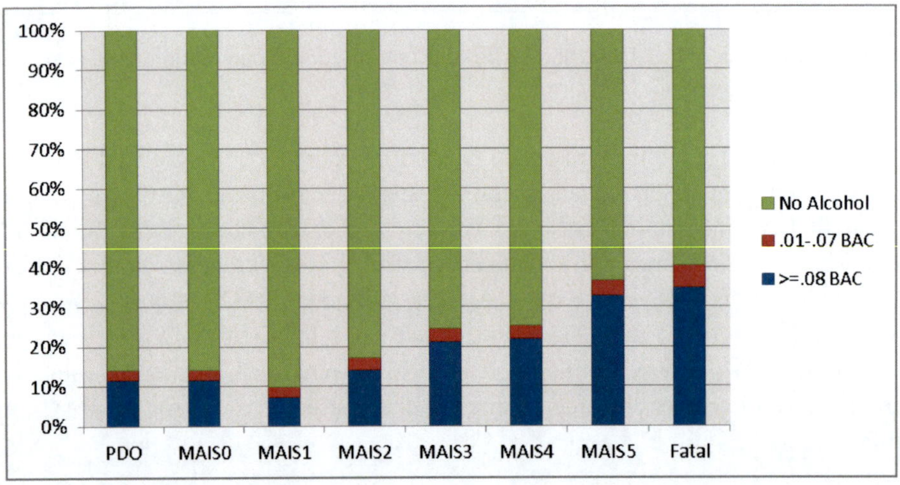

Figure 7-C. Relative Incidence of Impaired and Unimpaired BAC Levels to All Crashes.

The results illustrate the disproportionate impact that high BACs have on crash incidence. Less than 1 percent of overall miles are driven by impaired drivers (.08+ BAC), but they account for over 11 percent of all vehicle crashes, and over 80 percent of all alcohol related crashes, including 86 percent of all fatalities.

Figure 7-C illustrates the relative incidence of alcohol impaired and not impaired crashes to all crashes. Alcohol involved crashes account for 40 percent of all fatal crashes. There is a clear trend towards increased alcohol involvement as injury severity increases. This figure illustrates the fact that alcohol not only increases the likelihood of crashes, but their severity as well.

Figure 7-D illustrates the relative incidence of crashes at various BAC levels. The vast majority of all alcohol related crashes occur at legally impaired BAC levels of .08 and above.

Alcohol-Involved Crash Costs

The costs of alcohol-involved crashes tend to exceed those of non-alcohol-involved crashes due to a variety of factors. The first is a general tendency toward greater relative severity of alcohol-involved crashes. For all crashes, fatalities are approximately 0.8 percent of injured survivors. This rate nearly quadruples for crashes involving alcohol. Similarly, the rate for critical injuries (MAIS 5) triples for alcohol cases and for severe injuries (MAIS 4) it more than doubles. The more severe and expensive injuries represent a much higher portion of alcohol-involved cases. A second factor is demographics. Males are disproportionately represented in alcohol-involved crashes and this makes the cost for each alcohol-involved case higher. This occurs because males have higher earnings and participation in the work force than females; thus there is a higher lost productivity cost associated with these crashes. In non-alcohol-involved crashes, the gender distribution is more evenly distributed. In addition, the victims of alcohol-involved crashes tend to be of an age group where lost productivity is maximized by the discounting process.

Unit costs specific to alcohol-involved crashes were developed by extracting cases with police-reported alcohol from the previously discussed file based on 2008-2010 weights. As noted above, virtually all of these cases represent crashes with BACs of 0.5 or greater. Unit costs for these crashes were thus weighted by the relative incidence of 0.05 BAC+ cases within all positive BAC cases. The unit costs of cases with BACs of 0.0-0.04 were then derived as a function of the relative incidence and cost of the 0.05+BAC crashes and All Crashes as follows:

$$b=(cz-ax)/y$$

where: b=unit cost in crashes with BAC<0.05

c=average unit cost of all crashes

z=incidence of all crashes

a = unit cost of crashes with BAC>=0.05

x=incidence of crashes with BAC>=0.05

y = incidence of crashes with BAC<0.05

The results of this process are shown in Tables 7-7, 7-8, and 7-9

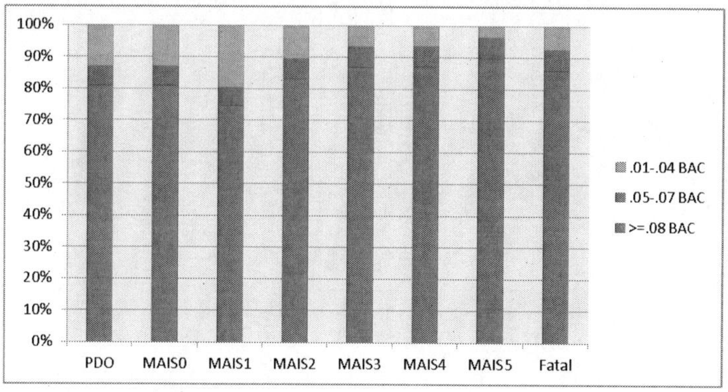

Figure 7-D. Relative Incidence of BAC Levels in Alcohol Involved Crashes by Injury Severity.

Table 7-7. Average Unit Costs, BAC>=.05 injuries, and BAC > .00 Fatalities (2010 Dollars)

	PDO	MAIS0	MAIS1	MAIS2	MAIS3	MAIS4	MAIS5	Fatal
Medical	$0	$0	$4,145	$26,138	$75,090	$201,540	$450,686	$11,317
Emergency Services	$28	$21	$89	$194	$416	$838	$855	$902
Market Productivity	$0	$0	$3,144	$41,582	$116,400	$185,948	$361,786	$1,156,859
Household Productivity	$60	$45	$1,050	$12,409	$35,861	$46,499	$110,424	$315,326
Insurance Admin.	$191	$143	$4,765	$11,152	$25,906	$39,551	$81,812	$28,322
Workplace Costs	$62	$46	$341	$2,644	$5,776	$6,361	$11,091	$11,783
Legal Costs	$0	$0	$1,759	$8,197	$20,955	$37,545	$95,220	$106,488
Injury Subtotal	$341	$255	$15,293	$102,316	$280,403	$518,283	$1,111,875	$1,630,997
Congestion Costs	$1,077	$760	$1,109	$1,197	$1,434	$1,511	$1,529	$5,720
Property Damage	$2,444	$1,828	$5,404	$5,778	$10,882	$16,328	$15,092	$11,212
Economic Subtotal	$3,862	$2,843	$21,806	$109,291	$292,719	$536,122	$1,128,496	$1,647,929
QALYs	$0	$0	$24,692	$365,054	$869,987	$2,117,774	$4,994,639	$8,495,097
Comprehensive Total	$3,862	$2,843	$46,498	$474,345	$1,162,706	$2,653,896	$6,123,135	$10,143,026

* Note: Unit costs are expressed on a per-person basis for all injury levels. PDO costs are expressed on a per-damaged-vehicle basis.

Table 7-8. Average Unit Costs, BAC=.00-0.04 injuries, and BAC = .00 Fatalities (2010 Dollars)

	PDO	MAIS0	MAIS1	MAIS2	MAIS3	MAIS4	MAIS5	Fatal
Medical	$0	$0	$3,771	$24,183	$69,663	$195,488	$436,661	$11,317
Emergency Services	$28	$21	$89	$194	$416	$838	$855	$902
Market Productivity	$0	$0	$2,833	$39,287	$108,220	$171,280	$346,925	$781,860
Household Productivity	$60	$45	$932	$11,833	$32,808	$44,007	$105,237	$272,700
Insurance Admin.	$191	$143	$3,864	$9,049	$23,887	$36,702	$78,958	$28,322
Workplace Costs	$62	$46	$341	$2,644	$5,776	$6,361	$11,091	$11,783
Legal Costs	$0	$0	$1,380	$6,477	$19,258	$34,619	$88,998	$106,488
Injury Subtotal	$341	$255	$13,210	$93,667	$260,028	$489,296	$1,068,724	$1,213,372

	PDO	MAIS0	MAIS1	MAIS2	MAIS3	MAIS4	MAIS5	Fatal
Congestion Costs	$1,077	$760	$1,109	$1,197	$1,434	$1,511	$1,529	$5,720
Property Damage	$2,444	$1,828	$5,404	$5,778	$10,882	$16,328	$15,092	$11,212
Economic Subtotal	$3,862	$2,843	$19,723	$100,642	$272,344	$507,135	$1,085,345	$1,230,304
QALYs	$0	$0	$23,116	$336,518	$786,674	$2,012,806	$4,351,042	$7,240,587
Comprehensive Total	$3,862	$2,843	$42,839	$437,159	$1,059,018	$2,519,941	$5,436,387	$8,470,891

* Note: Unit costs are expressed on a per-person basis for all injury levels. PDO costs are expressed on a per-damaged-vehicle basis.

Table 7-9. Average Unit Costs, All Positive BAC Injuries and Fatalities (2010 Dollars)

	PDO	MAIS0	MAIS1	MAIS2	MAIS3	MAIS4	MAIS5	Fatal
Medical	$0	$0	$4,071	$25,933	$74,729	$201,152	$450,168	$11,317
Emergency Services	$28	$21	$89	$194	$416	$838	$855	$902
Market Productivity	$0	$0	$3,083	$41,342	$115,857	$185,008	$361,237	$1,156,859
Household Productivity	$60	$45	$1,027	$12,349	$35,658	$46,339	$110,232	$315,326
Insurance Admin.	$191	$143	$4,588	$10,932	$25,772	$39,369	$81,707	$28,322
Workplace Costs	$62	$46	$341	$2,644	$5,776	$6,361	$11,091	$11,783
Legal Costs	$0	$0	$1,685	$8,017	$20,842	$37,358	$94,990	$106,488
Injury Subtotal	$341	$255	$14,883	$101,411	$279,050	$516,425	$1,110,280	$1,630,997
Congestion Costs	$1,077	$760	$1,109	$1,197	$1,434	$1,511	$1,529	$5,720
Property Damage	$2,444	$1,828	$5,404	$5,778	$10,882	$16,328	$15,092	$11,212
Economic Subtotal	$3,862	$2,843	$21,396	$108,386	$291,366	$534,264	$1,126,901	$1,647,929
QALYs	$0	$0	$24,382	$362,068	$864,455	$2,111,048	$4,970,847	$8,495,097
Comprehensive Total	$3,862	$2,843	$45,778	$470,453	$1,155,821	$2,645,312	$6,097,748	$10,143,026

* Note: Unit costs are expressed on a per-person basis for all injury levels. PDO costs are expressed on a per-damaged-vehicle basis.

Table 7-10. Summary of Total Economic Costs by BAC Level (Millions of 2010 Dollars)

	BAC= 0	BAC= .01-.04	BAC= .05-07	BAC>= .08	BAC= .01+	Total
PDO	$61,325	$1,318	$628	$8,209	$10,155	$71,480
MAIS0	$11,179	$240	$114	$1,496	$1,851	$13,030
MAIS1	$61,503	$1,322	$424	$5,547	$7,294	$68,797
MAIS2	$28,280	$608	$401	$5,248	$6,257	$34,537
MAIS3	$20,726	$446	$478	$6,255	$7,179	$27,905
MAIS4	$6,488	$139	$153	$2,000	$2,293	$8,781
MAIS5	$3,950	$85	$163	$2,131	$2,378	$6,327
Fatal	$24,207	$1,651	$1,416	$18,889	$21,955	$46,163
Total	$217,659	$5,810	$3,778	$49,774	$59,362	$277,020
% Total Alcohol Costs	NA	9.79%	6.36%	83.85%	100.00%	NA
% Total	78.57%	2.10%	1.36%	17.97%	21.43%	100.00%

Table 7-11. Percent of Economic Injury Costs by Alcohol Involvement Rate

	BAC= 0	BAC=.01-.04	BAC=.05-.07	BAC>=.08	BAC=.01+	Total
PDO	85.79%	1.84%	0.88%	11.48%	14.21%	100.00%
MAIS0	85.79%	1.84%	0.88%	11.48%	14.21%	100.00%
MAIS1	89.40%	1.92%	0.62%	8.06%	10.60%	100.00%
MAIS2	81.88%	1.76%	1.16%	15.19%	18.12%	100.00%
MAIS3	74.27%	1.60%	1.71%	22.42%	25.73%	100.00%
MAIS4	73.89%	1.59%	1.74%	22.78%	26.11%	100.00%
MAIS5	62.42%	1.34%	2.57%	33.67%	37.58%	100.00%
Fatal	52.44%	3.58%	3.07%	40.92%	47.56%	100.00%
Total	78.57%	2.10%	1.36%	17.97%	21.43%	100.00%

Table 7-10 lists the aggregate 2010 costs of alcohol related crashes, and Table 7-11 lists the proportion of total economic crash costs that each BAC level represents. Alcohol is involved in crashes that account for 14 percent of the costs of PDO crashes, 17 percent of the costs that result from nonfatal injuries and 48 percent of the costs that result from fatalities. Overall, these crashes are responsible for 21 percent of total economic costs. The impact of alcohol-involved crashes on overall costs is thus higher than would be indicated by the alcohol-involved incidence rates. Overall, alcohol involved crashes cost $59 billion in economic costs in 2010, with 84 percent of this or $50 billion, occurring in crashes where the highest BAC was >=.08.

Table 7-12 lists the aggregate 2010 comprehensive costs of alcohol related crashes, and Table 7-13 lists the proportion of total comprehensive crash costs that each BAC level represents. Alcohol is involved in crashes that account for 14 percent of the societal harm of PDO crashes, 20 percent of the harm that result from nonfatal injuries, and 45 percent of the harm that result from fatalities. All alcohol involved crashes are responsible for 28 percent of total societal harm from motor vehicle crashes, but crashes with BAC>=.08 are responsible for 85 percent of this or 24 percent. The impact of alcohol-involved crashes on overall costs is thus higher than would be indicated by the alcohol-involved incidence rates. Overall, alcohol involved crashes cost $243 billion in comprehensive societal costs in 2010, with 85 percent of this or $207 billion, occurring in crashes where the highest BAC was >=.08.

Alcohol Crash Causation

Inebriated drivers often experience impaired perceptions that can lead to risky behavior such as speeding, reckless driving, and failure to wear seat belts. They also experience reduced reaction times, which can make it more difficult for them to perform defensive safety maneuvers. As a result, there is a general tendency to equate the presence of alcohol with crash causation. However, there are clearly some instances in which crashes would occur regardless of whether the driver had consumed alcohol. For example, if a distracted texting driver were to run into a driver with a positive BAC who was stopped at a red light, a police investigation or medical records might record that the struck driver had a positive BAC, even though that driver was not at fault. In this case, the crash would be recorded as alcohol-involved, even though alcohol was not a causative factor.

Table 7-12. Total Comprehensive Costs by BAC Level (Millions of 2010 Dollars)

	BAC= 0	BAC= .01-.04	BAC= .05-.07	BAC>= .08	BAC= .01+	Total
PDO	$61,325	$1,318	$628	$8,209	$10,155	$71,480
MAIS0	$11,179	$240	$114	$1,496	$1,851	$13,030
MAIS1	$133,586	$2,872	$905	$11,829	$15,606	$149,192
MAIS2	$122,842	$2,641	$1,742	$22,776	$27,158	$150,001
MAIS3	$80,593	$1,733	$1,900	$24,845	$28,477	$109,070
MAIS4	$32,241	$693	$758	$9,902	$11,353	$43,594
MAIS5	$19,783	$424	$882	$11,560	$12,866	$32,649
Fatal	$166,673	$10,163	$8,713	$116,259	$135,136	$301,809
Total	$628,223	$20,084	$15,642	$206,876	$242,602	$870,826
% Total Alcohol Costs	NA	8.28%	6.45%	85.27%	100.00%	NA
% Total	72.14%	2.31%	1.80%	23.76%	27.86%	100.00%

Table 7-13. Percent of Comprehensive Injury Costs by Alcohol Involvement Rate

	BAC= 0	BAC=.01-.04	BAC=.05-.07	BAC>=.08	BAC=.01+	Total
PDO	85.79%	1.84%	0.88%	11.48%	14.21%	100.00%
MAIS0	85.79%	1.84%	0.88%	11.48%	14.21%	100.00%
MAIS1	89.54%	1.92%	0.61%	7.93%	10.46%	100.00%
MAIS2	81.89%	1.76%	1.16%	15.18%	18.11%	100.00%
MAIS3	73.89%	1.59%	1.74%	22.78%	26.11%	100.00%
MAIS4	73.96%	1.59%	1.74%	22.71%	26.04%	100.00%
MAIS5	60.59%	1.30%	2.70%	35.41%	39.41%	100.00%
Fatal	55.22%	3.37%	2.89%	38.52%	44.78%	100.00%
Total	72.14%	2.31%	1.80%	23.76%	27.86%	100.00%

Miller, Spicer and Levy (1999) estimated the percentages of alcohol-related crashes that are actually attributable to alcohol. In this study they examined the probability of crash involvement for drivers based on their BAC level and then removed the normal risk of crash involvement without alcohol from the overall risk found for drivers with positive BACs. Their study found that 94 percent of crashes at BACs of .10 or higher, and 31 percent of crashes with positive BACs less than .10, were actually caused by alcohol. The remaining crashes were due to bad weather, poor road conditions, non-drinking drivers, etc. Currently .08 BAC is considered to be the definition of "illegal *per se*" alcohol impairment rather than 0.10. More recently, Blomberg et al. (2005) examined the relative crash risk of drinking and non- drinking drivers. The methods and results of this study were discussed previously (see Figure B above). Table 6 displayed the relative risk for various BAC categories that were derived from Blomberg and colleagues' BAC specific risk factors. These factors can be used to estimate the incidence of crashes where alcohol consumption actually contributed to the crash occurrence across the various BAC groupings examined in this report. These proportions were estimated as the ratio of the added risk in an alcohol involved crash to the total risk in this crash. Specifically:

$$y=(r-1)/r$$

where: y = proportion of BAC + crashes that are attributable to alcohol.
r= relative risk ratio of specific BAC category

Table 7-14 and Figures 7-E and 7-F illustrate the results of this process. The second to the last row in Table 7-14 lists the relative risk calculated from data in Dunlop, while the last row lists the proportion of injuries in each BAC category that are attributable to alcohol. Roughly 6 percent of BAC = .01-.04 injuries, 40 percent of BA = .05-.07 injuries, and 94 percent of BAC>= .08 injuries are attributable to alcohol. The increasing proportions are expected since higher BAC levels cause more inebriation, with its associated reduction in awareness and motor skills. Overall, about 79 percent of injuries from crashes recorded as alcohol-involved can be attributed to alcohol as a causative factor. This is roughly the same percentage calculated in Blincoe et al., 2002 (80.8 percent), which was based on the earlier Miller, Spicer, and Levy analysis. Alcohol thus appears to be a causative factor in roughly 80 percent of cases coded as alcohol-involved, but is irrelevant to crash causation in the other 20 percent of cases.

Table 7-14. Injuries Attributable to Alcohol Use by BAC Level

Injury severity	BAC=.01-.04	BAC=.05-.079	BAC>=.08	BAC=.01+
PDO	20,688	64,529	2,007,336	2,092,553
MAIS0	5,123	15,977	497,007	518,107
MAIS1	4,063	7,723	240,257	252,043
MAIS2	366	1,458	45,346	47,169
MAIS3	99	649	20,180	20,928
MAIS4	17	113	3,524	3,654
MAIS5	5	57	1,783	1,845
Fatal	61	341	10,825	11,226

Injury severity	BAC=.01-.04	BAC=.05-.079	BAC>=.08	BAC=.01+
Total	30,421	90,846	2,826,257	2,947,525
Relative Risk	1.0645	1.6581	17.9870	4.7477
% Attributable to Alcohol	6.06%	39.69%	94.44%	79.16%

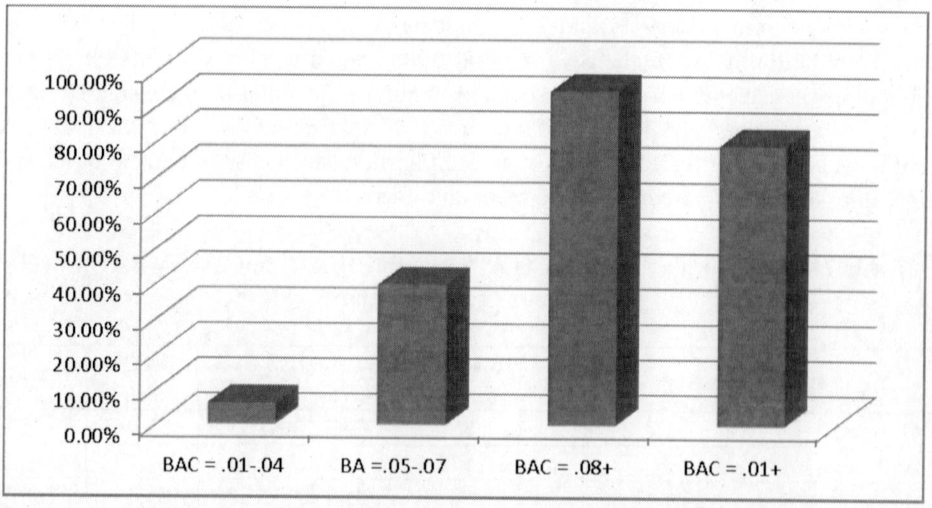

Figure 7-E. Percent of Positive BAC Crashes Attributable to Alcohol by BAC Level.

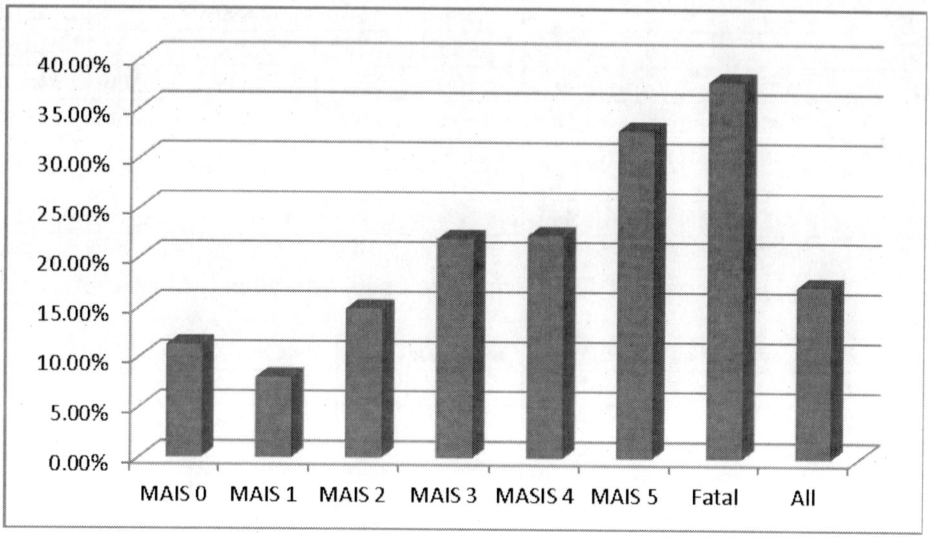

Figure 7-F. Percent of Injuries Attributable to Alcohol by Injury Severity Level.

To estimate the economic cost of crashes actually attributable to alcohol, the incidence from Table 7-14 was combined with the unit costs from Tables 7-7 and 7-8. The results, summarized in Table 7-15, indicate that alcohol causes crashes that result in roughly $49 billion in economic costs annually. This accounts for 82 percent of the crash costs associated with crashes that are considered alcohol-involved.

It represents 18 percent of all crash costs (including those without alcohol involvement), accounting for 11 percent of PDO costs, 14 percent of nonfatal injury costs, and 40 percent of fatality costs.

To estimate the comprehensive cost of crashes actually attributable to alcohol, the incidence from Table 7-14 was combined with the unit costs from Tables 7-7 and 7-8. The results, summarized in Table 7-17 and 7-18, indicate that alcohol causes crashes that result in roughly $199 billion in comprehensive societal costs annually. This accounts for 82 percent of the comprehensive crash costs associated with crashes that are considered alcohol-involved. It represents 23 percent of all crash costs (including those without alcohol involvement, accounting for 11 percent of societal harm from PDOs, 16 percent of harm from nonfatal injuries, and 37 percent of harm from fatalities.

Table 7-15. Economic Crash Costs Attributable to Alcohol Use by BAC Level (Millions of 2010 Dollars)

Injury severity	BAC= .01-04	BAC= .05-079	BAC>= .08	BAC= .01+	Total
PDO	$80	$249	$7,752	$8,081	$71,480
MAIS0	$15	$45	$1,413	$1,473	$13,030
MAIS1	$80	$168	$5,239	$5,488	$68,797
MAIS2	$37	$159	$4,956	$5,152	$34,537
MAIS3	$27	$190	$5,907	$6,124	$27,905
MAIS4	$8	$61	$1,889	$1,958	$8,781
MAIS5	$5	$65	$2,012	$2,082	$6,327
Fatal	$100	$562	$17,838	$18,500	$46,163
Total	$352	$1,499	$47,007	$48,858	$277,020
% of Total Alcohol Involved Costs Attributable to Alcohol	6.06%	39.69%	94.44%	82.31%	
% of Total Costs Attributable to Alcohol	0.13%	0.54%	16.97%	17.64%	

Table 7-16. Percent of Total Economic Costs Attributable to Alcohol

Injury severity	BAC=.01-.04	BAC=.05-.079	BAC>=.08	BAC=.01+
PDO	0.11%	0.35%	10.85%	11.31%
MAIS0	0.11%	0.35%	10.84%	11.30%
MAIS1	0.12%	0.24%	7.62%	7.98%
MAIS2	0.11%	0.46%	14.35%	14.92%
MAIS3	0.10%	0.68%	21.17%	21.95%
MAIS4	0.10%	0.69%	21.51%	22.30%
MAIS5	0.08%	1.02%	31.80%	32.90%
Fatal	0.22%	1.22%	38.64%	40.08%
Total	0.13%	0.54%	16.97%	17.64%

Table 7-17. Comprehensive Crash costs Attributable to Alcohol Use by BAC Level (Millions of 2010 Dollars)

Injury severity	BAC= .01-04	BAC= .05-079	BAC>= .08	BAC= .01+	Total
PDO	$80	$249	$7,752	$8,081	$71,480
MAIS0	$15	$45	$1,413	$1,473	$13,030
MAIS1	$174	$359	$11,171	$11,705	$149,192
MAIS2	$160	$691	$21,509	$22,361	$150,001
MAIS3	$105	$754	$23,463	$24,323	$109,070
MAIS4	$42	$301	$9,351	$9,694	$43,594
MAIS5	$26	$350	$10,918	$11,293	$32,649
Fatal	$616	$4	$109,796	$110,416	$301,809
Total	$1,217	$2,754	$195,374	$199,346	$870,826
% of Total Alcohol Involved Costs Attributable to Alcohol	6.06%	17.61%	94.44%	82.17%	
% of Total Costs Attributable to Alcohol	0.14%	0.32%	22.44%	22.89%	

Table 7-18. Percent of Total Comprehensive Costs Attributable to Alcohol

Injury severity	BAC=.01-.04	BAC=.05-.079	BAC>=.08	BAC=.01+
PDO	0.11%	0.35%	10.85%	11.31%
MAIS0	0.11%	0.35%	10.84%	11.30%
MAIS1	0.12%	0.24%	7.49%	7.85%
MAIS2	0.11%	0.46%	14.34%	14.91%
MAIS3	0.10%	0.69%	21.51%	22.30%
MAIS4	0.10%	0.69%	21.45%	22.24%
MAIS5	0.08%	1.07%	33.44%	34.59%
Fatal	0.20%	0.00%	36.38%	36.58%
Total	0.14%	0.32%	22.44%	22.89%

8. SPEEDING

Excess speed can contribute to both the frequency and severity of motor vehicle crashes. At higher speeds, additional time is required to stop a vehicle and more distance is traveled before corrective maneuvers can be implemented. Speeding reduces a driver's ability to react to emergencies created by driver inattention; by unsafe maneuvers of other vehicles; by roadway hazards; by vehicle system failures (such as tire blowouts); or by hazardous weather conditions. The fact that a vehicle was exceeding the speed limit does not necessarily mean that this was the cause of the crash, but the probability of avoiding the crash would likely be greater had the driver or drivers been traveling at slower speeds.

A speed-related crash is defined as any crash in which the police indicate that one or more drivers involved was exceeding the posted speed limit, driving too fast for conditions, driving at a speed greater than reasonable or prudent, exceeding a special speed limit or zone, or racing. FARS data indicate that in 2010, a total of 10,536 fatalities, representing 32 percent of all motor vehicle fatalities, occurred in speed-related crashes.

To estimate the cost of these crashes, we examined the relative incidence of each injury severity level that was represented by crashes that were speed related. These estimates reflect the relative proportions of specific injury severities that occur under each scenario. GES was used for each non-fatal case, while FARS was used for each fatal case. Each case in FARS contained information regarding speeding status, so the proportion of fatalities that occurred under each scenario was obtained directly from the FARS database. For nonfatal injuries and PDOs, GES data was queried to determine whether the case fell under the scenario or not. However, GES data is only recorded using the KABCO severity system, whereas this report is based on the Abbreviated Injury Scale. To translate GES data to an MAIS basis, we used a variety of KABCO/MAIS translators. For CDS equivalent crashes, we used a current translator derived from 2000-2008 CDS data. Since this data is relatively recent, they reflect roughly current levels of seat belt usage. For non-CDS cases, the only available data from which to develop translators were contained in the 1982-1986 NASS files. Seat belt use has increased dramatically since this time. Observed belt use during this period ranged from roughly 10-37 percent as public awareness of the importance of belt use and belt use laws were just beginning to take hold in 1986. Belt use has since risen dramatically, and has been between 80 and 85 percent since 2004. Belt use can influence injury reporting significantly in a number of ways. It changes the nature of injuries by preventing many more visible injuries (such as head/face contact with the windshield) but replaces them with often less visible (and also typically less serious) abdominal injuries such as bruising caused by pressure from the belt across the torso. This can influence the relationship between the KABCO reported injury severity and the corresponding MAIS injury level. For this reason, separate translators were developed from the 1982-86 NASS data for non-CDS cases where the victim was belted, unbelted, unknown belted status, and for nonoccupants/motorcyclists. These translators are presented in Tables 13-1 through 13-5 in Chapter 13.

2010 GES KABCO incidence counts were obtained both for speed involved and uninvolved cases. Consistent with NHTSA publication practice, cases where speed involvement was unknown were grouped with the uninvolved cases. Thus, one set of incidence counts was obtained for speed involved, and another for all other crashes. Each of this data sets was run through its corresponding translator to produce a set of MAIS based injury counts. These counts from each grouping (CDS equivalent cases, belted non-CDS cases, unbelted non-CDS cases, unknown belt status non-CDS cases, and nonoccupant/motorcycle cases) were added together to produce a total MAIS injury profile for each scenario. The percentage of each MAIS injury incidence that was appropriate to each scenario was then calculated as:

$$x=a/(a+b)$$

where x is the percentage of incidence attributable to speed related crashes a = the incidence of speed related crashes

b = the incidence of crashes not related to speed, including those where the speed related variable was coded unknown

The speed attributable portion of each MAIS level was then multiplied by the total cost of all 2010 crashes for that MAIS level and the MAIS level results were summed to produce the total cost of each crash scenario. MAIS0 portions were calculated using the same procedure described elsewhere in this report for Urban/Rural crashes, based on the relative incidence of MAIS 0 cases in injury crashes. The PDO portion was based on a direct count of PDO vehicles from each crash scenario compared to those not in that scenario.

Table 8-1. Economic Costs of Speed Related Crashes (Millions of 2010 Dollars)

	% Speed Related	Incidence		Total Economic Crash Costs		
		Total	Speed related	Total	Speed Related	Other
PDO Vehicles	16.28%	18,508,632	3,013,887	$71,480	$11,640	$59,841
MAIS0	20.54%	4,583,265	941,619	$13,030	$2,677	$10,353
MAIS1	20.51%	3,459,200	709,566	$68,797	$14,112	$54,685
MAIS2	20.00%	338,730	67,733	$34,537	$6,906	$27,631
MAIS3	20.08%	100,740	20,234	$27,905	$5,605	$22,300
MAIS4	22.34%	17,086	3,816	$8,781	$1,961	$6,820
MAIS5	22.72%	5,749	1,306	$6,327	$1,438	$4,890
Fatalities	31.93%	32,999	10,536	$46,163	$14,740	$31,423
Total	17.63%	27,046,402	4,768,697	$277,020	$59,078	$217,943
Percent of Total		100.00%	17.63%	100.00%	21.33%	78.67%

Table 8-2. Comprehensive Costs of Speed Related Crashes (Millions of 2010 Dollars)

	% Speed Related	Incidence		Total Comprehensive Crash Costs		
		Total	Speed related	Total	Speed Related	Other
PDO Vehicles	16.28%	18,508,632	3,013,887	$71,480	$11,640	$59,841
MAIS0	20.54%	4,583,265	941,619	$13,030	$2,677	$10,353
MAIS1	20.51%	3,459,200	709,566	$149,192	$30,603	$118,589
MAIS2	20.00%	338,730	67,733	$150,001	$29,994	$120,006
MAIS3	20.08%	100,740	20,234	$109,070	$21,907	$87,164
MAIS4	22.34%	17,086	3,816	$43,594	$9,737	$33,857
MAIS5	22.72%	5,749	1,306	$32,649	$7,418	$25,231
Fatalities	31.93%	32,999	10,536	$301,809	$96,366	$205,443
Total	17.63%	27,046,402	4,768,697	$870,826	$210,342	$660,484
Percent of Total		100.00%	17.63%	100.00%	24.15%	75.85%

The results of this process are summarized in Tables 8-1 and 8-2 for economic and comprehensive costs. Speed related crashes resulted in 10,536 fatalities, over 800,000 nonfatal injuries, and over 3 million PDO damaged vehicles in 2010. This represents 32 percent of all fatalities and roughly 20 percent of all nonfatal crashes (including both nonfatal injury and PDO). Speed related crashes caused $59 billion in economic costs and $210 billion in comprehensive costs, accounting for 21 percent of all economic costs and 24 percent of all societal harm (measured as comprehensive costs) from motor vehicle crashes.

One note of caution is in order when using these estimates - there is a significant overlap between alcohol involvement and speed. Many speed-related crashes involved alcohol and vice-versa. These two estimates should not be added together in order to account for the portion of costs that represent the combined factors of speed and alcohol. This same caveat applies to many of the other scenarios examined in this report, as multiple factors can be involved in any given crash.

9. DISTRACTED DRIVING

Driver error has long been recognized as the primary cause of motor vehicle crashes. In a landmark 1979 Tri-Level study by the University of Indiana (Teat et al., 1999), human factors such as speeding, inattention, distraction, and performance errors were found to be a factor in 92.6 percent of all crashes. The Tri-Level study found that inattention was a definite crash cause in roughly 9.8 percent and a probable cause in 15.0 percent of crashes. It also found that "internal distraction" was a definite cause in 5.7 percent of crashes and a probable cause in 9.0 percent. More recently, the National Motor Vehicle Crash Causation Survey (NMVCCS, NHTSA, 2008, July) sponsored by NHTSA found that driver related factors were the primary cause in 95.4 percent of crashes. Driver factors include both performance errors and errors related to non-driving activities, which typically involve distraction, inattention, inadequate surveillance, etc. Distraction, including interior distraction, exterior distraction, and inattention, was involved in about 17.7 percent of all cases where the critical pre-crash event was attributed to drivers. With vehicles traveling and interacting with other vehicles at high speeds, even momentary distraction can result in a crash.

For the National databases, FARS and GES, NHTSA essentially defines distraction to include both interior and exterior sources of distraction including inattentive driving. Types of distraction include talking on cell phones, texting, talking to other passengers, adjusting interior devices such as radios or mirrors, eating or drinking, diverting your attention to an exterior object, person , or event, or being lost in thought. All of these activities can potentially distract drivers from the task of safely driving an automobile. Data indicate that distracted driving is playing a substantial role in motor vehicle crashes:[45]

- In 2010, about 10 percent of fatal crashes were reported as distraction-affected crashes.
- Eighteen percent of injury crashes in 2010 were reported as distraction-affected crashes.

- In 2010, there were 3,267 people killed in crashes involving distracted drivers and an estimated additional 735,000 were injured in motor vehicle crashes involving distracted drivers.
- Of those people killed in distraction-affected crashes, 419 occurred in crashes in which at least one of the drivers was using a cell phone (13 percent of fatalities in distraction-affected crashes) at the time of the crash. Use of a cell phone includes talking/listening to a cell phone, dialing/texting a cell phone, or other cell-phone-related activities.[46]
- Of those injured in distraction-affected crashes, an estimated 27,000 were injured in crashes that involved the use of cell phones at the time of the crashes (5 percent of injured people in distraction-affected crashes).
- Eleven percent of all drivers under age 20 involved in fatal crashes were reported as distracted at the time of the crashes. This age group has the largest proportion of drivers who were distracted (NHTSA, 2012, September).
- For drivers under age 20 involved in fatal crashes, 19 percent of the distracted drivers were distracted by the use of cell phones (NHTSA, 2012, September).

To estimate the cost of distracted driving crashes, we examined the relative incidence of each injury severity level that was represented by distraction affected crashes. These incidence estimates reflect the relative proportions of specific injury severities that occur in crashes involving distraction. FARS was used for each fatal case. For nonfatal injuries, the rate of distraction involvement is taken from GES and applied to the total incidence estimates previously derived. Application of these rates rather than direct counts is required to cover the various incidence cases not covered by GES, including both the adjustment for GES undercounting of police-reported crashes as well as unreported crashes.

GES data is only recorded using the KABCO severity system, whereas this report is based on the Abbreviated Injury Scale. To translate GES data to an MAIS basis, we used a variety of KABCO/MAIS translators. For CDS equivalent crashes, we used a current translator derived from 2000-2008 CDS data. Since this data is relatively recent, they reflect roughly current levels of seat belt usage. For non-CDS cases, the only available data from which to develop translators were contained in the 1982-1986 NASS files. Seat belt use has increased dramatically since this time. Observed belt use during this period ranged from roughly 10-37 percent as public awareness of the importance of belt use and belt use laws were just beginning to take hold in 1986. Belt use has since risen dramatically, and has been between 80 and 85 percent since 2004. Belt use can influence injury reporting significantly in a number of ways. It changes the nature of injuries by preventing many more visible injuries (such as head/face contact with the windshield) but replaces them with often less visible (and also typically less serious) shoulder and abdominal injuries such as bruising caused by pressure from the belt across the torso. This can influence the relationship between the KABCO reported injury severity and the corresponding MAIS injury level. For this reason, separate translators were developed from the 1982-86 NASS data for non-CDS cases where the victim was belted, unbelted, unknown belted status, and for nonoccupants/motorcyclists. These translators are presented in Tables 13-1 through 13-5 in Chapter 13.

2010 GES KABCO incidence counts were obtained for each distraction status (distracted, not distracted, unknown if distracted). Each of these data sets was divided according to belt use and occupancy status and run through its corresponding translator to produce a set of MAIS based injury counts. These counts from each grouping (CDS equivalent cases, belted non-CDS cases, unbelted non-CDS cases, unknown belt status non-CDS cases, and nonoccupant/motorcycle cases) were added together to produce a total MAIS injury profile for each distraction scenario. The percentage of each MAIS injury incidence that resulted from a distraction-affected crash was then calculated as:

$$x=a/(a+b)$$

where x = the percentage of incidence attributable to a distraction-affected crash at each injury severity level

a = the incidence of distraction injuries

b = the incidence of injuries that were not specifically coded as being distraction related (includes "not distracted" and Unknown).

The distraction-attributable portion of each injury severity level was then multiplied by the total cost of all 2010 crashes for that severity level and the results were summed to produce the total cost of distraction-affected crashes. MAIS0 portions were calculated using the same procedure described previously for Urban/Rural crashes, based on the relative incidence of MAIS 0 cases in injury crashes. The PDO portion was based on a direct count of PDO vehicles from the 2010 GES crashes involving distraction compared to those that did not.

The results of this analysis are summarized in Tables 9-1 and 9-2 for economic and comprehensive societal costs. Distracted driving is identified as a factor for roughly 10 percent of all fatalities and 18 percent of all crashes overall. In 2010 distraction-affected crashes caused $46 billion in economic costs and are responsible for 17 percent of all economic impacts from motor vehicle crashes. They caused $129 billion in societal harm (as measured by comprehensive costs), representing roughly 15 percent of total harm caused by motor vehicle crashes.

These estimates are almost certainly conservative because they are based only on identified distraction cases. Police records frequently fail to identify whether or not distraction was involved in the crash. Roughly 21 percent of all fatal crashes and 7 percent of all nonfatal crashes were coded in GES as "Distraction Unknown."[47] Although it is likely that a portion of these cases could involve distraction, none of them are distributed to distraction in this analysis.

In previous publications NHTSA has noted that there are limitations to the collection and reporting of FARS and GES data with regard to driver distraction (NHTSA, 2012, September). The data for FARS and GES are based on PARs and investigations conducted after the crash has occurred. One significant challenge for collection of distracted driving data is the PAR itself. Police accident reports vary across jurisdictions, thus creating potential inconsistencies in reporting. Many variables on the police accident report are nearly universal, but distraction is not one of those variables. Some police accident reports identify distraction as a distinct reporting field, while others do not have such a field and identification of distraction is based upon the narrative portion of the report. The variation in reporting forms contributes to variation in the reported number of distraction-affected crashes. Any National or State count of distraction-affected crashes should be interpreted with this limitation in mind due to potential under-reporting in some States/primary sampling units and over-reporting in others.

There are several potential reasons for underreporting of distraction-affected crashes.

- There are negative implications associated with distracted driving—especially in conjunction with a crash. Survey research shows that self-reporting of negative behavior is lower than actual occurrence of that negative behavior. There is no reason to believe that self-reporting of distracted driving to a law enforcement officer would differ. The inference is that the reported driver distraction during crashes is lower than the actual occurrence.
- If a driver fatality occurs in the crash, law enforcement must rely on the crash investigation in order to report on whether driver distraction was involved. Law enforcement may not have information to indicate distraction. For example, some forms of distraction such as cognitive distraction (lost in thought) are impossible to identify. These investigations often rely on witness account and these accounts are often not available, especially in fatal crashes.

Another concern is the speed at which technologies are changing and the difficulty in updating the PAR to accommodate these changes. Without broad-sweeping changes to the PAR to incorporate new technologies and features of technologies, it is difficult to capture the data that involve interaction with these devices.

In the reporting of distraction-affected crashes, oftentimes external distractions are identified as a distinct type of distraction. Some of the scenarios captured under external distractions might actually be related to the task of driving (e.g., looking at a street sign). However, the crash reports may not differentiate these driving-related tasks from other external distractions (looking at previous crash or billboard). Currently, the category of external distractions is included in the counts of distraction- affected crashes.

Table 9-1. Economic Cost of Identified Distracted Driving Crashes (Millions of 2010 Dollars)

	% Distracted	Incidence		Total Economic Crash Costs		
		Total	Distracted	Total	Distracted	Other
PDO Vehicles	17.81%	18,508,632	3,295,716	$71,480	$12,728	$58,752
MAIS0	18.66%	4,583,265	855,361	$13,030	$2,432	$10,598
MAIS1	18.96%	3,459,200	656,014	$68,797	$13,047	$55,750
MAIS2	17.20%	338,730	58,272	$34,537	$5,941	$28,596
MAIS3	16.46%	100,740	16,586	$27,905	$4,594	$23,310
MAIS4	16.66%	17,086	2,847	$8,781	$1,463	$7,318
MAIS5	15.60%	5,749	897	$6,327	$987	$5,341
Fatalities	9.90%	32,999	3,267	$46,163	$4,570	$41,593
Total	18.08%	27,046,402	4,888,960	$277,020	$45,763	$231,258
Percent of Total		100.00%	18.08%	100.00%	16.52%	83.48%

Table 9-2. Comprehensive Cost of Identified Distracted Driving Crashes (Millions of 2010 Dollars)

	% Distracted	Incidence		Total Comprehensive Crash Costs		
		Total	Distracted	Total	Distracted	Other
PDO Vehicles	17.81%	18,508,632	3,295,716	$71,480	$12,728	$58,752
MAIS0	18.66%	4,583,265	855,361	$13,030	$2,432	$10,598
MAIS1	18.96%	3,459,200	656,014	$149,192	$28,293	$120,899
MAIS2	17.20%	338,730	58,272	$150,001	$25,805	$124,196
MAIS3	16.46%	100,740	16,586	$109,070	$17,957	$91,113
MAIS4	16.66%	17,086	2,847	$43,594	$7,265	$36,329
MAIS5	15.60%	5,749	897	$32,649	$5,092	$27,557
Fatalities	9.90%	32,999	3,267	$301,809	$29,880	$271,929
Total	18.08%	27,046,402	4,888,960	$870,826	$129,452	$741,373
Percent of Total		100.00%	18.08%	100.00%	14.87%	85.13%

10. MOTORCYCLE CRASHES

Motorcycles are the most hazardous form of motor vehicle transportation. The lack of external protection provided by vehicle structure, the lack of internal protection provided by seat belts and air bags, their speed capability, the propensity for riders to become airborne through ejection, and the relative instability inherent with riding a two-wheeled vehicle all contribute to making the motorcycle the most risky passenger vehicle. In 2010, 4,518 motorcyclists were killed and 96,000[48] were injured in police-reported crashes on our Nation's roadways. This represents 14 percent of all traffic fatalities and 3 percent of all police-reported injuries. Motorcycles accounted for only 0.6 percent of all vehicle miles traveled in 2010. Per vehicle mile traveled in 2010, a motorcyclist was about 30 times more likely than a passenger car occupant to die in a motor vehicle traffic crash and 5 times more likely to be injured. The difference in these proportions reflects the more severe injury profile that results from motorcycle crashes.

Over the past several decades motorcycle fatalities and injuries have generally increased relative to those in other vehicle types. Figure 1 shows the fatality rate/100,000 registered vehicles by vehicle type.[49] The rates for passenger cars, light trucks, and heavy trucks declined steadily from 1995 through 2006.[50] The recession that occurred in 2007 caused a dramatic decline in fatality rates for heavy trucks, and a less severe but still noticeable decline in the rates for passenger cars and light trucks. The heavy truck rate began increasing in 2010 as the economy began to rebound. By contrast, the motorcycle fatality rate climbed steadily from the mid-1990s through 2006 as middle-aged baby boomers showed increased interest in motorcycle riding (Blincoe & Shankar, 2007). Motorcycle fatality rates were also affected by the recession and declined sharply from 2007 through 2009, but have since stabilized. What is most apparent from Figure 10-A is the magnitude of the fatality rate for motorcycles when compared with other vehicles.

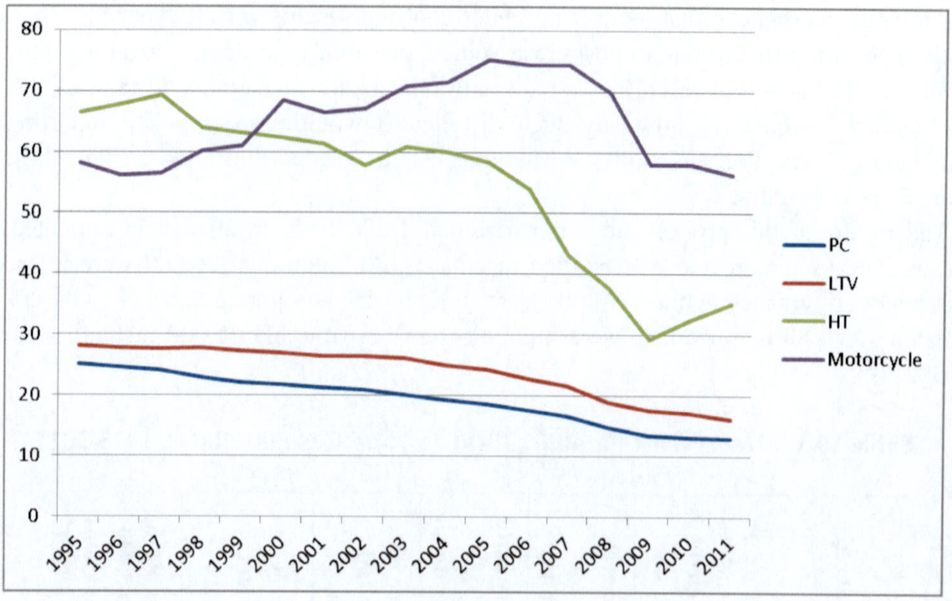

Figure 10-A. Fatality Rates per 100,000 Registered Vehicles by Vehicle Type.

Figure 10-2. Portions of Occupant Fatalities Represented by Vehicle Type.

Figure 10-B illustrates the percentage of occupant fatalities by vehicle type. The portions of fatalities represented by motorcycles and light trucks have been increasing since 1995, while the portions represented by passenger cars and heavy trucks have declined. The light truck increase is explained by the increasing sales of these vehicles relative to other types. However, as shown in Table 10-1, the fatality rate for these vehicles has actually been declining while for motorcycles it has increased overall. Light trucks have benefitted from a variety of occupant protection safety standards such as air bags, increased seat belt use, and side door beams that cannot be installed in motorcycles. The increase in the portion of fatalities represented by motorcycles thus represents both their increased popularity and the relative safety improvements made in other vehicle types, but not in motorcycles. If these trends continue, motorcycle riders will make up an increasing share of occupant fatalities.

Table 10-B lists a history of motorcycle fatalities and injuries along with fatality and injury rates from 1975 through 2011. Fatalities are taken directly from NHTSA's FARS database while injury totals represent the sum of all A, B, and C injuries from NHTSA's GES system. As noted elsewhere, these KABCO based injury counts are not consistent with the Abbreviated Injury Scale used to stratify injury in this report. They must therefore be adjusted using a KABCO/MAIS translator. This translator was derived from motorcycle crashes contained in the 1982-1986 NASS – the only database containing both KABCO and MAIS information that also has motorcycle crashes.

Nonfatal injuries were further adjusted to reflect the undercounting of police-reported crashes inherent in the GES database. This 10.7 percent adjustment represents the difference between total State police- reported crashes and the GES total. It is discussed in detail in the Incidence chapter.

A final adjustment was made to nonfatal injuries to represent unreported crashes. We know of no studies that indicate the extent to which motorcycle crashes go unreported, but we have no reason to believe that there is no underreporting for this vehicle type.[51] It's possible that the rates are different due to post-crash vehicle drivability, insurance coverage rates, the prevalence of single-vehicle crashes, or the different nature of motorcycle injuries, but we have no data to quantify how any such differences would impact police reporting for motorcycle crashes. For this study, we assume that underreporting rates are the same for motorcycles as for all vehicles.

The results of this process are summarized in Table 10-2. In 2010 it is estimated that there were 4519 motorcycle riders killed in crashes. An additional 96,000 were injured in police-reported crashes while 27,000 were injured in unreported crashes. Overall, an estimated 123,000 motorcyclists were injured in crashes, roughly 41,000 of them seriously (MAIS2-5).

Table 10-1. Motorcyclist Fatalities, Injuries, and Casualty Rates, 1975-2011

Year	Registered Motorcycles	Vehicle Miles Traveled - millions	Motorcycle Rider Fatalities	Fatality Rate per 100,000 Registrations	Fatality Rate per 100 Million VMT	Motorcycle Riders Injured	Injury Rate per 100,000 Registrations	Injury Rate per 100Million VMT
1975	4,964,070	5,629	3,189	64.24	56.65	-	-	-
1976	4,933,332	6,003	3,312	67.14	55.17	-	-	-
1977	4,933,256	6,349	4,104	83.19	64.64	-	-	-
1978	4,867,855	7,158	4,577	94.02	63.94	-	-	-
1979	5,422,132	8,637	4,894	90.26	56.66	-	-	-
1980	5,693,940	10,214	5,144	90.34	50.36	-	-	-
1981	5,831,132	10,690	4,906	84.13	45.89	-	-	-
1982	5,753,858	9,910	4,453	77.39	44.93	-	-	-
1983	5,585,112	8,760	4,265	76.36	48.69	-	-	-
1984	5,479,822	8,784	4,608	84.09	52.46	-	-	-
1985	5,444,404	9,086	4,564	83.83	50.23	-	-	-
1986	5,198,993	9,397	4,566	87.82	48.59	-	-	-
1987	4,885,772	9,506	4,036	82.61	42.46	-	-	-
1988	4,584,284	10,024	3,662	79.88	36.53	105,168	2,294	1,049
1989	4,420,420	10,371	3,141	71.06	30.29	83,435	1,888	805
1990	4,259,462	9,557	3,244	76.16	33.94	84,285	1,979	882
1991	4,177,365	9,178	2,806	67.17	30.57	80,435	1,925	876
1992	4,065,118	9,557	2,395	58.92	25.06	65,099	1,601	681
1993	3,977,856	9,906	2,449	61.57	24.72	59,436	1,494	600
1994	3,756,555	10,240	2,320	61.76	22.66	57,405	1,528	561
1995	3,897,191	9,797	2,227	57.14	22.73	57,480	1,475	587
1996	3,871,599	9,920	2,161	55.82	21.78	55,281	1,428	557
1997	3,826,373	10,081	2,116	55.3	20.99	52,574	1,374	522
1998	3,879,450	10,283	2,294	59.13	22.31	48,974	1,262	476
1999	4,152,433	10,584	2,483	59.8	23.46	49,986	1,204	472
2000	4,346,068	10,469	2,897	66.66	27.67	57,723	1,328	551

Year	Registered Motorcycles	Vehicle Miles Traveled - millions	Motorcycle Rider Fatalities	Fatality Rate per 100,000 Registrations	Fatality Rate per 100 Million VMT	Motorcycle Riders Injured	Injury Rate per 100,000 Registrations	Injury Rate per 100Million VMT
2001	4,903,056	9,633	3,197	65.2	33.19	60,236	1,229	625
2002	5,004,156	9,552	3,270	65.35	34.23	64,713	1,293	677
2003	5,370,035	9,576	3,714	69.16	38.78	67,103	1,250	701
2004	5,767,934	10,122	4,028	69.83	39.79	76,379	1,324	755
2005	6,227,146	10,454	4,576	73.48	43.77	87,335	1,402	835
2006	6,678,958	12,049	4,837	72.42	40.14	87,652	1,312	727
2007	7,138,476	21,396	5,174	72.48	24.18	102,994	1,443	481
2008	7,752,926	20,811	5,312	68.52	25.52	95,986	1,238	461
2009	7,929,724	20,822	4,469	56.36	21.46	89,607	1,130	430
2010	8,009,503	18,513	4,518	56.41	24.4	81,979	1,024	443
2011	8,437,502	18,500	4,612	54.66	24.93	81,399	965	440

Source: Traffic Safety Facts, 2011, Table 10, NHTSA, DOT HS 811 754.

Table 10-2. Motorcycle Riders, Incidence Summary, 2010

	GES Translated	Adjusted to State Total	% Unreported	# Unreported	Total
MAIS 0	2,954	3,270	53.14%	3,708	6,977
MAIS 1	55,075	60,970	25.45%	20,809	81,779
MAIS 2	19,065	21,105	19.95%	5,259	26,365
MAIS 3	10,978	12,153	4.31%	548	12,701
MAIS 4	808	895	0.00%	0	895
MAIS 5	746	826	0.00%	0	826
Nonfatal Injury Total	86,673	95,950	21.72%	26,616	122,566
Fatal	4519	4,519	0.00%	0	4,519
PDO Vehicle	9,518	10,536	59.72%	15,623	26,160

The crash environment faced by motorcyclists results in an injury profile skewed more towards serious injuries than the typical crash in a passenger car or light truck. Minor injuries (MAIS1) represent over 87 percent of injuries for the general crash population, but they represent only 67 percent of motorcyclist injuries. The more serious MAIS 2-5 injuries represent 33 percent of motorcyclist injuries, compared to only 13 percent for the general crash population. In addition, within each MAIS category, the type of injuries typically received is different for motorcyclists. For example, motorcyclists, especially those who do not wear helmets, are more likely to receive head injuries than are their counterparts in regular passenger vehicles. Lower extremity injuries are also more likely since a crashed motorcycle is likely to fall over and crush lower limbs. These differences produce different average injury costs within each MAIS category. To assess these injuries, we isolated crash records for motorcycle occupants on the crash file described in the Incidence chapter. The resulting average unit costs are shown in Table 3. Also in Table 3, these motorcycle occupant specific costs are combined with injury incidence from Table 2 to estimate the total costs associated with motorcycle crashes.

In 2010, motorcycle crashes cost $13.5 billion in economic impacts, and $66 billion in societal harm as measured by comprehensive costs. Compared to other motor vehicle crashes, these costs are disproportionately caused by fatalities and serious injuries.

Impacts of Helmet Use

Motorcycle helmet usage is the most important action that motorcycle riders can take to protect themselves in the event of a crash. Helmets reduce the chance of fatal injury by 37 percent for motorcycle operators and by 41 percent for passengers (Deutermann, 2004). They reduce the chance of nonfatal serious injury by 13 percent and minor injury by 8 percent (Blincoe, 1988). Unfortunately, only about two-thirds of motorcycle riders currently wear helmets. This causes unnecessary loss of life and critical injury, as well as considerable preventable economic loss to society. Figure 3 illustrates the historical trend in motorcycle rider helmet use from 1994 through 2011.

Helmet use peaked in 2000, but then declined after a number of States repealed their helmet use laws. It reached a nadir in 2005, but has since slowly increased due to a number of factors including public awareness and possibly shifting attitudes associated with the age of riders. Note that there was a noticeable decline in observed use in 2010, but this was followed by a return to 2009 levels in 2011. The 2010 drop was not mirrored by a similar drop in use in crash data, raising the possibility that the observation survey recorded a less representative sample that year. In any case, calculations used in this analysis are based on police-reported use in crashes, not the NOPUS survey.

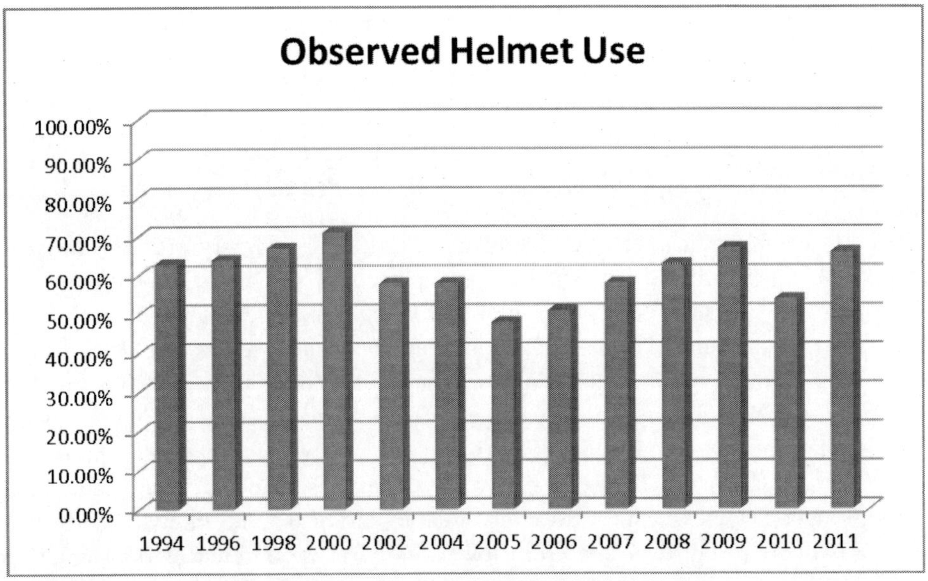

Source: National Occupant Protection Use Survey (NOPUS), 1994-2011 data

Figure 10-C. Observed Helmet Use.

Table 10-3. Motorcycle Rider Unit Costs and Total Costs (2010 Dollars)

	Incidence	Unit Economic Costs	Total Economic Costs (Millions)	Unit Comprehensive Costs	Total Comprehensive Costs (Millions)
MAIS 0	6,977	$2,975	$21	$2,975	$21
MAIS 1	81,779	$14,022	$1,147	$27,214	$2,226
MAIS 2	26,365	$65,398	$1,724	$239,636	$6,318
MAIS 3	12,701	$241,453	$3,067	$817,326	$10,381
MAIS 4	895	$421,628	$377	$1,840,343	$1,646
MAIS 5	826	$1,055,816	$872	$5,593,880	$4,622
Fatal	4,519	$1,381,645	$6,244	$9,090,622	$41,081
PDO	26,160	$3,272	$86	$3,272	$86
Total			$13,537		$66,380

NHTSA has published historical estimates of lives that have been saved by helmets, as well as those that could have been saved, but were instead lost due to helmet nonuse. These estimates are shown in Table 10-4. To determine the cost impact of these savings, similar estimates must be derived for nonfatal injuries. The methods used to estimate savings from helmet use have been established in a number of studies.[52] Different approaches apply depending on the type of data available. For this study we based our calculations on methods used in the NHTSA 2011 Research Note because we were able to develop separate incidence for helmeted and unhelmeted riders and it contains the most up to date effectiveness estimates.

A first step in this process was to develop separate helmeted and unhelmeted incidence profiles using the same methods and translators as were used for the overall incidence estimate, but specific to cases with known helmet status. We then distributed cases with Unknown helmet status according to the known cases. The results are shown in Tables 10-4 and 10-5.

Table 10-4. Helmeted Motorcycle Injured Riders, Incidence Summary, 2010

	GES Translated	Adj to State Total	% Unreported	# Unreported	Total
MAIS 0	2,126	2,353	53.14%	2,668	5,021
MAIS 1	37,949	42,011	25.45%	14,338	56,350
MAIS 2	12,810	14,181	19.95%	3,534	17,715
MAIS 3	7,129	7,892	4.31%	356	8,247
MAIS 4	515	570	0.00%	0	570
MAIS 5	472	523	0.00%	0	523
Nonfatal Injury Total	58,875	65,176	21.72%	18,228	83,404
Fatal	2,636	2,636	0.00%	0	2,636
PDO	6,849	7,582	59.72%	11,243	18,825

Table 10-5. Unhelmeted Motorcycle Injured Riders, Incidence Summary, 2010

	GES Translated	Adj to State Total	% Unreported	# Unreported	Total
MAIS 0	828	917	53.14%	1,040	1,956
MAIS 1	17,126	18,959	25.45%	6,471	25,430
MAIS 2	6,255	6,924	19.95%	1,725	8,650
MAIS 3	3,850	4,262	4.31%	192	4,454
MAIS 4	293	325	0.00%	0	325
MAIS 5	274	304	0.00%	0	304
Nonfatal Injury Total	27,798	30,773	21.72%	8,388	39,162
Fatal	1,883	1,883	0.00%	0	1,883
PDO	2,669	2,954	59.72%	4,380	7,334

Table 10-6. Helmeted Motorcycle Rider Unit Costs and Total Costs (2010 Dollars)

	Incidence	Unit Economic Costs	Total Economic Costs (Millions)	Unit Comprehensive Costs	Total Comprehensive Costs (Millions)
MAIS 0	5,021	$2,975	$15	$2,975	$15
MAIS 1	56,350	$13,878	$782	$26,713	$1,505
MAIS 2	17,715	$64,730	$1,147	$237,124	$4,201
MAIS 3	8,247	$241,358	$1,991	$815,524	$6,726
MAIS 4	570	$414,695	$236	$1,787,272	$1,018
MAIS 5	523	$899,930	$470	$4,618,345	$2,414
Fatal	2,636	$1,381,645	$3,641	$9,090,622	$23,959
PDO Vehicle	18,825	$3,272	$62	$3,272	$62
Total			$8,344		$39,899

Unhelmeted motorcycle crash victims have more severe injuries than do those who wear helmets. These injuries are also more expensive to treat and result in more lost quality-of-life. To determine benefits from helmet use, we isolated crash records separately for helmeted and unhelmeted motorcycle occupants on the crash file described in section 2. The resulting average unit costs are shown in Tables 10-6 and 10-7. Also in these tables, these motorcycle occupant helmet status specific costs are combined with injury incidence from Tables 4 and 5 to estimate the total costs associated with helmeted and unhelmeted motorcycle riders. Unhelmeted riders make up 32 percent of all motorcycle injuries, but, due to their more serious injury profile, account for 38 percent of total economic costs and 40 percent of total comprehensive costs caused by these crashes.

Using methods described in NHTSA 2011, the lives saved, serious (MAIS 2-5) injuries and minor (MAIS 1) injuries avoided due to helmet use and non-use were calculated and combined with the unit costs from Table 10-7 to derive estimates of the economic impact of helmet use and non-use from 1975 through 2010. The results are summarized in Tables 8 and 9. Over this 36 year period, motorcycle helmets have saved over $62 billion in economic costs. Another $49 billion in potential economic savings was lost due to the refusal of some riders to wear helmets. Helmets are currently saving $2.8 billion in economic costs annually. As shown in Figures 4 and 5, the gap between potential benefits and achieved benefits has grown smaller over time, but there is still considerable progress to be made if all motorcycle riders can be persuaded to wear helmets.

Table 10-7. Unhelmeted Motorcycle Rider Unit Costs and Total Costs (2010 Dollars)

	Incidence	Unit Economic Costs	Total Economic Costs (Millions)	Unit Comprehensive Costs	Total Comprehensive Costs (Millions)
MAIS 0	1,956	$2,975	$6	$2,975	$6
MAIS 1	25,430	$14,341	$365	$28,327	$720
MAIS 2	8,650	$66,765	$578	$244,779	$2,117
MAIS 3	4,454	$241,630	$1,076	$820,664	$3,655
MAIS 4	325	$433,790	$141	$1,933,428	$628
MAIS 5	304	$1,324,105	$402	$7,272,831	$2,209
Fatal	1,883	$1,381,645	$2,602	$9,090,622	$17,122
PDO Vehicle	7,334	$3,272	$24	$3,272	$24
Total			$5,193		$26,481

Table 10-8. Economic Benefits of Helmet Use, 1975-2010

Year	Lives Saved by Helmets	Cost/ Fatality Current$	MAIS 2-5 Injuries Prevented	MAIS 2-5 Cost/Injury Current$	MAIS 1 Injuries Prevented	MAIS 1 Cost/Injury Current$	Current $ (Millions)	2010 Dollars (Millions)
1975	823	$336,380	3,088	$36,969	3,743	$1,794	$398	$1,612
1976	788	$355,763	2,955	$39,099	3,582	$1,897	$403	$1,543
1977	970	$378,897	3,648	$41,642	4,422	$2,020	$528	$1,901
1978	900	$407,658	3,382	$44,802	4,100	$2,174	$527	$1,764
1979	885	$453,926	3,318	$49,887	4,022	$2,421	$577	$1,733
1980	871	$515,200	3,289	$56,621	3,986	$2,747	$646	$1,709
1981	843	$568,345	3,171	$62,462	3,843	$3,031	$689	$1,652
1982	816	$603,359	3,071	$66,310	3,722	$3,217	$708	$1,600
1983	735	$622,741	2,765	$68,441	3,351	$3,321	$658	$1,441
1984	813	$649,627	3,066	$71,395	3,716	$3,464	$760	$1,595
1985	788	$672,761	2,960	$73,938	3,588	$3,587	$762	$1,544
1986	807	$685,266	3,056	$75,312	3,704	$3,654	$797	$1,585
1987	667	$710,275	2,521	$78,061	3,055	$3,787	$682	$1,309
1988	622	$739,662	2,283	$81,290	2,767	$3,944	$657	$1,210
1989	561	$775,301	2,006	$85,207	2,431	$4,134	$616	$1,083
1990	655	$817,192	2,271	$89,811	2,753	$4,358	$751	$1,253

Table 10-8. (Continued)

Year	Lives Saved by Helmets	Cost/ Fatality Current$	MAIS 2-5 Injuries Prevented	MAIS 2-5 Cost/Injury Current$	MAIS 1 Injuries Prevented	MAIS 1 Cost/Injury Current$	Current $ (Millions)	2010 Dollars (Millions)
1991	595	$851,580	2,018	$93,590	2,446	$4,541	$707	$1,131
1992	641	$877,215	2,118	$96,408	2,567	$4,678	$778	$1,210
1993	671	$903,475	2,156	$99,294	2,614	$4,818	$833	$1,257
1994	625	$926,609	1,960	$101,836	2,376	$4,941	$791	$1,163
1995	624	$952,869	1,913	$104,722	2,319	$5,081	$807	$1,154
1996	617	$981,005	1,846	$107,814	2,237	$5,231	$816	$1,134
1997	627	$1,003,514	1,837	$110,288	2,226	$5,351	$844	$1,146
1998	660	$1,019,145	1,888	$112,006	2,289	$5,434	$897	$1,199
1999	745	$1,041,654	2,090	$114,480	2,533	$5,554	$1,029	$1,347
2000	872	$1,076,667	2,390	$118,328	2,896	$5,741	$1,238	$1,568
2001	947	$1,107,304	2,547	$121,695	3,087	$5,905	$1,377	$1,695
2002	992	$1,124,811	2,615	$123,619	3,170	$5,998	$1,458	$1,767
2003	1,173	$1,150,446	3,022	$126,436	3,663	$6,135	$1,754	$2,079
2004	1,324	$1,181,083	3,415	$129,803	4,140	$6,298	$2,033	$2,347
2005	1,554	$1,221,098	4,013	$134,201	4,865	$6,511	$2,468	$2,755
2006	1,667	$1,260,489	4,313	$138,530	5,228	$6,721	$2,734	$2,957
2007	1,788	$1,296,390	4,627	$142,476	5,608	$6,913	$3,016	$3,172
2008	1,836	$1,346,166	4,748	$147,946	5,755	$7,178	$3,215	$3,256
2009	1,486	$1,341,376	3,844	$147,420	4,659	$7,153	$2,593	$2,636
2010	1,556	$1,363,378	4,010	$149,838	4,860	$7,270	$2,758	$2,758
Total	33,544		104,221		126,323		$42,303	$62,268

Table 10-9. Economic Benefits Forgone by Helmet Nonuse

Year	Lives Lost Due to Helmet Nonuse	Cost/Fatality Current$	MAIS 2-5 Injuries Caused by Helmet Nonuse	MAIS 2-5 Cost/Injury Current$	MAIS 1 Injuries Caused by Helmet Nonuse	MAIS 1 Cost/Injury Current$	Current $ (Millions)	2010 Dollars (Millions)
1975	1,164	$336,380	1,114	$36,969	1,270	$1,794	$435	$1,763
1976	1,189	$355,763	1,313	$39,099	1,496	$1,897	$477	$1,829
1977	1,472	$378,897	1,636	$41,642	1,864	$2,020	$630	$2,266
1978	1,588	$407,658	2,248	$44,802	2,562	$2,174	$754	$2,521
1979	1,676	$453,926	2,588	$49,887	2,950	$2,421	$897	$2,694
1980	1,744	$515,200	2,843	$56,621	3,240	$2,747	$1,068	$2,827

Year	Lives Lost Due to Helmet Nonuse	Cost/Fatality Current$	MAIS 2-5 Injuries Caused by Helmet Nonuse	MAIS 2-5 Cost/Injury Current$	MAIS 1 Injuries Caused by Helmet Nonuse	MAIS 1 Cost/Injury Current$	Current $ (Millions)	2010 Dollars (Millions)
1981	1,667	$568,345	2,691	$62,462	3,067	$3,031	$1,125	$2,698
1982	1,528	$603,359	2,323	$66,310	2,647	$3,217	$1,084	$2,451
1983	1,450	$622,741	2,334	$68,441	2,660	$3,321	$1,072	$2,346
1984	759	$649,627	2,473	$71,395	2,819	$3,464	$679	$1,426
1985	764	$672,761	2,497	$73,938	2,845	$3,587	$709	$1,436
1986	751	$685,266	2,439	$75,312	2,780	$3,654	$709	$1,410
1987	697	$710,275	2,268	$78,061	2,584	$3,787	$682	$1,309
1988	644	$739,662	2,060	$81,290	2,348	$3,944	$653	$1,204
1989	553	$775,301	1,738	$85,207	1,980	$4,134	$585	$1,029
1990	541	$817,192	1,671	$89,811	1,905	$4,358	$601	$1,002
1991	467	$851,580	1,412	$93,590	1,610	$4,541	$537	$860
1992	323	$877,215	961	$96,408	1,096	$4,678	$381	$592
1993	336	$903,475	989	$99,294	1,127	$4,818	$407	$614
1994	339	$926,609	988	$101,836	1,125	$4,941	$420	$618
1995	326	$952,869	928	$104,722	1,058	$5,081	$413	$591
1996	324	$981,005	908	$107,814	1,034	$5,231	$421	$585
1997	315	$1,003,514	871	$110,288	992	$5,351	$417	$567
1998	369	$1,019,145	1,007	$112,006	1,148	$5,434	$495	$662
1999	396	$1,041,654	1,062	$114,480	1,210	$5,554	$541	$708
2000	478	$1,076,667	1,269	$118,328	1,446	$5,741	$673	$852
2001	558	$1,107,304	1,457	$121,695	1,660	$5,905	$805	$991
2002	576	$1,124,811	1,483	$123,619	1,690	$5,998	$841	$1,020
2003	651	$1,150,446	1,653	$126,436	1,883	$6,135	$969	$1,149
2004	673	$1,181,083	1,707	$129,803	1,946	$6,298	$1,029	$1,188
2005	731	$1,221,098	1,857	$134,201	2,117	$6,511	$1,156	$1,290
2006	756	$1,260,489	1,919	$138,530	2,187	$6,721	$1,234	$1,334
2007	805	$1,296,390	2,045	$142,476	2,331	$6,913	$1,351	$1,421
2008	827	$1,346,166	2,101	$147,946	2,394	$7,178	$1,441	$1,460
2009	733	$1,341,376	1,861	$147,420	2,120	$7,153	$1,273	$1,294
2010	708	$1,363,378	1,804	$149,838	2,056	$7,270	$1,251	$1,251
Total	28,878		62,520		71,248		$28,215	$49,257

In Table 10-10, the societal impact of helmet use over this same time period is shown. Over $370 billion in societal harm, as measured by comprehensive costs, has been averted over the past 36 years due to motorcycle helmet use. Over this same period, an additional $301 billion in societal harm could have been prevented had all motorcycle riders worn helmets. Motorcycle helmets are currently preventing $17 billion in societal harm annually, but another $8 billion in harm could be prevented if all riders were to wear their helmets.

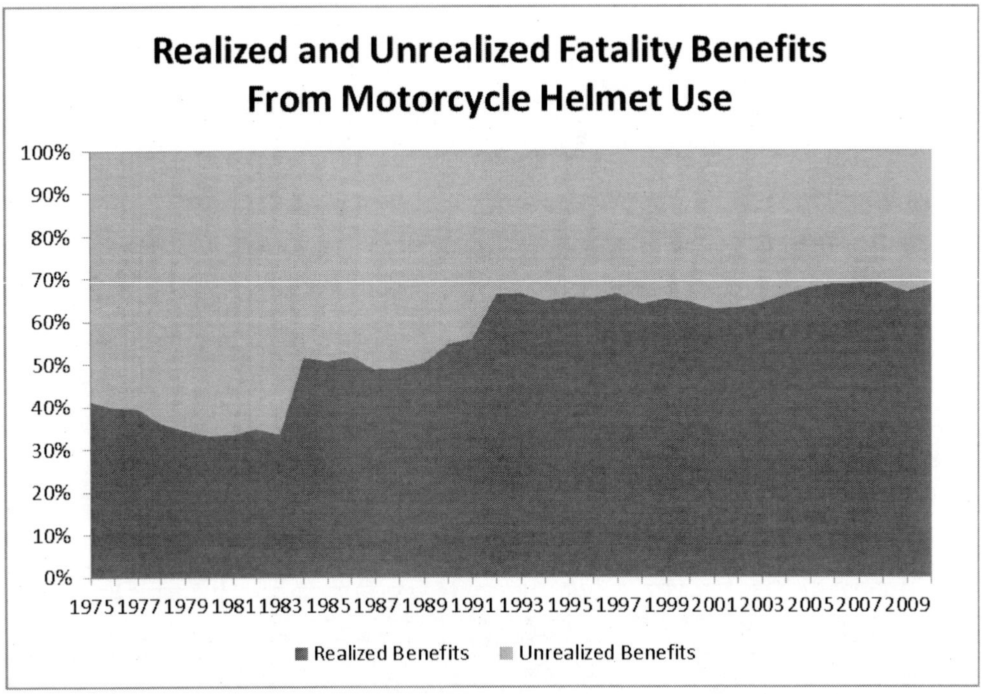

Figure 10-4. Realized and Unrealized Fatality Benefits from Motorcycle Helmet Use.

Figure 10-5. Realized and Unrealized Economic Benefits From Motorcycle Helmet Use

Table 10-10. Comprehensive Societal Benefits of Helmet Use, 1975-2010

Year	Lives Saved by Helmets	Cost/Fatality Current$	MAIS 2-5 Injuries Prevented	MAIS 2-5 Cost/Injury Current$	MAIS 1 Injuries Prevented	MAIS 1 Cost/ Injury Current$	Current $ (Millions)	2010 Dollars (Millions)
1975	823	$2,238,383	3,088	$152,178	3,743	$3,543	$2,325	$9,425
1976	788	$2,367,360	2,955	$160,947	3,582	$3,747	$2,355	$9,023
1977	970	$2,521,301	3,648	$171,413	4,422	$3,991	$3,089	$11,114
1978	900	$2,712,687	3,382	$184,424	4,100	$4,294	$3,083	$10,310
1979	885	$3,020,568	3,318	$205,356	4,022	$4,781	$3,374	$10,133
1980	871	$3,428,303	3,289	$233,076	3,986	$5,426	$3,774	$9,988
1981	843	$3,781,951	3,171	$257,119	3,843	$5,986	$4,026	$9,659
1982	816	$4,014,943	3,071	$272,959	3,722	$6,355	$4,138	$9,351
1983	735	$4,143,920	2,765	$281,728	3,351	$6,559	$3,847	$8,422
1984	813	$4,322,824	3,066	$293,890	3,716	$6,842	$4,441	$9,320
1985	788	$4,476,765	2,960	$304,356	3,588	$7,086	$4,454	$9,026
1986	807	$4,559,976	3,056	$310,013	3,704	$7,218	$4,654	$9,260
1987	667	$4,726,399	2,521	$321,328	3,055	$7,481	$3,985	$7,650
1988	622	$4,921,945	2,283	$334,622	2,767	$7,791	$3,847	$7,091
1989	561	$5,159,097	2,006	$350,745	2,431	$8,166	$3,618	$6,362
1990	655	$5,437,855	2,271	$369,697	2,753	$8,607	$4,425	$7,383
1991	595	$5,666,686	2,018	$385,254	2,446	$8,969	$4,171	$6,678
1992	641	$5,837,269	2,118	$396,851	2,567	$9,239	$4,606	$7,159
1993	671	$6,012,013	2,156	$408,731	2,614	$9,516	$4,940	$7,455
1994	625	$6,165,953	1,960	$419,197	2,376	$9,760	$4,699	$6,913
1995	624	$6,340,697	1,913	$431,077	2,319	$10,036	$4,805	$6,875
1996	617	$6,527,922	1,846	$443,806	2,237	$10,332	$4,870	$6,768
1997	627	$6,677,703	1,837	$453,989	2,226	$10,570	$5,044	$6,853
1998	660	$6,781,717	1,888	$461,060	2,289	$10,734	$5,371	$7,185
1999	745	$6,931,497	2,090	$471,243	2,533	$10,971	$6,177	$8,084
2000	872	$7,164,488	2,390	$487,083	2,896	$11,340	$7,444	$9,427
2001	947	$7,368,356	2,547	$500,943	3,087	$11,663	$8,290	$10,207
2002	992	$7,484,852	2,615	$508,863	3,170	$11,847	$8,793	$10,658
2003	1,173	$7,655,435	3,022	$520,461	3,663	$12,117	$10,597	$12,559
2004	1,324	$7,859,302	3,415	$534,321	4,140	$12,440	$12,282	$14,178
2005	1,554	$8,125,578	4,013	$552,424	4,865	$12,861	$14,907	$16,644
2006	1,667	$8,387,694	4,313	$570,244	5,228	$13,276	$16,511	$17,859
2007	1,788	$8,626,593	4,627	$586,485	5,608	$13,654	$18,214	$19,156
2008	1,836	$8,957,816	4,748	$609,004	5,755	$14,179	$19,420	$19,668
2009	1,486	$8,925,946	3,844	$606,837	4,659	$14,128	$15,662	$15,919
2010	1,556	$9,072,356	4,010	$616,791	4,860	$14,360	$16,660	$16,660
Total	33,544		104,221		126,323		$252,898	$370,420

Table 10-11. Comprehensive Benefits Forgone by Helmet Nonuse

Year	Lives Lost Due to Helmet Nonuse	Cost/Fatality Current$	MAIS 2-5 Injuries Caused by Helmet Nonuse	MAIS 2-5 Cost/ Injury Current$	MAIS 1 Injuries Caused by Helmet Nonuse	MAIS 1 Cost/Injury Current$	Current $ (Millions)	2010 Dollars (Millions)
1975	1,164	$2,238,383	1,114	$152,178	1,612	$3,543	$2,781	$11,271
1976	1,189	$2,367,360	1,313	$160,947	1,543	$3,747	$3,032	$11,619
1977	1,472	$2,521,301	1,636	$171,413	1,901	$3,991	$3,999	$14,391
1978	1,588	$2,712,687	2,248	$184,424	1,764	$4,294	$4,730	$15,819
1979	1,676	$3,020,568	2,588	$205,356	1,733	$4,781	$5,602	$16,827
1980	1,744	$3,428,303	2,843	$233,076	1,709	$5,426	$6,651	$17,600
1981	1,667	$3,781,951	2,691	$257,119	1,652	$5,986	$7,006	$16,807
1982	1,528	$4,014,943	2,323	$272,959	1,600	$6,355	$6,779	$15,318
1983	1,450	$4,143,920	2,334	$281,728	1,441	$6,559	$6,676	$14,615
1984	759	$4,322,824	2,473	$293,890	1,595	$6,842	$4,019	$8,434
1985	764	$4,476,765	2,497	$304,356	1,544	$7,086	$4,191	$8,493
1986	751	$4,559,976	2,439	$310,013	1,585	$7,218	$4,192	$8,341
1987	697	$4,726,399	2,268	$321,328	1,309	$7,481	$4,033	$7,741
1988	644	$4,921,945	2,060	$334,622	1,210	$7,791	$3,868	$7,131
1989	553	$5,159,097	1,738	$350,745	1,083	$8,166	$3,471	$6,104
1990	541	$5,437,855	1,671	$369,697	1,253	$8,607	$3,571	$5,957
1991	467	$5,666,686	1,412	$385,254	1,131	$8,969	$3,201	$5,124
1992	323	$5,837,269	961	$396,851	1,210	$9,239	$2,278	$3,541
1993	336	$6,012,013	989	$408,731	1,257	$9,516	$2,436	$3,676
1994	339	$6,165,953	988	$419,197	1,163	$9,760	$2,516	$3,701
1995	326	$6,340,697	928	$431,077	1,154	$10,036	$2,479	$3,547
1996	324	$6,527,922	908	$443,806	1,134	$10,332	$2,530	$3,516
1997	315	$6,677,703	871	$453,989	1,146	$10,570	$2,511	$3,411
1998	369	$6,781,717	1,007	$461,060	1,199	$10,734	$2,980	$3,986
1999	396	$6,931,497	1,062	$471,243	1,347	$10,971	$3,260	$4,267
2000	478	$7,164,488	1,269	$487,083	1,568	$11,340	$4,061	$5,142
2001	558	$7,368,356	1,457	$500,943	1,695	$11,663	$4,861	$5,985
2002	576	$7,484,852	1,483	$508,863	1,767	$11,847	$5,087	$6,166
2003	651	$7,655,435	1,653	$520,461	2,079	$12,117	$5,869	$6,955
2004	673	$7,859,302	1,707	$534,321	2,347	$12,440	$6,231	$7,193
2005	731	$8,125,578	1,857	$552,424	2,755	$12,861	$7,001	$7,817
2006	756	$8,387,694	1,919	$570,244	2,957	$13,276	$7,475	$8,085
2007	805	$8,626,593	2,045	$586,485	3,172	$13,654	$8,187	$8,610
2008	827	$8,957,816	2,101	$609,004	3,256	$14,179	$8,734	$8,845
2009	733	$8,925,946	1,861	$606,837	2,636	$14,128	$7,709	$7,835
2010	708	$9,072,356	1,804	$616,791	2,758	$14,360	$7,576	$7,576
Total	28,878		62,520		62,268		$171,581	$301,447

11. SEAT BELT USE

When properly fastened, seat belts provide significant protection to vehicle occupants involved in a crash. The simple act of buckling a seat belt can improve an occupant's chance of surviving a potentially fatal crash by from 44 to 73 percent, depending on the type of vehicle and seating position involved. They are also highly effective against serious nonfatal injuries. Belts reduce the chance of receiving an MAIS 2-5 injury (moderate to critical) by 49 to 78 percent.

The effectiveness of seat belts is a function of vehicle type, restraint type, and seat position. Table 11-1 shows the estimated effectiveness of seat belts for various seating positions for passenger cars and for light trucks, vans, and sports utility vehicles (LTVs).

Although all passenger vehicles have been equipped with seat belts since 1968, a sizable minority of vehicle occupants still neglect to use these devices. As of 2012, about 86 percent of occupants wear their seat belts. Usage has risen steadily throughout the last two decades, largely in response to public education programs sponsored by State and Federal safety agencies, as well as private consumer and safety advocacy groups. A major factor in this increase has been the passage of seat belt use laws. As of 2001, all States except New Hampshire had some form of adult usage law. These laws can take the form of either primary enforcement laws, under which police can stop drivers specifically for failing to wear seat belts, or secondary laws, under which fines can only be levied if a driver is stopped for some other offense. Primary enforcement laws are far more effective in increasing seat belt use. Experience in a number of States indicates that usage rates rise from 10-15 percentage points when primary laws are passed. For example, usage in California jumped from 70 percent to 82 percent when a primary law was passed in 1993. Similar impacts occurred in Louisiana where usage rose 18 points, in Georgia where usage rose 17 points, in Maryland where usage rose 13 points, and in the District of Columbia where usage rose 24 points when they combined a new primary enforcement law with penalty points. Overall, States with primary belt use laws have an average belt use rate that is 12 percentage points higher than States with only secondary enforcement (NHTSA, 2012, November). Figure 11-A illustrates the nationwide trend in seat belt use rates from 1983 through 2010.

By combining seat belt use rates with effectiveness rates and National injury counts, an estimate can be made of the impact of seat belts on fatality and casualty rates. The basic methods for these calculations are well documented (Partyka & Womble, 1989, Blincoe, 1994, Wang & Blincoe, 2001, Wang & Blincoe, 2003, Glassbrenner, 2007). The effect of increases in seat belt use on fatalities is curvilinear, i.e., the more the observed usage rate in the general population approaches 100 percent, the more lives are saved for each incremental point increase. This occurs because those who are most resistant to buckling up tend to be in high-risk groups such as impaired drivers or people who are risk takers in general. These people are more likely to be involved in serious crashes and are thus more likely to actually benefit from wearing their belts. Belt use by people involved in potentially fatal crashes (UPFC) tends to be lower than observed use for these same reasons. Figure 11-B illustrates the relationship between use in potentially fatal crashes as well as lives saved and increasing rates of observed seat belt usage.

Table 11-1. Effectiveness of Seat Belts against Fatalities and Serious Injuries

	Percent Effectiveness	
	Lap Belts	Lap/Shoulder Belts
Passenger Cars, Front Seat		
Fatalities	35	45
MAIS 2-5 Injuries	30	50
Passenger Cars, Rear Seat		
Fatalities	32	44
MAIS 2-5 Injuries	37	49
Light Trucks, Front Seat		
Fatalities	50	60
MAIS 2-5 Injuries	55	65
Light Trucks, Rear Seat		
Fatalities	63	73
MAIS 2-5 Injuries	68	78

Sources: Kahane, 2000; Morgan, 1999; NHTSA 1984.

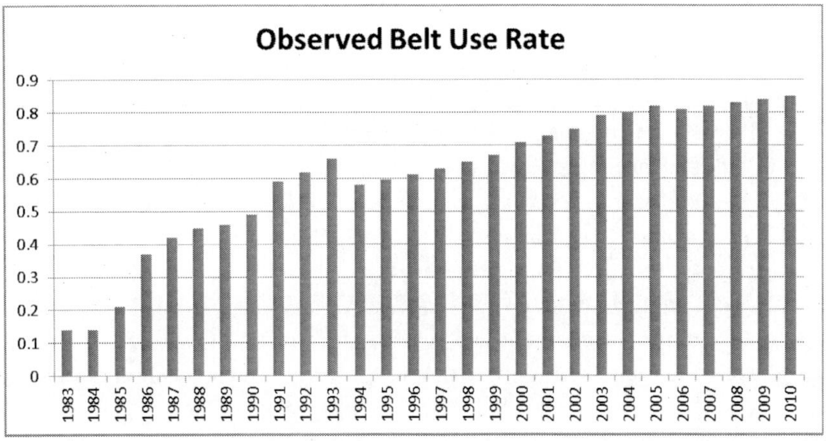

Figure 11-A. Observed Belt Use Rate.

Table 11-2 lists the historical and cumulative impact of seat belt use on motor vehicle casualties. Through 2010, seat belts have saved 280,000 lives and prevented 7.2 million serious nonfatal police- reported injuries. At current (2010) use rates, they are preventing 12,500 fatalities and 308,000 serious (MAIS 2-5) police-reported nonfatal injuries annually.

The failure of a large segment of the driving population to wear their belts also has significant safety implications. If all occupants had used seat belts properly, many more lives would have been saved. Table 11-2 also lists the potential safety benefits that could have been realized since 1975 had all occupants worn their seat belts. Over this period, passenger vehicles were equipped with devices that could have saved over 360,000 additional lives and prevented 5.8 million additional serious police- reported injuries[53] if all vehicle occupants had taken a few seconds to buckle their seat belts. At current (2010) belt use rates, an additional 3,353 fatalities and 54,000 serious injuries could be prevented every year if all passengers were to wear their seat belts. This represents an enormous lost opportunity for injury prevention.

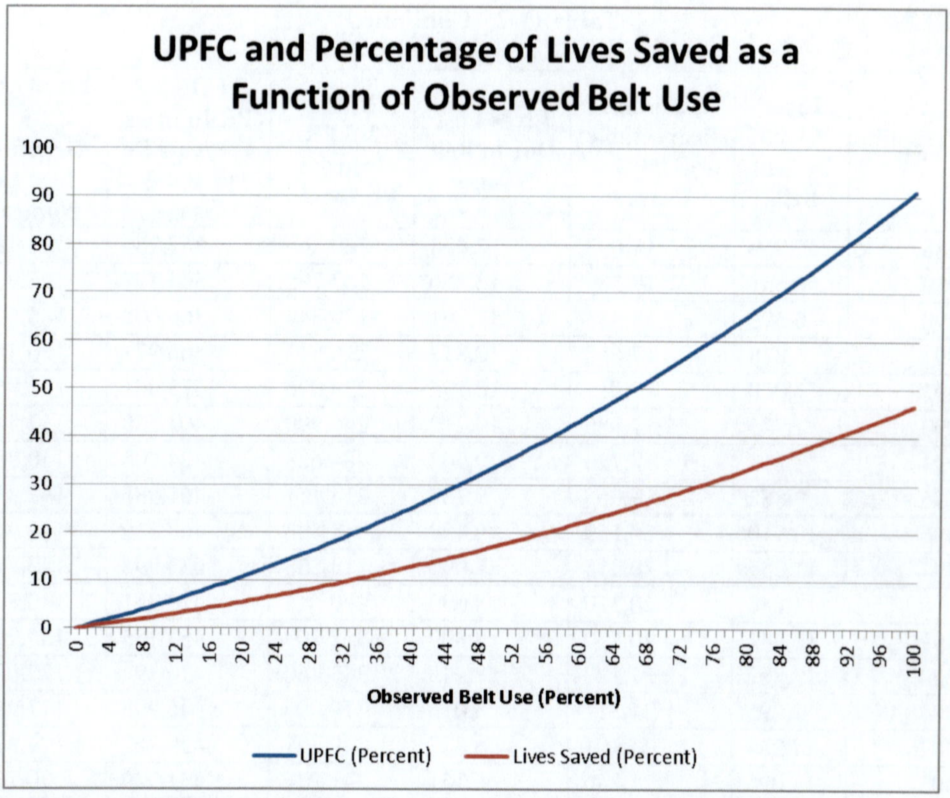

Figure 11-B. UPFC and Percentage of Lives Saved as a Function of Observed Belt Use.

Table 11-2. Achieved and Potential Impact of Seat belt Use on Fatalities and Serious Injuries, 1975-2010

Year	Lives Saved by Seat Belts	Fatalities Preventable @100% Usage*	Lives Lost Due to Belt Nonuse	MAIS 2-5 PR Injuries Prevented	MAIS 2-5 PR Injuries Preventable @ 100% Usage	PR MAIS 2-5 Benefits Lost to Nonuse
1975	978	14,279	13,301	35,118	259,709	224,591
1976	796	14,647	13,851	28,754	262,473	233,719
1977	682	15,142	14,460	25,311	274,272	248,961
1978	679	16,220	15,541	22,016	253,280	231,264
1979	594	16,320	15,726	22,254	294,178	271,924
1980	575	16,305	15,730	21,602	292,407	270,805
1981	548	15,770	15,222	20,596	281,042	260,446
1982	678	13,928	13,250	25,378	253,185	227,807
1983	809	13,722	12,913	35,448	253,197	217,750
1984	1,197	14,424	13,227	36,728	262,346	225,617
1985	2,435	14,943	12,508	56,778	270,373	213,595
1986	4,094	16,822	12,728	119,801	323,788	203,986
1987	5,141	17,819	12,678	146,489	348,782	202,294

Table 11-2. (Continued)

Year	Lives Saved by Seat Belts	Fatalities Preventable @100% Usage*	Lives Lost Due to Belt Nonuse	MAIS 2-5 PR Injuries Prevented	MAIS 2-5 PR Injuries Preventable @ 100% Usage	PR MAIS 2-5 Benefits Lost to Nonuse
1988	5,959	18,633	12,674	160,765	357,255	196,490
1989	6,333	18,589	12,256	167,795	364,771	196,976
1990	6,592	18,353	11,761	179,344	366,009	186,665
1991	6,838	17,650	10,812	216,513	366,971	150,458
1992	7,020	17,215	10,195	233,096	375,961	142,865
1993	7,773	17,985	10,212	266,990	404,530	137,540
1994	9,219	18,726	9,507	284,688	424,908	140,220
1995	9,882	19,663	9,781	314,151	461,986	147,836
1996	10,710	20,169	9,459	318,593	468,519	149,926
1997	11,259	20,355	9,096	313,258	453,998	140,739
1998	11,680	20,370	8,690	290,042	420,351	130,309
1999	11,941	20,750	8,809	317,209	453,155	135,947
2000	12,882	21,127	8,245	343,460	470,494	127,033
2001	13,295	21,311	8,016	319,006	436,995	117,989
2002	14,264	21,101	6,837	329,791	439,722	109,930
2003	15,095	21,246	6,151	341,224	431,929	90,705
2004	15,548	21,422	5,874	339,907	424,884	84,977
2005	15,688	21,355	5,667	348,798	425,363	76,565
2006	15,458	20,926	5,468	319,550	394,506	74,956
2007	15,223	20,271	5,048	318,772	388,747	69,974
2008	13,312	17,483	4,171	303,555	365,729	62,174
2009	12,763	16,463	3,700	300,994	358,326	57,332
2010	12,546	15,899	3,353	307,958	362,303	54,345
Total	280,486	647,403	366,917	7,231,732	13,046,444	5,814,711

Seat belt use has also had a significant economic impact. Table 11-3 lists the economic savings that have resulted from seat belt use over the 36 years. Since 1975, about $1.6 trillion in economic costs (2010 dollars) have been saved due to seat belt use. At 2010 usage rates, seat belts saved society an estimated $69 billion annually in medical care, lost productivity, and other injury related costs. Table 11-4 lists the potential economic savings that were lost due to nonuse. These lost savings could be viewed as costs of seat belt nonuse. Since 1975, nearly $1.5 trillion in unnecessary economic costs (2010 dollars) have been incurred due to seat belt nonuse. At current usage rates, the needless deaths and injuries that result from nonuse continue to cost society an estimated $14 billion annually in medical care, lost productivity, and other injury related costs.[54]

Figure 11-C compares the portion of potential seat belt fatality benefits to those that could be achieved if observed belt use rose to 100 percent. For nearly 2 decades, between 1975 and 1985, belt use was so low that less than 10 percent of potential safety benefits were actually achieved. However, belt use and its corresponding life saving benefits increased dramatically over the past 25 years, and by 2010, 79 percent of potential safety benefits were being realized.

Table 11-3. Impact of Historical Seat belt Use on Economic Costs

					Total Cost Savings (Millions)	
Year	Lives Saved by Seat Belts	Cost/ Fatality Current$	MAIS 2-5 Injuries Prevented	MAIS 2-5 Cost/ Injury Current$	Current $	2010 Dollars
1975	978	$340,971	35,118	$41,698	$1,798	$7,287
1976	796	$360,618	28,754	$44,101	$1,555	$5,960
1977	682	$384,068	25,311	$46,969	$1,451	$5,220
1978	679	$413,221	22,016	$50,534	$1,393	$4,659
1979	594	$460,121	22,254	$56,270	$1,526	$4,582
1980	575	$522,230	21,602	$63,865	$1,680	$4,445
1981	548	$576,101	20,596	$70,453	$1,767	$4,238
1982	678	$611,593	25,378	$74,794	$2,313	$5,226
1983	809	$631,240	35,448	$77,196	$3,247	$7,109
1984	1,197	$658,492	36,728	$80,529	$3,746	$7,862
1985	2,435	$681,942	56,778	$83,397	$6,396	$12,961
1986	4,094	$694,617	119,801	$84,947	$13,021	$25,905
1987	5,141	$719,968	146,489	$88,047	$16,599	$31,862
1988	5,959	$749,756	160,765	$91,690	$19,208	$35,406
1989	6,333	$785,881	167,795	$96,108	$21,103	$37,111
1990	6,592	$828,344	179,344	$101,301	$23,628	$39,421
1991	6,838	$863,201	216,513	$105,564	$28,758	$46,042
1992	7,020	$889,186	233,096	$108,741	$31,589	$49,096
1993	7,773	$915,805	266,990	$111,997	$37,021	$55,865
1994	9,219	$939,254	284,688	$114,864	$41,360	$60,855
1995	9,882	$965,873	314,151	$118,120	$46,652	$66,751
1996	10,710	$994,393	318,593	$121,607	$49,393	$68,646
1997	11,259	$1,017,209	313,258	$124,398	$50,421	$68,503
1998	11,680	$1,033,053	290,042	$126,335	$48,709	$65,161
1999	11,941	$1,055,869	317,209	$129,126	$53,568	$70,113
2000	12,882	$1,091,360	343,460	$133,466	$59,899	$75,850
2001	13,295	$1,122,415	319,006	$137,264	$58,711	$72,288
2002	14,264	$1,140,161	329,791	$139,434	$62,247	$75,450
2003	15,095	$1,166,146	341,224	$142,612	$66,265	$78,530
2004	15,548	$1,197,201	339,907	$146,409	$68,380	$78,934
2005	15,688	$1,237,762	348,798	$151,370	$72,216	$80,630
2006	15,458	$1,277,690	319,550	$156,253	$69,681	$75,369
2007	15,223	$1,314,081	318,772	$160,703	$71,232	$74,913
2008	13,312	$1,364,536	303,555	$166,873	$68,820	$69,700
2009	12,763	$1,359,681	300,994	$166,280	$67,403	$68,508
2010	12,546	$1,381,984	307,958	$169,007	$69,385	$69,385
Total	280,486		7,231,732		$1,242,140	$1,609,842

Table 11-4. Impact of Potential Seat belt Use on Economic Costs

					Cost Savings Forgone (Millions)	
Year	Lives Lost Due to Belt Nonuse	Cost/ Fatality Current$	MAIS 2-5 Injury Benefits Lost Due to Belt Nonuse	MAIS 2-5 Cost/ Injury Current$	Current $	2010 Dollars
1975	13,301	$340,971	224,591	$41,698	$13,900	$56,339
1976	13,851	$360,618	233,719	$44,101	$15,302	$58,642
1977	14,460	$384,068	248,961	$46,969	$17,247	$62,060
1978	15,541	$413,221	231,264	$50,534	$18,109	$60,563
1979	15,726	$460,121	271,924	$56,270	$22,537	$67,690
1980	15,730	$522,230	270,805	$63,865	$25,510	$67,507
1981	15,222	$576,101	260,446	$70,453	$27,119	$65,054
1982	13,250	$611,593	227,807	$74,794	$25,142	$56,812
1983	12,913	$631,240	217,750	$77,196	$24,961	$54,647
1984	13,227	$658,492	225,617	$80,529	$26,879	$56,410
1985	12,508	$681,942	213,595	$83,397	$26,343	$53,385
1986	12,728	$694,617	203,986	$84,947	$26,169	$52,065
1987	12,678	$719,968	202,294	$88,047	$26,939	$51,710
1988	12,674	$749,756	196,490	$91,690	$27,519	$50,724
1989	12,256	$785,881	196,976	$96,108	$28,563	$50,228
1990	11,761	$828,344	186,665	$101,301	$28,651	$47,801
1991	10,812	$863,201	150,458	$105,564	$25,216	$40,371
1992	10,195	$889,186	142,865	$108,741	$24,601	$38,235
1993	10,212	$915,805	137,540	$111,997	$24,756	$37,358
1994	9,507	$939,254	140,220	$114,864	$25,036	$36,837
1995	9,781	$965,873	147,836	$118,120	$26,909	$38,502
1996	9,459	$994,393	149,926	$121,607	$27,638	$38,411
1997	9,096	$1,017,209	140,739	$124,398	$26,760	$36,356
1998	8,690	$1,033,053	130,309	$126,335	$25,440	$34,033
1999	8,809	$1,055,869	135,947	$129,126	$26,855	$35,150
2000	8,245	$1,091,360	127,033	$133,466	$25,953	$32,864
2001	8,016	$1,122,415	117,989	$137,264	$25,193	$31,019
2002	6,837	$1,140,161	109,930	$139,434	$23,123	$28,028
2003	6,151	$1,166,146	90,705	$142,612	$20,109	$23,830
2004	5,874	$1,197,201	84,977	$146,409	$19,474	$22,479
2005	5,667	$1,237,762	76,565	$151,370	$18,604	$20,772
2006	5,468	$1,277,690	74,956	$156,253	$18,699	$20,225
2007	5,048	$1,314,081	69,974	$160,703	$17,879	$18,802
2008	4,171	$1,364,536	62,174	$166,873	$16,067	$16,272
2009	3,700	$1,359,681	57,332	$166,280	$14,564	$14,803
2010	3,353	$1,381,984	54,345	$169,007	$13,819	$13,819
Total	366,917		5,814,711		$827,582	$1,489,802

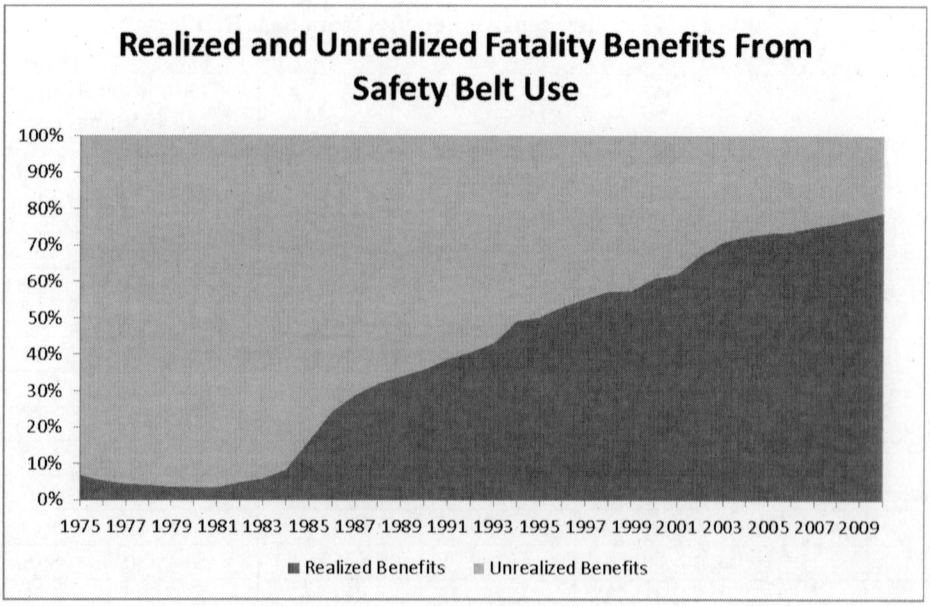

Figure 11-C. Realized and Unrealized Fatality Benefits From Safety Belt Use.

Figure 11-D compares the achieved economic benefits from seat belt use to those that could have been achieved if observed belt use was 100 percent. Cost impacts (which include impacts to both fatalities and nonfatal injuries) roughly parallel the pattern seen for fatalities, with less than 10 percent of potential economic benefits being realized between 1975 and 1985, but with significant growth in later years due to increases in belt use. By 2010, some 83 percent of potential economic benefits were being realized.

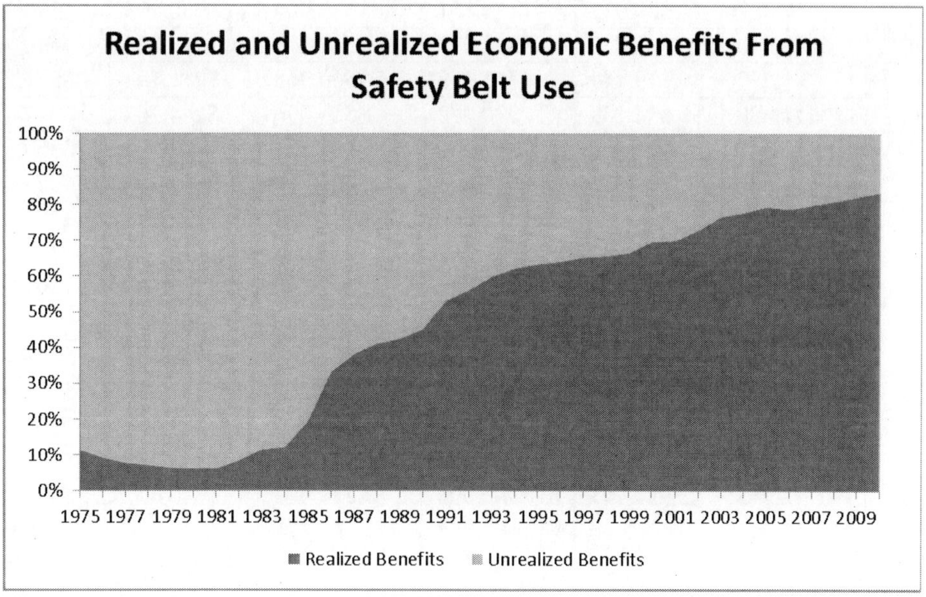

Figure 11-D. Realized and Unrealized Economic Benefits From Safety Belt Use.

Table 11-5. Comprehensive Benefits from Seat Belt Use

					Total Cost Savings (Millions)	
Year	Lives Saved by Seat Belts	Cost/ Fatality Current$	MAIS 2-5 Injuries Prevented	MAIS 2-5 Cost/Injury Current$	Current $	2010 Dollars
1975	978	$2,252,374	35,118	$187,836	$8,799	$35,664
1976	796	$2,382,158	28,754	$198,659	$7,608	$29,158
1977	682	$2,537,061	25,311	$211,577	$7,086	$25,496
1978	679	$2,729,643	22,016	$227,637	$6,865	$22,960
1979	594	$3,039,449	22,254	$253,474	$7,446	$22,365
1980	575	$3,449,733	21,602	$287,689	$8,198	$21,695
1981	548	$3,805,592	20,596	$317,366	$8,622	$20,683
1982	678	$4,040,040	25,378	$336,917	$11,289	$25,510
1983	809	$4,169,823	35,448	$347,741	$15,700	$34,372
1984	1,197	$4,349,846	36,728	$362,753	$18,530	$38,889
1985	2,435	$4,504,749	56,778	$375,672	$32,299	$65,455
1986	4,094	$4,588,480	119,801	$382,654	$64,628	$128,581
1987	5,141	$4,755,943	146,489	$396,620	$82,551	$158,456
1988	5,959	$4,952,712	160,765	$413,029	$95,914	$176,792
1989	6,333	$5,191,346	167,795	$432,930	$105,520	$185,559
1990	6,592	$5,471,846	179,344	$456,322	$117,909	$196,716
1991	6,838	$5,702,108	216,513	$475,525	$141,948	$227,259
1992	7,020	$5,873,757	233,096	$489,839	$155,413	$241,545
1993	7,773	$6,049,593	266,990	$504,503	$181,721	$274,223
1994	9,219	$6,204,496	284,688	$517,421	$204,503	$300,898
1995	9,882	$6,380,332	314,151	$532,085	$230,205	$329,381
1996	10,710	$6,568,728	318,593	$547,796	$244,875	$340,322
1997	11,259	$6,719,444	313,258	$560,365	$251,193	$341,272
1998	11,680	$6,824,108	290,042	$569,094	$244,767	$327,441
1999	11,941	$6,974,825	317,209	$581,662	$267,795	$350,506
2000	12,882	$7,209,273	343,460	$601,214	$299,363	$379,082
2001	13,295	$7,414,415	319,006	$618,322	$295,823	$364,235
2002	14,264	$7,531,639	329,791	$628,098	$314,572	$381,292
2003	15,095	$7,703,288	341,224	$642,412	$335,488	$397,582
2004	15,548	$7,908,430	339,907	$659,520	$347,136	$400,715
2005	15,688	$8,176,370	348,798	$681,865	$366,104	$408,762
2006	15,458	$8,440,124	319,550	$703,860	$355,386	$384,395
2007	15,223	$8,680,517	318,772	$723,908	$362,905	$381,658
2008	13,312	$9,013,810	303,555	$751,703	$348,175	$352,627
2009	12,763	$8,981,741	300,994	$749,028	$340,087	$345,666
2010	12,546	$9,129,066	307,958	$761,314	$348,986	$348,986
Total	280,486		7,231,732		$6,235,410	$8,066,198

Table 11-5 lists the comprehensive impact of seat belt use. This table reflects the combined impact of both economic cost savings and valuations for lost quality-of-life (see Chapter 4). The comprehensive societal benefits from seat belt use are enormous. From 1975 to 2010, seat belts have prevented $8.1 trillion in societal harm as measured by comprehensive costs, and they are currently preventing $349 billion in societal harm annually.

Table 11-6 lists the unnecessary societal harm (as measured by comprehensive costs) that resulted from failure of occupants to wear seat belts. These lost potential savings can be viewed as the societal cost of seat belt nonuse. Since 1975, some $7.8 trillion in unnecessary societal harm (2010 Dollars) has been incurred due to seat belt nonuse. At current usage rates, the needless deaths and injuries that result from nonuse continue to cost society an estimated $72 billion annually in lost quality-of-life, medical care, lost productivity, and other injury related costs.

Table 11-6. Impact of Potential Seat Belt Use on Societal Harm

					Cost Savings Forgone (Millions)	
Year	**Lives Lost Due to Belt Nonuse**	**Cost/ Fatality Current$**	**MAIS 2-5 Injury Benefits Lost Due to Belt Nonuse**	**MAIS 2-5 Cost/ Injury Current$**	**Current $**	**2010 Dollars**
1975	13,301	$2,252,374	224,591	$187,836	$72,145	$292,410
1976	13,851	$2,382,158	233,719	$198,659	$79,426	$304,380
1977	14,460	$2,537,061	248,961	$211,577	$89,360	$321,544
1978	15,541	$2,729,643	231,264	$227,637	$95,066	$317,940
1979	15,726	$3,039,449	271,924	$253,474	$116,724	$350,583
1980	15,730	$3,449,733	270,805	$287,689	$132,172	$349,768
1981	15,222	$3,805,592	260,446	$317,366	$140,585	$337,244
1982	13,250	$4,040,040	227,807	$336,917	$130,283	$294,393
1983	12,913	$4,169,823	217,750	$347,741	$129,565	$283,659
1984	13,227	$4,349,846	225,617	$362,753	$139,379	$292,516
1985	12,508	$4,504,749	213,595	$375,672	$136,587	$276,799
1986	12,728	$4,588,480	203,986	$382,654	$136,458	$271,492
1987	12,678	$4,755,943	202,294	$396,620	$140,530	$269,747
1988	12,674	$4,952,712	196,490	$413,029	$143,927	$265,292
1989	12,256	$5,191,346	196,976	$432,930	$148,902	$261,847
1990	11,761	$5,471,846	186,665	$456,322	$149,534	$249,477
1991	10,812	$5,702,108	150,458	$475,525	$133,198	$213,249
1992	10,195	$5,873,757	142,865	$489,839	$129,864	$201,836
1993	10,212	$6,049,593	137,540	$504,503	$131,168	$197,937
1994	9,507	$6,204,496	140,220	$517,421	$131,539	$193,541
1995	9,781	$6,380,332	147,836	$532,085	$141,067	$201,841
1996	9,459	$6,568,728	149,926	$547,796	$144,263	$200,493

Table 11-6. (Continued)

					Cost Savings Forgone (Millions)	
Year	**Lives Lost Due to Belt Nonuse**	**Cost/ Fatality Current$**	**MAIS 2-5 Injury Benefits Lost Due to Belt Nonuse**	**MAIS 2-5 Cost/ Injury Current$**	**Current $**	**2010 Dollars**
1997	9,096	$6,719,444	140,739	$560,365	$139,985	$190,185
1998	8,690	$6,824,108	130,309	$569,094	$133,459	$178,538
1999	8,809	$6,974,825	135,947	$581,662	$140,516	$183,916
2000	8,245	$7,209,273	127,033	$601,214	$135,815	$171,981
2001	8,016	$7,414,415	117,989	$618,322	$132,389	$163,005
2002	6,837	$7,531,639	109,930	$628,098	$120,541	$146,107
2003	6,151	$7,703,288	90,705	$642,412	$105,653	$125,208
2004	5,874	$7,908,430	84,977	$659,520	$102,498	$118,318
2005	5,667	$8,176,370	76,565	$681,865	$98,543	$110,025
2006	5,468	$8,440,124	74,956	$703,860	$98,909	$106,983
2007	5,048	$8,680,517	69,974	$723,908	$94,474	$99,356
2008	4,171	$9,013,810	62,174	$751,703	$84,333	$85,411
2009	3,700	$8,981,741	57,332	$749,028	$76,176	$77,425
2010	3,353	$9,129,066	54,345	$761,314	$71,984	$71,984
Total	366,917		5,814,711		$4,327,016	$7,776,433

12. Crashes by Roadway Location

Urban roadway environments are characterized by high population densities. This typically produces higher traffic volumes and, on average, lower average travel speeds than are found in more rural areas. These conditions affect crash impacts in a variety of ways. Slower travel speeds can reduce the severity of crashes when they occur, but higher traffic volume creates more opportunities for exposure to distracted or alcohol impaired drivers, as well as more complex driving interactions in general. Higher traffic volume also means that when crashes do occur, they will have more impact on uninvolved drivers and cause more aggregate travel delay and pollution. By contrast, the higher speeds typically encountered on less congested rural roadways can lead to more serious injury outcomes in the event of a crash.

The categorization of any specific crash locale as urban or rural is a function of the definition that is assumed for these designations. According to the Washington Post, within the U.S. government there are at least 15 different official definitions of the word "rural (Fahrenthold, 2013)." The Department of Agriculture alone has 11 different definitions depending on the specific program that the definition relates to. Most definitions seem to be based on absolute population size: for example, "fewer than 50,000 inhabitants and not located next to an urban area" or "20,000 or fewer inhabitants," or "10,000 or fewer inhabitants," or "5,000 or fewer inhabitants." In some cases, these definitions are based on area as well, such as "less than 20 people per square mile." The Federal Highway Administration (FHWA) is a primary user of crash cost information related to roadway systems. FHWA uses this data to allocate resources towards improving safety on the U.S. roadways. The definition adopted for this study is that used by the FHWA to define the Nation's roadway system. FHWA's roadway designations were designed to be consistent with designations used by the U.S. Census Bureau. Urban areas are defined in the Federal aid highway law (Section 101 of Title 23, U.S. Code) as follows:

> "The term 'urban area' means an urbanized area or, in the case of an urbanized area encompassing more than one State, that part of the urbanized area in each such State, or an urban place as designated by the Bureau of the Census having a population of five thousand or more and not within any urbanized area, within boundaries to be fixed by responsible State and local officials in cooperation with each other, subject to approval by the Secretary. Such boundaries shall, as a minimum, encompass the entire urban place designated by the Bureau of the Census… .
>
> Small urban areas are those urban places, as designated by the Bureau of the Census having a population of five thousand (5,000) or more and not within any urbanized area.
>
> Urbanized areas are designated as such by the Bureau of the Census.
>
> Rural areas comprise the areas outside the boundaries of small urban and urbanized areas, as defined above."

NHTSA's FARS system collects geospatial coordinates which permit the exact identification of crashes and allow for the overlay of theses crashes on the roadway land use designation map defined by FHWA. The urban/rural breakout of fatal crashes can thus be derived directly from the FARS database. Table 12-1 below lists this profile over the past 14 years. Over this period, there has been a very gradual decline in the portion of fatalities that occur in rural jurisdictions, from roughly 61 percent to 55 percent.

Table 12-1. Traffic Fatalities with Known Urban/Rural Designation

	Urban	% Urban	Rural	% Rural
1998	16219	39.2%	25185	60.8%
1999	16058	38.6%	25548	61.4%
2000	16113	39.3%	24838	60.7%
2001	16988	40.3%	25150	59.7%
2002	17013	39.6%	25896	60.4%
2003	17783	41.6%	24957	58.4%
2004	17581	41.1%	25179	58.9%
2005	18627	43.1%	24587	56.9%
2006	18791	44.3%	23646	55.7%
2007	17908	43.5%	23254	56.5%
2008	16218	43.6%	20987	56.4%
2009	14,501	42.9%	19,323	57.1%
2010	14,659	44.8%	18,089	55.2%
2011	14,464	44.9%	17,762	55.1%

Urban/rural designations for nonfatal crashes are more elusive. There are no definitive sources or surveys designed specifically to produce a nationally representative break out of urban and rural crashes. Until it was discontinued in 1997, the GES included a specific urban/rural variable. However, this variable was not directly linked to the crash itself. Rather, it represented the pre-determined urban/rural proportion of the general population that was covered by the primary sampling unit (PSU) from which each case was drawn. Thus, use of this variable assumes that crashes occur proportionally according to the population spread.

A second possible indicator of urban/rural status is the Land Use Variable that has been collected in GES since 1988 (except 2009). This variable categorizes land use based on population size within the specific police jurisdictions from which crash records are drawn. Each PSU has multiple police jurisdictions. Therefore, this variable represents a finer definition than the Urban/Rural variable which reflected populations at the PSU level. The categories included under land use are:

Within an area of population 25,000-50,000;
Within an area of population 50,000-100,000;
Within an area of population 100,000 +;
Other area; and
Unknown area.

This variable quantifies populations, but fails to define urban/rural. A small area with a population of 25,000 might be considered urban whereas a large area with the same population might be considered rural. Generally, since these areas are all specific police jurisdictions, the size of the area is somewhat limited and, although there would be exceptions, we might expect that most police jurisdictions with 25,000 or more population would be considered urban. By the same logic, the "Other Area" category is more likely to represent rural roadways, but may also include some urban areas as defined by FHWA.

A third possible source for insight into the urban/rural breakdown of nonfatal crashes is the National Highway System variable collected in GES from 1995-1998. This variable has 20 possible selections based on the type of roadway on which the crash occurred and whether that roadway was urban or rural. The urban/rural designation for these roadways reflects the characteristics of the surrounding land use area. The selections for this variable for roadways with known urban/rural characteristics are as follows:

National Highway System Variables
Urban
1 = Eisenhower Interstate (EIS) 2= Congressional High Priority Route 3 = STRAHNET Route 4 = STRAHNET Major Connector 5 = Other NHS Route 9 = Unknown Urban Route
Rural
11 = Eisenhower Interstate (EIS) 12 = Congressional High Priority Route 13 = STRAHNET Route 14 = STRAHNET Major Connector 15 = Other NHS Route 19 = Unknown Rural Route

Only about 15 percent of the crashes in the 1994-1998 GES occurred on roadways in the NHS, so this data represents only a sample of all crashes and use of these variables to represent nationwide distributions assumes that the urban/rural distribution for crashes on all roadways is similar to that on NHS roadways.

A fourth possible source of urban/rural designations is State data files collected in NHTSA's State Data System (SDS). Many States do not collect urban/rural information, but a subsample of the 34 States in the SDS system do have urban/rural indicators. NHTSA found 10 States with urban/rural indicators within their data sets in 2008 or later. Table 12-2 summarizes the average rural portion of crashes in each State by injury severity based on the designations found in State files for all available years within each State from 2008-2011. Table 12-3 presents the relative rates of rural proportions according to injury severity for each State and the average across all States.

Table 12-2. Portion of Injuries Occurring in Rural Jurisdictions, by Injury Severity

State	Injury Severity					Total
	K	A	B	C	O	
Arkansas	67.26%	57.20%	31.74%	18.94%	14.22%	16.88%
Florida	61.81%	56.63%	46.77%	44.97%	39.44%	42.08%
Illinois	49.90%	32.21%	21.20%	9.19%	11.36%	12.11%
Kansas	69.17%	53.82%	40.31%	25.86%	27.62%	28.69%
Minnesota	76.98%	62.18%	48.26%	33.93%	24.50%	26.98%
Missouri	73.40%	64.55%	46.56%	28.15%	30.74%	32.13%
Nebraska	91.34%	66.07%	51.94%	35.52%	37.15%	38.35%
Texas	28.77%	20.60%	14.12%	7.93%	9.13%	9.44%
Washington	66.73%	54.57%	48.83%	23.02%	21.75%	23.74%
Wisconsin	78.57%	65.10%	52.74%	35.24%	36.66%	37.74%
Average	66.39%	53.29%	40.25%	26.27%	25.26%	26.81%

Table 12-3. Rural Proportions Relative to Fatal Proportions by Injury Severity

State	Injury Severity					Total
	K	A	B	C	O	
Arkansas	1.0000	0.8504	0.4720	0.2816	0.2114	0.2510
Florida	1.0000	0.9161	0.7566	0.7274	0.6380	0.6808
Illinois	1.0000	0.6455	0.4250	0.1841	0.2277	0.2428
Kansas	1.0000	0.7781	0.5828	0.3739	0.3993	0.4148
Minnesota	1.0000	0.8077	0.6269	0.4408	0.3183	0.3505
Missouri	1.0000	0.8795	0.6343	0.3835	0.4188	0.4378
Nebraska	1.0000	0.7233	0.5686	0.3888	0.4067	0.4199
Texas	1.0000	0.7161	0.4909	0.2756	0.3173	0.3282
Washington	1.0000	0.8177	0.7317	0.3449	0.3259	0.3557
Wisconsin	1.0000	0.8285	0.6712	0.4486	0.4666	0.4803
Average	1.0000	0.7963	0.5960	0.3849	0.3730	0.3962

A final source for urban/rural definitions for injuries is the National Motor Vehicle Crash Causation Survey (NMVCSS). In 2008 NHTSA's National Center for Statistics and Analysis completed a nationwide survey of crashes involving light passenger vehicles, with a focus on the factors related to pre-crash events. This Nationally representative sample of crashes was investigated from 2005 to 2007. NMVCCS investigated a total of 6,950 crashes during the 3-year period from January 2005 to December 2007. However, the final report was based on a nationally representative sample of 5,471 crashes that were investigated during a 2 ½- year period from July 3, 2005, to December 31, 2007. The remaining 1,479 crashes were investigated but were not used because (1) these crashes were investigated during the transition period from January 1, 2005, to July 2, 2005, when the data collection effort was being phased in, or (2) these crashes were investigated after the phase-in period, but ultimately determined not to meet the requisite sample selection criteria. Each investigated crash involved at least one light passenger vehicle that was towed due to damage. Data was collected on at least 600 data elements to capture information related to the drivers, vehicles, roadways, and environment. In addition, the NMVCCS database includes crash narratives, photographs, schematic diagrams, vehicle information, as well as event data recorder (EDR) data, when available. An important feature of NMVCCS relevant to this study is the fact that each crash location was recorded using geo-spatial devices. For this study, these coordinates were overlain on the previously discussed roadway land use designation map developed by FHWA to produce urban and rural designations for each crash.

As noted above, the higher average speeds encountered in less congested rural areas result in generally more severe crash outcomes. Based on this, we would expect to see a higher rural proportion of more serious crashes. To estimate the urban/rural portions for nonfatal crashes, we examined the relative proportions of these factors by injury severity level across the 5 sources cited above. As expected, the rural proportion of crashes was highest in the most serious crashes and declined fairly steadily as crash severity diminished. However, the absolute portions of fatal crashes that were rural in these databases differed significantly from the rural rates that were found in FARS. It is uncertain why these differences occur, but it is possible that whatever is biasing the fatality number in these sources is also biasing the nonfatal injuries as well. This would seem to be the case given that the value of all nonfatal injury severity levels seems consistent with the absolute value of the fatal injury proportions measured in each data source. Generally speaking, the sources with rural fatality portion that are higher also have nonfatal injury rural portions that are higher. For this analysis we adopted this assumption (that both fatal and injury biases are similar) and normalize the results of each source to the known rural portion from FARS. That is, we assume the FARS urban/rural distribution is correct, but accept the relative ratios among injury severity from the 5 sources. The results are displayed in Table 12-4.

A limitation common to all 5 sources is that injury severity is only coded in the KABCO system. As previously mentioned, this report stratifies injury severity using the more precise MAIS basis. Previously the derivation and use of KABCO-MAIS translators was discussed. In order to derive urban/rural proportions under MAIS, a reverse translator was applied to the urban and rural KABCO injury distributions (see Table 12-5 below). This translator was derived from the same historical data bases as the previously discussed translators. However, since it will be applied to known nonfatal injuries only, it was normalized to remove categories of Unknowns and fatalities. Table 12-6 shows the resulting KABCO/MAIS matrix. Table 12-7 shows the estimated rural incidence counts for each MAIS level derived by applying the KABCO injury- severity- specific rural percentage (from Table 12-4 above) to the corresponding incidence counts in Table 12-6. The resulting MAIS totals were then used to obtain rates for urban and rural crashes for each MAIS severity level, and these rates were applied to the nationwide 2010 incidence data previously derived in the Incidence chapter to estimate total nationwide urban and rural crash incidence.

Table 12-4. Derivation of Estimated Urban/Rural Proportions of Crashes by Injury Severity

	No Injury (O)	Possible Injury (C)	Non-incapacitating Injury (B)	Incapacitating Injury (A)	Fatal Injury (K)
Estimated Percentage Rural					
1994-1996 GES Urban/Rural	23.8%	21.3%	25.8%	29.5%	33.0%
2010-2011 GES Land Use	29.9%	33.2%	41.1%	42.7%	50.0%
1995-1998 National Highway System	39.0%	30.2%	46.2%	66.2%	70.2%
SDS States w/Urban/Rural (10 States)	25.3%	26.3%	40.2%	53.3%	66.4%
NMVCCS	29.7%	18.6%	27.6%	42.8%	49.2%
Ratio/Fatal					
1994-1996 GES Urban/Rural	0.7202	0.6454	0.7827	0.8955	1.0000
2010-2011 GES Land Use	0.5985	0.6630	0.8208	0.8543	1.0000
1995-1998 National Highway System	0.5553	0.4295	0.6588	0.9433	1.0000
SDS States w/Urban/Rural (10 States)	0.3730	0.3849	0.5960	0.7963	1.0000
NMVCCS	0.6030	0.3776	0.5598	0.8693	1.0000
Normalized to 2010 FARS					
1994-1996 GES Urban/Rural	39.8%	35.6%	43.2%	49.5%	55.2%
2010-2011 GES Land Use	33.1%	36.6%	45.3%	47.2%	55.2%
1995-1998 National Highway System	30.7%	23.7%	36.4%	52.1%	55.2%
SDS States w/Urban/Rural (10 States)	21.0%	21.9%	33.5%	44.3%	55.2%
NMVCCS	33.3%	20.9%	30.9%	48.0%	55.2%
Average of 5 Methods	31.6%	27.7%	37.9%	48.2%	55.2%

Table 12-5. Reverse Translator for MAIS to KABCO Application to Injured Survivors

MAIS	O No Injury	C Possible Injury	B Non-Incapacitating	A Incapacitating	Total
0	0.96507	0.02827	0.00538	0.00128	1.00000
1	0.33042	0.36314	0.21622	0.09022	1.00000
2	0.08384	0.31360	0.28566	0.31690	1.00000
3	0.00984	0.14665	0.23328	0.61024	1.00000
4	0.00190	0.08278	0.19400	0.72132	1.00000
5	0.03750	0.02134	0.08451	0.85665	1.00000

Table 12-6. KABCO Incidence Counts from MAIS Injured Survivors

MAIS	O	C	B	A	Total
	No Injury	Possible Injury	Non-Incapacitating	Incapacitating	
0	4,423,168	129,580	24,650	5,867	4,583,265
1	1,142,986	1,256,178	747,956	312,080	3,459,200
2	28,400	106,224	96,761	107,344	338,730
3	991	14,773	23,500	61,476	100,740
4	32	1,414	3,315	12,324	17,086
5	216	123	486	4,925	5,749

Table 12-7. Rural KABCO Incidence Counts from MAIS Injured Survivors

MAIS	O	C	B	A	Total Rural	% Rural
	No Injury	Possible Injury	Non-Incapacitating	Incapacitating		
0	1,502,706	38,689	10,047	3,045	1,554,487	33.92%
1	360,812	348,498	283,270	150,490	1,143,070	33.04%
2	8,965	29,470	36,646	51,763	126,844	37.45%
3	313	4,098	8,900	29,645	42,956	42.64%
4	10	392	1,255	5,943	7,601	44.49%
5	68	34	184	2,375	2,661	46.29%

Cases designated as O-Uninjured in KABCO records are likely to be predominately PDO crashes. In addition, they would include cases where uninjured people were involved in crashes that did produce injury, which are categorized as MAIS 0 in this study. Since PDOs are counted separately, the rural portion for MAIS 0 injuries should equal the weighted average rural portion of MAIS 0 incidence in injury crashes. To estimate rural MAIS 0 incidence we examined the frequency of uninjured occupants in injury crashes by MAIS level. Data from the 2009-2011 CDS and 1982-86 NASS were examined to determine ratios of uninjured occupant frequencies. These are the only two databases with MAIS stratification. Neither database is ideal – CDS represents tow-away crashes for light vehicles while the NASS data, which include all crash types, are old. However, both sources gave very similar frequencies. These are shown in Table 12-8. Using these frequencies, we calculated a weighted average rural portion across all injury severity categories. We then assumed that the frequency of PDOs versus MAIS 0 was similar to the relative frequency of these cases nationwide. This would imply that 80.1 percent of these cases were PDOs and 19.9 percent were MAIS0s. Using these weights, we derived the rural portion of MAIS0s. We then imputed the PDO portion. The results, which are shown in the lower half of Table 12-8, are nearly identical for either data source. We chose to base our estimate on the NASS data because it includes all crash types, but the difference if we used CDS would be insignificant.

As noted previously, for fatalities, the urban/rural designation contained in the FARS files is used directly. The resulting urban/rural incidence counts are illustrated in Figure 12-A and shown in Table 12-9 for each injury severity category.

Table 12-8. Derivation of Rural Incidence Percentage for MAIS 0 and PDO

	2009-11 CDS	1982-86 NASS
Distribution of MAIS 0 by Crash Severity		
MAIS1	0.8662	0.8722
MAIS2	0.1029	0.0875
MAIS3	0.0230	0.0284
MAIS4	0.0044	0.0033
MAIS5	0.0006	0.0015
Fatal	0.0029	0.0070
Imputed % Rural		
All KABCO O Injuries	31.57%	31.57%
MAIS 0 in Injury Crashes	33.84%	33.92%
PDO	31.00%	30.99%

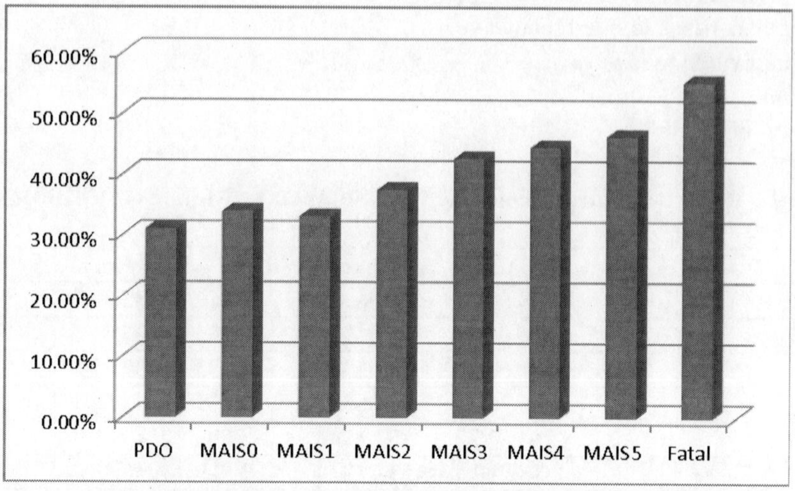

Figure 12-A. Rural Percentage of Motor Vehicle Injury by Injury Severity.

Table 12-9. Urban/Rural Incidence Summary

	Urban	% Urban	Rural	% Rural	Total
MAIS0	3,028,778	66.08%	1,554,487	33.92%	4,583,265
MAIS1	2,316,130	66.96%	1,143,070	33.04%	3,459,200
MAIS2	211,886	62.55%	126,844	37.45%	338,730
MAIS3	57,784	57.36%	42,956	42.64%	100,740
MAIS4	9,485	55.51%	7,601	44.49%	17,086
MAIS5	3,088	53.71%	2,661	46.29%	5,749
Fatal	14,771	44.76%	18,228	55.24%	32,999
PDO	12,773,589	69.01%	5,735,043	30.99%	18,508,632

In Table 12-10, the incidence from Table 12-9 is combined with the per-unit economic costs of crashes from Table 2-15 in chapter 2. In Table 12-11, incidence is combined with the per-unit comprehensive costs from Table 4-2 in chapter 4. The results indicate that urban crashes cost an estimated $171 billion and rural crashes cost $106 billion in 2010. Roughly 62 percent of all economic crash costs thus occur in urban areas while 38 percent occur in rural areas. Comparing Tables 12-9, 12-10, and 12-11, the rural portion of incidence was 32 percent, but the rural portion rises to 38 percent for economic costs and 44 percent for comprehensive costs. This is a reflection of the more severe injury profile associated with rural crashes, which have larger proportions of more costly and debilitating injuries. Nonetheless, the higher frequency of crashes in urban areas results in urban crashes causing the majority of all injury incidence, economic costs, and comprehensive costs.

A further breakdown of these crash cost estimates was made by roadway classification. Data on roadway crash costs is useful for highway safety planning and allocation of limited roadway construction funds. For this analysis, roadways were divided into the following 5 classifications.

4-lane divided roadways
Greater than 4-lane divided roadways
2-lane undivided roadways
Multi-lane undivided roadways
All other roadways

Table 12-10. Urban/Rural Economic Cost Summary, (Millions of 2010 Dollars)

	Urban	% Urban	Rural	% Rural	Total
MAIS0	$8,611	66.08%	$4,419	33.92%	$13,030
MAIS1	$46,063	66.96%	$22,733	33.04%	$68,797
MAIS2	$21,604	62.55%	$12,933	37.45%	$34,537
MAIS3	$16,006	57.36%	$11,899	42.64%	$27,905
MAIS4	$4,875	55.51%	$3,907	44.49%	$8,781
MAIS5	$3,399	53.71%	$2,929	46.29%	$6,327
Fatal	$20,664	44.76%	$25,499	55.24%	$46,163
PDO	$49,332	69.01%	$22,149	30.99%	$71,480
Total Economic Costs	$170,553	61.57%	$106,467	38.43%	$277,020

Table 12-11. Urban/Rural Comprehensive Cost Summary (Millions of 2010 Dollars)

	Urban	% Urban	Rural	% Rural	Total
MAIS0	$8,611	66.08%	$4,419	33.92%	$13,030
MAIS1	$99,892	66.96%	$49,299	33.04%	$149,192
MAIS2	$93,830	62.55%	$56,171	37.45%	$150,001
MAIS3	$62,562	57.36%	$46,508	42.64%	$109,070
MAIS4	$24,200	55.51%	$19,394	44.49%	$43,594
MAIS5	$17,537	53.71%	$15,112	46.29%	$32,649
Fatal	$135,099	44.76%	$166,710	55.24%	$301,809
PDO	$49,332	69.01%	$22,149	30.99%	$71,480
Total Comprehensive Costs	$491,064	56.39%	$379,762	43.61%	$870,826

Table 12-12. NMVCCS Cases, Percentages Urban Versus Rural by Roadway Type

A Injuries	Rural	Urban	Total
Greater than 4 lanes divided	13.11%	86.89%	100.00%
Multi-lane undivided	26.63%	73.37%	100.00%
Four lanes divided	64.59%	35.41%	100.00%
Two-lane undivided	67.79%	32.21%	100.00%
Other	37.20%	62.80%	100.00%
Total	42.88%	57.12%	100.00%
B Injuries			
Greater than 4 lanes divided	14.14%	85.86%	100.00%
Multi-lane undivided	13.14%	86.86%	100.00%
Four lanes divided	43.27%	56.73%	100.00%
Two-lane undivided	43.67%	56.33%	100.00%
Other	14.30%	85.70%	100.00%
Total	27.82%	72.18%	100.00%
C Injuries			
Greater than 4 lanes divided	9.42%	90.58%	100.00%
Multi-lane undivided	12.70%	87.30%	100.00%
Four lanes divided	24.34%	75.66%	100.00%
Two-lane undivided	30.74%	69.26%	100.00%
Other	10.11%	89.89%	100.00%
Total	18.77%	81.23%	100.00%
Uninjured			
Greater than 4 lanes divided	23.14%	76.86%	100.00%
Multi-lane undivided	14.82%	85.18%	100.00%
Four lanes divided	67.71%	32.29%	100.00%
Two-lane undivided	34.91%	65.09%	100.00%
Other	20.19%	79.81%	100.00%
Total	29.90%	70.10%	100.00%
Injured Severity Unknown			
Greater than 4 lanes divided	4.98%	95.02%	100.00%
Multi-lane undivided	5.11%	94.89%	100.00%
Four lanes divided	31.20%	68.80%	100.00%
Two-lane undivided	0.79%	99.21%	100.00%
Other	0.00%	100.00%	100.00%
Total	4.72%	95.28%	100.00%

The lane count designations in the above categories include lanes in both directions. Thus, for example, 4 lane divided roadways would include 2 lanes in each direction. These categories and designations were selected based on discussions with FHWA staff regarding the most useful categories for planning purposes.

As previously noted, NHTSA's FARS system collects geospatial coordinates which permit the exact identification of crashes and allow for the overlay of these crashes on the roadway land use designation maps defined by FHWA, or, for some roadway designations, individual States. The urban/rural and roadway designation breakout of fatal crashes can thus be derived directly from the FARS database.

For nonfatal injuries and PDOs, roadway designations are available within NHTSA's GES data system, but as noted earlier, urban and rural designations for these roadways are not collected in GES. To stratify these nonfatal impacts by roadway category, we first examined the roadway designation proportions within GES under the 5 categories discussed above. All cases were stratified by their coded roadway type and lumped under the appropriate category. The approach we used involved determining proportions of cases that occurred under each roadway type from the data files and then applying these proportions to the total urban and rural costs already derived. Cases where the roadway designation was unknown were thus ignored, because redistributing these cases across known roadway cases would not alter the proportions assigned to that roadway type. In other words, we used the police-reported cases with known roadway types to determine the proportions of crashes that occurred on each roadway type, and then applied that proportion to the total costs of urban and rural crashes.

Because GES is stratified only by KABCO, data were organized into the same 5 categories noted in previous sections of this report to be run through KABCO/MAIS translators to produce an MAIS based injury profile. Those categories are CDS equivalent cases, Unbelted Non-CDS cases, Belted Non-CDS cases, Unknown Belt Use Non-CDS cases, and Motorcycle/nonoccupant cases. The MAIS injury totals from each of these cases were then combined to form a full MAIS injury profile representing all 5 translator scenarios. This was done separately for each of the 5 roadway types.

Because GES records do not include an urban/rural designation, this feature was derived from the database we created from NMVCCS discussed above. NMVCCS cases were stratified within one of the 5 roadway designations for both urban and rural crashes. The proportions of each Roadway Type that were Urban and Rural were calculated within each KABCO injury severity level. Table 12-12 lists the NMVCCS case distributions that resulted from this process. These proportions were then applied to each translated MAIS case total that was derived from the corresponding KABCO distribution.

This process produced separate tables that define the proportions of all crashes that occur on different roadway types by injury severity for urban and rural locations. Tables 12-13 and 12-14 summarize these results.

Table 12-13. Proportions of Fatalities, Injuries and PDOV by Roadway Designation in Rural Crashes

	Divided Roadways		Undivided Roadways			
	4 Lanes	**>4 Lanes**	**2 Lanes**	**>2 Lanes**	**All Other**	**Total**
MAIS0	31.47%	13.29%	39.42%	10.96%	4.87%	100.00%
MAIS1	23.45%	8.26%	54.16%	10.58%	3.55%	100.00%
MAIS2	22.13%	5.94%	59.17%	9.22%	3.53%	100.00%
MAIS3	21.68%	4.89%	60.65%	9.07%	3.72%	100.00%
MAIS4	22.99%	4.75%	60.18%	8.81%	3.28%	100.00%
MAIS5	22.57%	4.14%	60.98%	8.73%	3.58%	100.00%
Fatal	15.74%	7.09%	71.96%	4.17%	1.03%	100.00%
PDOV	32.35%	10.53%	41.50%	9.80%	5.82%	100.00%

Table 12-14. Proportions of Fatalities, Injuries and PDOV by Roadway Designation in Urban Crashes

	Divided Roadways		Undivided Roadways			
	4 Lanes	>4 Lanes	2 Lanes	>2 Lanes	All Other	Total
MAIS0	8.98%	21.74%	32.24%	27.84%	9.20%	100.00%
MAIS1	13.80%	20.52%	33.15%	24.21%	8.32%	100.00%
MAIS2	13.86%	19.62%	35.00%	23.11%	8.40%	100.00%
MAIS3	13.58%	20.14%	33.92%	23.99%	8.38%	100.00%
MAIS4	13.99%	21.72%	33.21%	24.06%	7.02%	100.00%
MAIS5	12.23%	20.98%	35.57%	23.73%	7.48%	100.00%
Fatal	17.21%	22.66%	37.09%	18.21%	4.83%	100.00%
PDOV	9.58%	17.89%	35.25%	25.87%	11.42%	100.00%

The resulting estimates indicate significant differences in the proportions of crashes that occur on various roadway types in rural versus urban settings. In rural settings over half of all injuries and 40 percent of PDOs occur on 2-lane undivided roadways, and over 20 percent of injuries and 30 percent of PDOs occur on 4-lane divided roadways. These two roadway types account for roughly 80 percent of all injuries and over 70 percent of all PDOs in rural settings. By contrast, in urban settings injury incidence is spread more evenly, with over 30 percent occurring on 2-lane undivided roadways, about 24 percent on undivided roadways with more than 2 lanes, about 20 percent on divided roadways with more than 4 lanes, and about 14 percent on 4-lane divided roadways. This might be expected given that the roadway infrastructure in rural areas is typically designed for a lower population density. Rural occupants travel exposure is more likely to occur on roadways with fewer lanes.

Table 12-15. Estimated Economic Cost of Crashes in Rural Areas by Roadway Designation (Millions 2010 Dollars)

	Divided Highways		Undivided Highways			
	4 Lanes	>4 Lanes	2 Lanes	>2 Lanes	All Other	Total
MAIS0	$1,391	$587	$1,742	$484	$215	$4,419
MAIS1	$5,331	$1,877	$12,312	$2,406	$807	$22,733
MAIS2	$2,863	$769	$7,653	$1,192	$456	$12,933
MAIS3	$2,579	$581	$7,217	$1,079	$442	$11,899
MAIS4	$898	$186	$2,351	$344	$128	$3,907
MAIS5	$661	$121	$1,786	$256	$105	$2,929
Fatal	$4,014	$1,809	$18,350	$1,062	$263	$25,499
PDOV	$7,164	$2,332	$9,192	$2,172	$1,289	$22,149
Total	$24,901	$8,262	$60,603	$8,995	$3,706	$106,467
% Rural	23.39%	7.76%	56.92%	8.45%	3.48%	100.00%
% All	8.99%	2.98%	21.88%	3.25%	1.34%	38.43%

The proportions in Table 12-13 were then applied to the total economic costs of rural crashes from Table 12-10, and the proportions in Table 12-14 were applied to the total economic costs of urban crashes in Table 12-10 to distribute these costs by roadway type. This same process was then repeated for the comprehensive costs in Table 12-11. The results are shown in Tables 12-15 and 12-16 for economic costs, and Tables 12-17 and 12-18 for comprehensive costs. Figures 12-B through 12-G illustrate the distribution of costs across roadways.

Table 12-16. Estimated Economic Cost of Crashes in Urban Areas by Roadway Designation (Millions 2010 Dollars)

	Divided Highways		Undivided Highways			
	4 Lanes	**>4 Lanes**	**2 Lanes**	**>2 Lanes**	**All Other**	**Total**
MAIS0	$773	$1,872	$2,777	$2,397	$792	$8,611
MAIS1	$6,357	$9,453	$15,271	$11,150	$3,832	$46,063
MAIS2	$2,995	$4,239	$7,561	$4,993	$1,815	$21,604
MAIS3	$2,173	$3,223	$5,429	$3,839	$1,341	$16,006
MAIS4	$682	$1,059	$1,619	$1,173	$342	$4,875
MAIS5	$416	$713	$1,209	$806	$254	$3,399
Fatal	$3,556	$4,683	$7,665	$3,762	$997	$20,664
PDOV	$4,725	$8,823	$17,388	$12,761	$5,634	$49,332
Total	$21,677	$34,065	$58,919	$40,882	$15,010	$170,553
% Rural	12.71%	19.97%	34.55%	23.97%	8.80%	100.00%
% All	7.82%	12.30%	21.27%	14.76%	5.42%	61.57%

Table 12-17. Estimated Comprehensive Cost of Crashes in Rural Areas by Roadway Designation (Millions 2010 Dollars)

	Divided Highways		Undivided Highways			
	4 Lanes	**>4 Lanes**	**2 Lanes**	**>2Lanes**	**All Other**	**Total**
MAIS0	$1,391	$587	$1,742	$484	$215	$4,419
MAIS1	$11,560	$4,071	$26,700	$5,217	$1,751	$49,299
MAIS2	$12,433	$3,339	$33,237	$5,179	$1,982	$56,171
MAIS3	$10,081	$2,272	$28,207	$4,219	$1,729	$46,508
MAIS4	$4,458	$921	$11,672	$1,708	$635	$19,394
MAIS5	$3,411	$625	$9,216	$1,319	$540	$15,112
Fatal	$26,246	$11,826	$119,969	$6,946	$1,723	$166,710
PDOV	$7,164	$2,332	$9,192	$2,172	$1,289	$22,149
Total	$76,744	$25,974	$239,936	$27,243	$9,864	$379,762
% Rural	20.21%	6.84%	63.18%	7.17%	2.60%	100.00%
% All	8.81%	2.98%	27.55%	3.13%	1.13%	43.61%

Table 12-18. Estimated Comprehensive Cost of Crashes in Urban Areas by Roadway Designation (Millions 2010 Dollars)

	Divided Highways		Undivided Highways			
	4 Lanes	>4 Lanes	2 Lanes	>2 Lanes	All Other	Total
MAIS0	$773	$1,872	$2,777	$2,397	$792	$8,611
MAIS1	$13,785	$20,500	$33,117	$24,179	$8,311	$99,892
MAIS2	$13,009	$18,409	$32,840	$21,687	$7,885	$93,830
MAIS3	$8,493	$12,598	$21,221	$15,007	$5,243	$62,562
MAIS4	$3,384	$5,255	$8,037	$5,824	$1,700	$24,200
MAIS5	$2,145	$3,680	$6,239	$4,161	$1,312	$17,537
Fatal	$23,251	$30,615	$50,114	$24,598	$6,521	$135,099
PDOV	$4,725	$8,823	$17,388	$12,761	$5,634	$49,332
Total	$69,566	$101,753	$171,732	$110,614	$37,399	$491,064
% Rural	14.17%	20.72%	34.97%	22.53%	7.62%	100.00%
% All	7.99%	11.68%	19.72%	12.70%	4.29%	56.39%

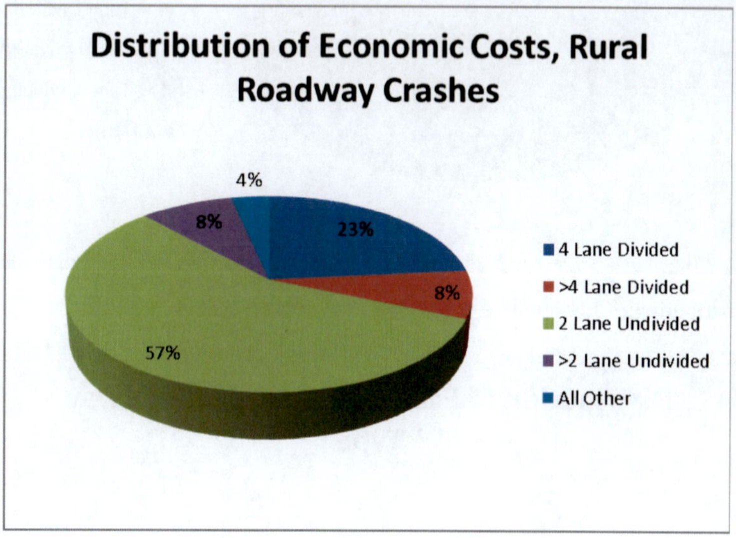

Figure 12-B. Distribution of Economic Costs, Rural Roadway Crashes.

This data indicates that the largest impact on society from motor vehicle crashes, whether measured by economic costs or by the societal harm implied in comprehensive costs, occurs on 2-lane undivided roadways in both rural and urban settings. In rural areas, this is likely a function of both exposure, since these roadways are the most common type, and the relatively serious injury profile that occurs on rural 2-lane undivided roadways, which lack dividers to separate vehicles traveling at relatively high speeds compared to those in urban areas where congestion slows down traffic. In urban areas, exposure is the primary cause of this disproportionate impact. Figures 12-H and 12-I show the relative portion of all motor vehicle injuries that are in the most serious injury categories – MAIS3, MAIS4, MAIS5, and Fatal. In rural crashes, this portion is significantly higher on 2-Lane undivided roadways than on other types, and it is higher for all rural roadways than for any urban roadways – an indication of the impact that higher travel speeds have on injury profiles.

Significant economic impact also occurs on urban divided highways (both lane count categories), and in urban undivided highways with more than two lanes. These impacts are primarily exposure driven, although high speed travel on urban divided roadways does contribute to a relatively severe injury profile as well.

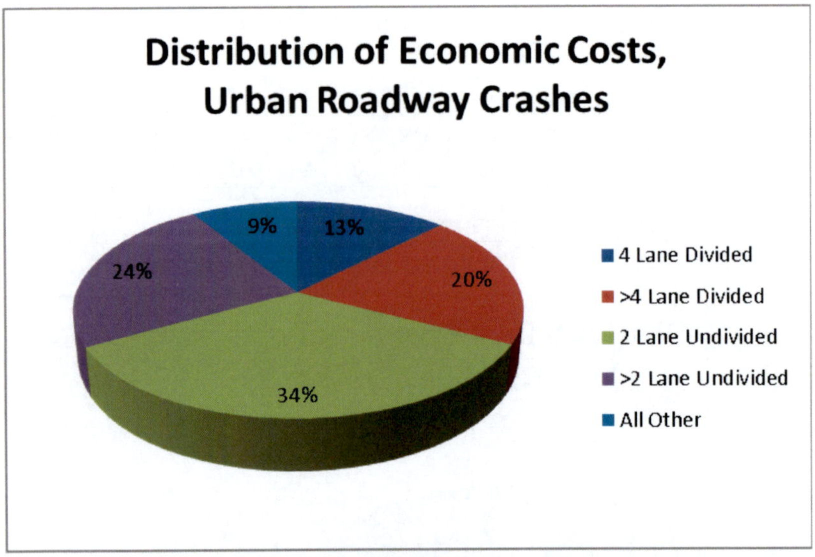

Figure 12-C. Distribution of Economic Costs, Urban Roadway Crashes.

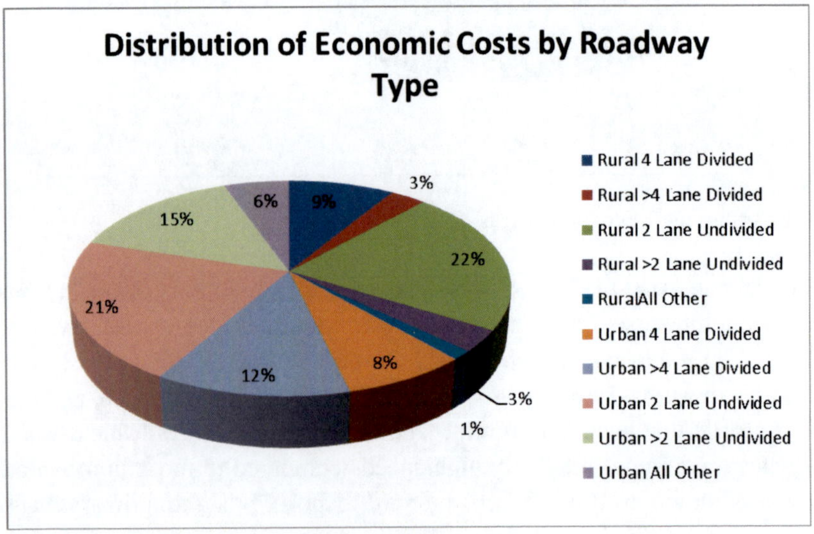

Figure 12-D. Distribution of Economic Costs by Roadway Type.

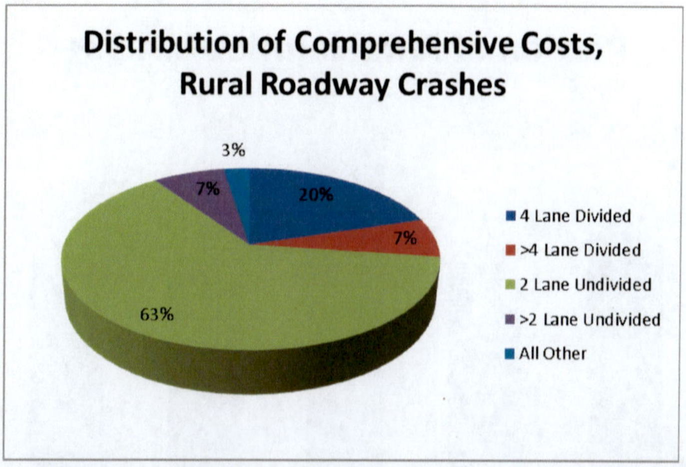

Figure 12-E. Distribution of Comprehensive Costs, Rural Roadway Crashes.

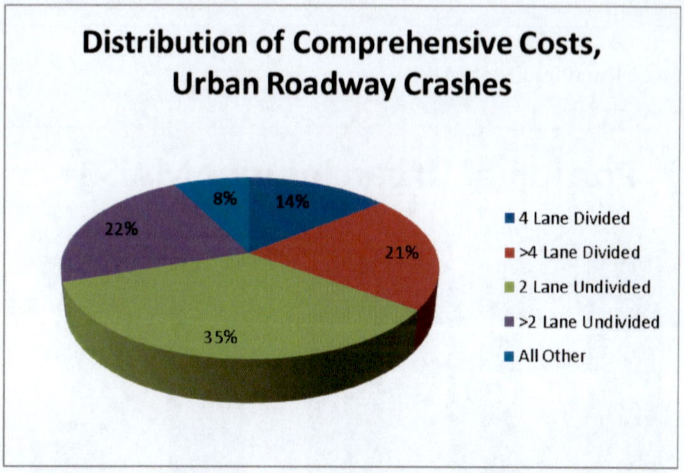

Figure 12-F. Distribution of Comprehensive Costs, Urban Roadway Crashes.

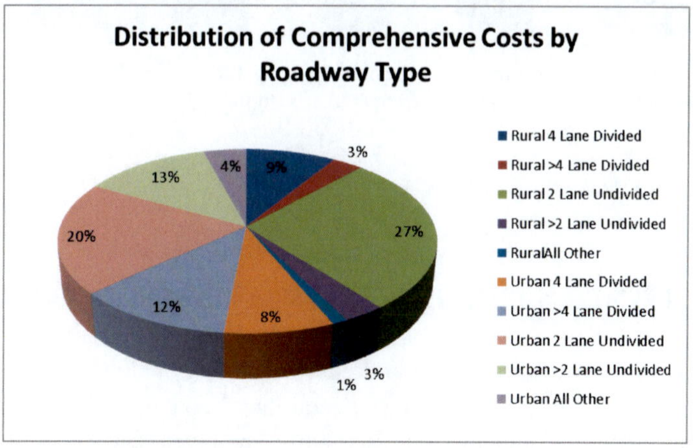

Figure 12-G. Distribution of Comprehensive Costs by Roadway Type.

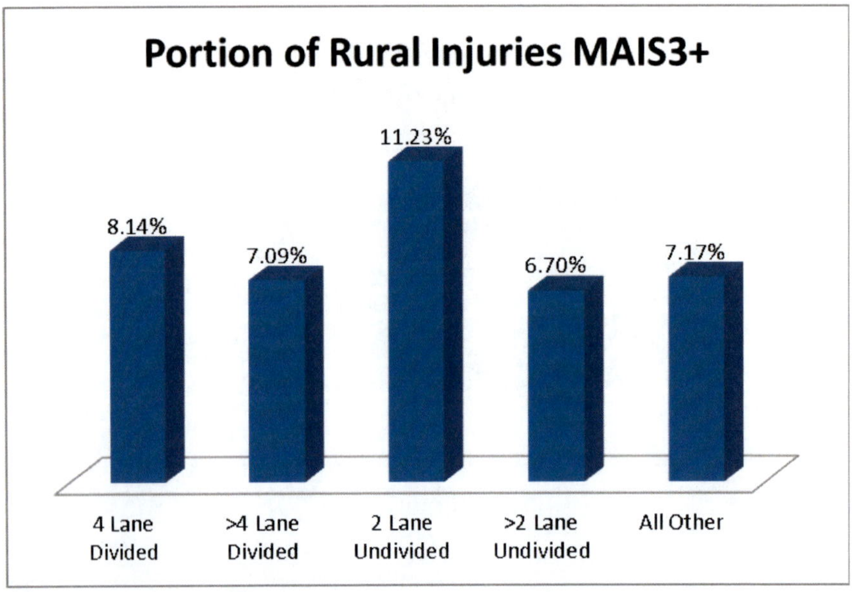

Figure 12-H. Portion of Rural Injuries MAIS3+.

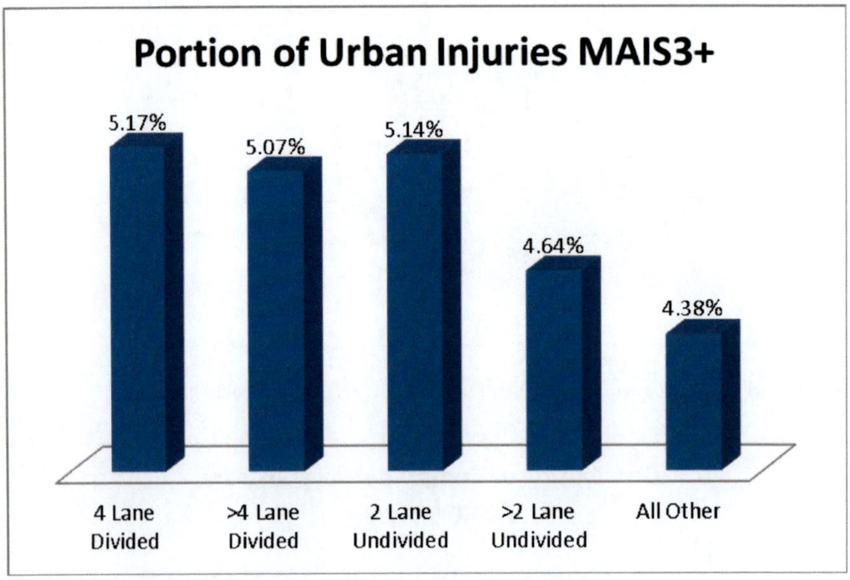

Figure 12-I. Portion of Urban Injuries MAIS3+.

13. OTHER SPECIAL INTEREST CRASH SCENARIOS

Motor vehicle crashes cost society hundreds of billions of dollars in medical care, lost productivity, legal costs, congestion, and other economic impacts. The cost of crashes is even higher when pain, suffering, and lost quality-of-life are taken into account. Federal, State, local, and private organizations are constantly striving to reduce these impacts through motor vehicle safety regulations, behavioral programs such as alcohol and seat belt laws, roadway improvements, traffic control measures, and public information and educational programs. Efforts to address the impacts of motor vehicle crashes are normally focused on specific types of crashes, based on locality, roadway type, crash causation, or victim characteristics. For this report we have estimated crash costs for a number of specific crash scenarios which are commonly of interest to organizations interested in improving vehicle safety. These include seat belt use, impaired driving, speeding, and distracted driving, urban/rural crashes, and State- specific costs, each of which is examined in a separate chapter due to the complexity of the methodology required. However, a number of additional crash scenarios were estimated based on a fairly straightforward examination of the data in NHTSA's two primary databases, FARS and GES. These include, crashes on interstate highways, crashes at intersections, single-vehicle crashes, roadway departure crashes, and pedestrian/bicyclist crashes.

To estimate the cost of these crashes, we examined the relative incidence of each injury severity level that was represented by crashes that matched each scenario. These estimates reflect the relative proportions of specific injury severities that occur under each scenario. GES was used for each non-fatal case, while FARS was used for each fatal case. Each case in FARS contained information regarding the status of the specific scenario, so the proportion of fatalities that occurred under each scenario was obtained directly from the FARS database. For nonfatal injuries and PDOs, GES data was queried to determine whether the case fell under the scenario or not. However, GES data is only recorded using the KABCO severity system, whereas this report is based on the Abbreviated Injury Scale. To translate GES data to an MAIS basis, we used a variety of KABCO/MAIS translators. For CDS equivalent crashes, we used a current translator derived from 2000-2008 CDS data. Since this data is relatively recent, it reflects roughly current levels of seat belt usage. For non-CDS cases, the only available data from which to develop translators were contained in the 1982-1986 NASS files. Seat belt use has increased dramatically since this time. Observed belt use during this period ranged from roughly 10-37 percent as public awareness of the importance of belt use and belt use laws were just beginning to take hold in 1986. Belt use has since risen dramatically, and has been between 80 and 85 percent since 2004. Belt use can influence injury reporting significantly in a number of ways. It changes the nature of injuries by preventing many more visible injuries (such as head/face contact with the windshield) but replaces them with often less visible (and also typically less serious) abdominal injuries such as bruising caused by pressure from the belt across the torso. This can influence the relationship between the KABCO reported injury severity and the corresponding MAIS injury level. For this reason, separate translators were developed from the 1982-86 NASS data for non-CDS cases where the victim was belted, unbelted, unknown belted status, and for nonoccupants/motorcyclists. These translators are presented in Tables 13-1 through 13-5.

2010 GES KABCO incidence counts were obtained for each scenario, both for the cases that met the scenario definition, and for all other cases. So, for example, one set of incidence counts was obtained for intersection crashes, and another for all other crashes. Each of these data sets was run through its corresponding translator to produce a set of MAIS based injury counts. These counts from each grouping (CDS equivalent cases, belted non-CDS cases, unbelted non-CDS cases, unknown belt status non-CDS cases, and nonoccupant/motorcycle cases) were added together to produce a total MAIS injury profile for each scenario. The process was repeated for each "Other" category (e.g., all non-intersection crashes). The percentage of each MAIS injury incidence that was appropriate to each scenario was then calculated as:

$$x=a/(a+b)$$

where x is the percentage of incidence attributable to the specific crash scenario
a = the incidence of the specific crash scenario
b = the incidence of each case not attributable to the specific crash scenario.

The attributable portion of each MAIS level was then multiplied by the total cost of all 2010 crashes for that MAIS level and the MAIS level results were summed to produce the total cost of each crash scenario. MAIS0 portions were calculated using the same procedure described previously for Urban/Rural crashes, based on the relative incidence of MAIS 0 cases in injury crashes. The PDO portion was based on a direct count of PDO vehicles from each crash scenario compared to those not in that scenario. For the interstate highway crash scenario, congestion costs were modified based on data in Chapter 3 to reflect congestion impacts specific to interstate highways, which have far more serious congestion impacts. This data indicates that crashes on interstates cause roughly three times the average congestion costs across all roadways.

The results of this process are summarized for each scenario in Tables 6 to 15 for both economic costs and comprehensive costs. Note that these categories are not exclusive or additive, since some crashes qualify under more than one category.

Intersection Crashes: Intersection crashes resulted in 8,682 fatalities, over 2.2 million injuries, and over 10 million PDO damaged vehicles in 2010.[55] This represents 26 percent of all fatalities and roughly 55 percent of all nonfatal crashes (including both nonfatal injury and PDO). Intersection crashes caused $139 billion in economic costs and $391 billion in comprehensive costs, accounting for 50 percent of all economic costs and 45 percent of all societal harm (measured as comprehensive costs) from motor vehicle crashes.

Interststate Highway Crashes: Crashes on interstate highways resulted in 4,288 fatalities, over 300,000 injuries, and 1.4 million PDO damaged vehicles in 2010. This represents 13 percent of all fatalities and roughly 8 percent of all nonfatal crashes (including both nonfatal injury and PDO). Interstate highway crashes caused $28 billion in economic costs and $88 billion in comprehensive costs, accounting for roughly 10 percent of both economic costs and societal harm (measured as comprehensive costs).

Single-Vehicle Crashes: Single-vehicle crashes resulted in 19,241fatalities, 962,000 injuries, and nearly 3.2 million PDO damaged vehicles in 2010. This represents 58 percent of all fatalities and roughly 20 percent of all nonfatal crashes (including both nonfatal injuries and PDO). Single-vehicle crashes caused $89 billion in economic costs and $357 billion in comprehensive costs, accounting for 32 percent of all economic costs, and 41 percent of all societal harm (measured as comprehensive costs).

Roadway Departure Crashes: Roadway departure crashes resulted in 18,850 fatalities, 795,000 injuries, and over 2.4 million PDO damaged vehicles in 2010.[56] This represents 57 percent of all fatalities and roughly 16 percent of all nonfatal crashes (including both nonfatal injury and PDO). Roadway departure crashes caused $73 billion in economic costs and $307 billion in comprehensive costs, accounting for 26 percent of all economic costs, and 35 percent of all societal harm (measured as comprehensive costs).

Pedestrian/bicyclist Crashes: Pedestrian/bicyclist crashes resulted in 5,123 fatalities, 189,000 injuries, and over 13,000 PDO damaged vehicles in 2010. [57] This represents 16 percent of all fatalities and roughly 2 percent of all nonfatal crashes (including both nonfatal injury and PDO). These crashes caused $19 billion in economic costs and $90 billion in comprehensive costs, accounting for 7 percent of all economic costs, and 10 percent of all societal harm (measured as comprehensive costs).

Pedestrian Crashes: Pedestrian crashes resulted in 4,372 fatalities, 110,000 injuries, and 4,370 PDO damaged vehicles in 2010.[58] This represents 13 percent of all fatalities and roughly 1 percent of all nonfatal crashes (including both nonfatal injury and PDO). These crashes caused $14 billion in economic costs and $67 billion in comprehensive costs, accounting for 5 percent of all economic costs, and 8 percent of all societal harm (measured as comprehensive costs).

Bicyclist Crashes: Bicyclist crashes resulted in 632 fatalities, 79,000 injuries, and 9.078 PDO damaged vehicles in 2010.[59] This represents 2 percent of all fatalities and roughly 1 percent of all nonfatal crashes (including both nonfatal injury and PDO). These crashes caused $6 billion in economic costs and $23 billion in comprehensive costs, accounting for 2 percent of all economic costs, and 3 percent of all societal harm (measured as comprehensive costs).

Table 13-1. KABCO/MAIS Translator for CDS Equivalent Cases

MAIS	O No Injury	C Possible Injury	B Non-incapacitating	A Incapacitating	K Killed	Injured Severity Unknown	Unknown if Injured
0	0.8191	0.2188	0.0906	0.0376	0.0032	0.2429	0.5935
1	0.1759	0.7014	0.7518	0.5782	0.0110	0.5961	0.3751
2	0.0047	0.0674	0.1113	0.1924	0.0019	0.1039	0.0208
3	0.0002	0.0101	0.0348	0.1259	0.0041	0.0406	0.0091
4	0.0000	0.0021	0.0085	0.0444	0.0027	0.0047	0.0008
5	0.0001	0.0001	0.0014	0.0171	0.0007	0.0117	0.0006
Fatal	0.0000	0.0001	0.0015	0.0043	0.9765	0.0000	0.0001
Total	1.000	1.000	1.000	1.000	1.000	1.000	1.000

Table 13-2. KABCO/MAIS Translator for Non-CDS Equivalent Cases, Unbelted

MAIS	O	C	B	A	Injured Severity Unknown	Unknown if Injured
	No Injury	Possible Injury	Non-incapacitating	Incapacitating		
0	0.9591	0.2655	0.0777	0.0394	0.1082	0.9196
1	0.0388	0.4360	0.6667	0.5033	0.3931	0.0681
2	0.0011	0.0389	0.0640	0.1517	0.0345	0.0004
3	0.0000	0.0045	0.0041	0.0512	0.0000	0.0013
4	0.0000	0.0000	0.0008	0.0116	0.0000	0.0000
5	0.0000	0.0000	0.0000	0.0061	0.0000	0.0000
7	0.0010	0.2545	0.1863	0.2347	0.4615	0.0106
Fatality	0.0000	0.0004	0.0005	0.0020	0.0027	0.0000
Total	1.0000	1.0000	1.0000	1.0000	1.0000	1.0000

Table 13-3. KABCO/MAIS Translator for Non-CDS Equivalent Cases, Belted

AIS	O	C	B	A	Injured Severity Unknown	Unknown if Injured
	No Injury	Possible Injury	Non-incapacitating	Incapacitating		
0	0.9491	0.3002	0.1558	0.0962	0.5115	0.9164
1	0.0494	0.4888	0.6544	0.5418	0.1948	0.0598
2	0.0006	0.0154	0.0192	0.1966	0.0655	0.0000
3	0.0001	0.0124	0.0082	0.0505	0.0179	0.0000
4	0.0000	0.0001	0.0012	0.0073	0.0000	0.0000
5	0.0000	0.0000	0.0000	0.0000	0.0000	0.0000
7	0.0009	0.1831	0.1611	0.1076	0.2104	0.0239
Fatality	0.0000	0.0000	0.0000	0.0000	0.0000	0.0000
Total	1.0000	1.0000	1.0000	1.0000	1.0000	1.0000

Table 13-4. KABCO/MAIS Translator for Non-CDS Equivalent Cases, Unknown Belt Use

MAIS	O	C	B	A	Injured Severity Unknown	Unknown if Injured
	No Injury	Possible Injury	Non-incapacitating	Incapacitating		
0	0.9960	0.1078	0.0908	0.0252	0.0413	0.9476
1	0.0001	0.1161	0.4971	0.2687	0.1866	0.0050
2	0.0000	0.0057	0.0067	0.1491	0.0067	0.0000
3	0.0000	0.0000	0.0000	0.1305	0.0000	0.0000
4	0.0000	0.0000	0.0000	0.0000	0.0000	0.0000
5	0.0000	0.0000	0.0000	0.0000	0.0000	0.0000
7	0.0038	0.7704	0.4053	0.4265	0.7654	0.0475
Fatality	0.0000	0.0000	0.0000	0.0000	0.0000	0.0000
Total	1.0000	1.0000	1.0000	1.0000	1.0000	1.0000

Table 13-5. KABCO/MAIS Translator for Nonoccupants and Motorcyclists

MAIS	O	C	B	A	Injured Severity Unknown	Unknown if Injured
	No Injury	Possible Injury	Non-incapacitating	Incapacitating		
0	0.6403	0.0429	0.0221	0.0050	0.0324	0.2164
1	0.3079	0.5724	0.6180	0.3013	0.4332	0.4686
2	0.0435	0.1636	0.1577	0.2776	0.0847	0.1172
3	0.0035	0.0397	0.0593	0.2701	0.0422	0.0639
4	0.0005	0.0012	0.0021	0.0248	0.0045	0.0083
5	0.0000	0.0014	0.0003	0.0263	0.0000	0.0037
7	0.0044	0.1788	0.1402	0.0890	0.4029	0.0981
Fatality	0.0000	0.0000	0.0002	0.0058	0.0000	0.0238
Total	1.0000	1.0000	1.0000	1.0000	1.0000	1.0000

Table 13-6. Economic Costs of Intersection Crashes, (Millions of 2010 Dollars)

	% Intersection	Incidence		Total Economic Crash Costs		
		Total	Intersection	Total	Intersection	Other
PDO Vehicles	54.72%	18,508,632	10,127,014	$71,480	$39,111	$32,370
MAIS0	56.68%	4,583,265	2,597,665	$13,030	$7,385	$5,645
MAIS1	57.19%	3,459,200	1,978,467	$68,797	$39,348	$29,449
MAIS2	55.45%	338,730	187,814	$34,537	$19,150	$15,388
MAIS3	53.24%	100,740	53,632	$27,905	$14,856	$13,049
MAIS4	50.28%	17,086	8,591	$8,781	$4,415	$4,366
MAIS5	49.39%	5,749	2,839	$6,327	$3,125	$3,202
Fatalities	26.31%	32,999	8,682	$46,163	$12,145	$34,017
Total	55.33%	27,046,402	14,964,704	$277,020	$139,535	$137,486
Percent of Total		100.00%	55.33%	100.00%	50.37%	49.63%

Table 13-7. Comprehensive Costs of Intersection Crashes, (Millions of 2010 Dollars)

	% Intersection	Incidence		Total Comprehensive Crash Costs		
		Total	Intersection	Total	Intersection	Other
PDO Vehicles	54.72%	18,508,632	10,127,014	$71,480	$39,111	$32,370
MAIS0	56.68%	4,583,265	2,597,665	$13,030	$7,385	$5,645
MAIS1	57.19%	3,459,200	1,978,467	$149,192	$85,329	$63,863
MAIS2	55.45%	338,730	187,814	$150,001	$83,170	$66,831
MAIS3	53.24%	100,740	53,632	$109,070	$58,067	$51,003
MAIS4	50.28%	17,086	8,591	$43,594	$21,919	$21,674
MAIS5	49.39%	5,749	2,839	$32,649	$16,125	$16,525
Fatalities	26.31%	32,999	8,682	$301,809	$79,406	$222,403
Total	55.33%	27,046,402	14,964,704	$870,826	$390,512	$480,313
Percent of Total		100.00%	55.33%	100.00%	44.84%	55.16%

Table 13-8. Economic Costs of Interstate Highway Crashes, (Millions of 2010 Dollars)

	% Interstate	Incidence		Interstate Unit Costs	Total Economic Crash Costs		
		Total	Interstate		Total	Interstate	Other
PDO Vehicles	7.62%	18,508,632	1,411,077	$5,517	$71,480	$7,784	$63,696
MAIS0	8.11%	4,583,265	371,519	$4,404	$13,030	$1,636	$11,394
MAIS1	8.11%	3,459,200	280,704	$22,388	$68,797	$6,284	$62,512
MAIS2	7.74%	338,730	26,203	$104,692	$34,537	$2,743	$31,794
MAIS3	7.66%	100,740	7,713	$280,350	$27,905	$2,162	$25,742
MAIS4	8.94%	17,086	1,527	$517,504	$8,781	$790	$7,991
MAIS5	8.44%	5,749	485	$1,104,193	$6,327	$536	$5,791
Fatalities	12.99%	32,999	4,288	$1,410,277	$46,163	$6,047	$40,116
Total	7.78%	27,046,402	2,103,517		$277,020	$27,983	$249,037
Percent of Total		100.00%	7.78%		100.00%	10.10%	89.90%

Table 13-9. Comprehensive Costs of Interstate Highway Crashes, (millions of 2010$)

	% Interstate	Incidence		Interstate Unit Costs	Total Comprehensive Crash Costs		
		Total	Interstate		Total	Interstate	Other
PDO Vehicles	7.62%	18,508,632	1,411,077	$5,517	$71,480	$7,784	$63,696
MAIS0	8.11%	4,583,265	371,519	$4,404	$13,030	$1,636	$11,394
MAIS1	8.11%	3,459,200	280,704	$45,629	$149,192	$12,808	$136,384
MAIS2	7.74%	338,730	26,203	$445,564	$150,001	$11,675	$138,326
MAIS3	7.66%	100,740	7,713	$1,086,047	$109,070	$8,377	$100,694
MAIS4	8.94%	17,086	1,527	$2,554,987	$43,594	$3,901	$39,692
MAIS5	8.44%	5,749	485	$5,682,718	$32,649	$2,759	$29,891
Fatalities	12.99%	32,999	4,288	$9,157,359	$301,809	$39,263	$262,546
Total	7.78%	27,046,402	2,103,517		$870,826	$88,203	$782,622
Percent of Total		100.00%	7.78%		100.00%	10.13%	89.87%

Table 10. Economic Costs of Single-Vehicle Crashes, (Millions of 2010 Dollars)

	% Single-Vehicle	Incidence		Total Economic Crash Costs		
		Total	Single-Vehicle	Total	Single-Vehicle	Other
PDO Vehicles	17.22%	18,508,632	3,186,682	$71,480	$12,307	$59,173
MAIS0	24.83%	4,583,265	1,137,803	$13,030	$3,235	$9,795
MAIS1	22.82%	3,459,200	789,271	$68,797	$15,697	$53,100
MAIS2	35.41%	338,730	119,945	$34,537	$12,230	$22,308
MAIS3	42.87%	100,740	43,187	$27,905	$11,963	$15,942
MAIS4	38.94%	17,086	6,653	$8,781	$3,420	$5,362
MAIS5	44.85%	5,749	2,578	$6,327	$2,838	$3,490
Fatalities	58.31%	32,999	19,241	$46,163	$26,917	$19,246
Total	19.62%	27,046,402	5,305,360	$277,020	$88,605	$188,416
Percent of Total		100.00%	19.62%	100.00%	31.98%	68.02%

Table 13-11. Comprehensive Costs of Single-Vehicle Crashes, (Millions of 2010 Dollars)

	% Single-Vehicle	Incidence		Total Comprehensive Crash Costs		
		Total	Single-Vehicle	Total	Single-Vehicle	Other
PDO Vehicles	17.22%	18,508,632	3,186,682	$71,480	$12,307	$59,173
MAIS0	24.83%	4,583,265	1,137,803	$13,030	$3,235	$9,795
MAIS1	22.82%	3,459,200	789,271	$149,192	$34,040	$115,151
MAIS2	35.41%	338,730	119,945	$150,001	$53,115	$96,885
MAIS3	42.87%	100,740	43,187	$109,070	$46,758	$62,312
MAIS4	38.94%	17,086	6,653	$43,594	$16,976	$26,618
MAIS5	44.85%	5,749	2,578	$32,649	$14,642	$18,007
Fatalities	58.31%	32,999	19,241	$301,809	$175,978	$125,831
Total	19.62%	27,046,402	5,305,360	$870,826	$357,052	$513,773
Percent of Total		100.00%	19.62%	100.00%	41.00%	59.00%

Table 13-12. Economic Costs of Roadway Departure Crashes, (Millions of 2010 Dollars)

	% Roadway Departure	Incidence		Total Economic Crash Costs		
		Total	Roadway Departure	Total	Roadway Departure	Other
PDO Vehicles	13.17%	18,508,632	2,437,564	$71,480	$9,414	$62,066
MAIS0	20.54%	4,583,265	941,521	$13,030	$2,677	$10,353
MAIS1	19.47%	3,459,200	673,382	$68,797	$13,392	$55,404
MAIS2	25.45%	338,730	86,193	$34,537	$8,788	$25,749
MAIS3	27.16%	100,740	27,366	$27,905	$7,580	$20,324
MAIS4	34.13%	17,086	5,831	$8,781	$2,997	$5,784
MAIS5	33.11%	5,749	1,904	$6,327	$2,095	$4,232
Fatalities	57.12%	32,999	18,850	$46,163	$26,370	$19,793
Total	15.50%	27,046,402	4,192,611	$277,020	$73,313	$203,707
Percent of Total		100.00%	15.50%	100.00%	26.46%	73.54%

Table 13-13. Comprehensive Costs of Roadway Departure Crashes, (Millions of 2010 Dollars)

	% Roadway Departure	Incidence		Total Comprehensive Crash Costs		
		Total	Roadway Departure	Total	Roadway Departure	Other
PDO Vehicles	13.17%	18,508,632	2,437,564	$71,480	$9,414	$62,066
MAIS0	20.54%	4,583,265	941,521	$13,030	$2,677	$10,353
MAIS1	19.47%	3,459,200	673,382	$149,192	$29,042	$120,150
MAIS2	25.45%	338,730	86,193	$150,001	$38,169	$111,832

Table 13-13. (Continued)

	% Roadway Departure	Incidence		Total Comprehensive Crash Costs		
		Total	Roadway Departure	Total	Roadway Departure	Other
MAIS3	27.16%	100,740	27,366	$109,070	$29,629	$79,442
MAIS4	34.13%	17,086	5,831	$43,594	$14,879	$28,715
MAIS5	33.11%	5,749	1,904	$32,649	$10,811	$21,839
Fatalities	57.12%	32,999	18,850	$301,809	$172,402	$129,407
Total	15.50%	27,046,402	4,192,611	$870,826	$307,022	$563,804
Percent of Total		100.00%	15.50%	100.00%	35.26%	64.74%

Table 13-14. Economic Costs of Pedestrian/Bicyclist Crashes, (Millions of 2010 Dollars)

	% Pedestrian/ Bicyclist	Incidence		Total Economic Crash Costs		
		Total	Pedestrian/ Bicyclist	Total	Pedestrian/ Bicyclist	Other
PDO Vehicles	0.07%	18,508,632	13,449	$71,480	$52	$71,428
MAIS0	4.93%	4,583,265	226,076	$13,030	$643	$12,387
MAIS1	3.99%	3,459,200	137,926	$68,797	$2,743	$66,053
MAIS2	10.39%	338,730	35,180	$34,537	$3,587	$30,950
MAIS3	14.04%	100,740	14,147	$27,905	$3,919	$23,986
MAIS4	7.06%	17,086	1,207	$8,781	$620	$8,161
MAIS5	11.26%	5,749	648	$6,327	$713	$5,615
Fatalities	15.15%	32,999	5,000	$46,163	$6,995	$39,168
Total	1.60%	27,046,402	433,632	$277,020	$19,271	$257,750
Percent of Total		100.00%	1.60%	100.00%	6.96%	93.04%

Table 13-15. Comprehensive Costs of Pedestrian/Bicyclist Crashes, (Millions of 2010 Dollars)

	% Pedestrian/ Bicyclist	Incidence		Total Comprehensive Crash Costs		
		Total	Pedestrian/ Bicyclist	Total	Pedestrian/ Bicyclist	Other
PDO Vehicles	0.07%	18,508,632	13,449	$71,480	$52	$71,428
MAIS0	4.93%	4,583,265	226,076	$13,030	$643	$12,387
MAIS1	3.99%	3,459,200	137,926	$149,192	$5,949	$143,243
MAIS2	10.39%	338,730	35,180	$150,001	$15,579	$134,422
MAIS3	14.04%	100,740	14,147	$109,070	$15,316	$93,754
MAIS4	7.06%	17,086	1,207	$43,594	$3,079	$40,515

	% Pedestrian/ Bicyclist	Incidence		Total Comprehensive Crash Costs		
		Total	Pedestrian/ Bicyclist	Total	Pedestrian/ Bicyclist	Other
MAIS5	11.26%	5,749	648	$32,649	$3,677	$28,972
Fatalities	15.15%	32,999	5,000	$301,809	$45,730	$256,079
Total	1.60%	27,046,402	433,632	$870,826	$90,025	$780,801
Percent of Total		100.00%	1.60%	100.00%	10.34%	89.66%

Table 13-16. Economic Cost of Pedestrian Crashes, (Millions of 2010 Dollars)

	% Pedestrian	Incidence		Total Economic Crash Costs		
		Total	Pedestrian	Total	Pedestrian	Other
PDO Vehicles	0.02%	18,508,632	4,370	$71,480	$17	$71,463
MAIS0	2.91%	4,583,265	133,154	$13,030	$379	$12,652
MAIS1	2.29%	3,459,200	79,183	$68,797	$1,575	$67,222
MAIS2	6.16%	338,730	20,863	$34,537	$2,127	$32,410
MAIS3	8.78%	100,740	8,844	$27,905	$2,450	$25,455
MAIS4	4.65%	17,086	794	$8,781	$408	$8,373
MAIS5	7.71%	5,749	443	$6,327	$488	$5,839
Fatalities	13.25%	32,999	4,372	$46,163	$6,116	$40,047
Total	0.93%	27,046,402	252,024	$277,020	$13,559	$263,461
Percent of Total		100.00%	0.93%	100.00%	4.89%	95.11%

Table 13-17. Comprehensive Cost of Pedestrian Crashes, (Millions of 2010 Dollars)

	% Pedestrian	Incidence		Total Comprehensive Crash Costs		
		Total	Pedestrian	Total	Pedestrian	Other
PDO Vehicles	0.02%	18,508,632	4,370	$71,480	$17	$71,463
MAIS0	2.91%	4,583,265	133,154	$13,030	$379	$12,652
MAIS1	2.29%	3,459,200	79,183	$149,192	$3,415	$145,777
MAIS2	6.16%	338,730	20,863	$150,001	$9,239	$140,762
MAIS3	8.78%	100,740	8,844	$109,070	$9,576	$99,495
MAIS4	4.65%	17,086	794	$43,594	$2,025	$41,569
MAIS5	7.71%	5,749	443	$32,649	$2,518	$30,131
Fatalities	13.25%	32,999	4,372	$301,809	$39,986	$261,822
Total	0.93%	27,046,402	252,024	$870,826	$67,154	$803,671
Percent of Total		100.00%	0.93%	100.00%	7.71%	92.29%

Table 13-18. Economic Cost of Bicyclist Crashes, (Millions of 2010 Dollars)

	% Bicyclist	Incidence		Total Economic Crash Costs		
		Total	Bicyclist	Total	Bicyclist	Other
PDO Vehicles	0.05%	18,508,632	9,078	$71,480	$35	$71,445
MAIS0	2.04%	4,583,265	93,390	$13,030	$266	$12,765
MAIS1	1.70%	3,459,200	58,920	$68,797	$1,172	$67,625
MAIS2	4.27%	338,730	14,465	$34,537	$1,475	$33,062
MAIS3	5.32%	100,740	5,363	$27,905	$1,486	$26,419
MAIS4	2.44%	17,086	417	$8,781	$214	$8,567
MAIS5	3.60%	5,749	207	$6,327	$228	$6,099
Fatalities	1.92%	32,999	632	$46,163	$884	$45,279
Total	0.67%	27,046,402	182,472	$277,020	$5,759	$271,261
Percent of Total		100.00%	0.67%	100.00%	2.08%	97.92%

Table 13-19. Comprehensive Cost of Bicyclist Crashes, (Millions of 2010 Dollars)

	% Bicyclist	Incidence		Total Comprehensive Crash Costs		
		Total	Bicyclist	Total	Bicyclist	Other
PDO Vehicles	0.05%	18,508,632	9,078	$71,480	$35	$71,445
MAIS0	2.04%	4,583,265	93,390	$13,030	$266	$12,765
MAIS1	1.70%	3,459,200	58,920	$149,192	$2,541	$146,651
MAIS2	4.27%	338,730	14,465	$150,001	$6,405	$143,595
MAIS3	5.32%	100,740	5,363	$109,070	$5,807	$103,264
MAIS4	2.44%	17,086	417	$43,594	$1,063	$42,531
MAIS5	3.60%	5,749	207	$32,649	$1,177	$31,473
Fatalities	1.92%	32,999	632	$301,809	$5,780	$296,029
Total	0.67%	27,046,402	182,472	$870,826	$23,074	$847,752
Percent of Total		100.00%	0.67%	100.00%	2.65%	97.35%

14. Source of Payment

The economic toll of motor vehicle crashes is borne by society through a variety of payment mechanisms. The most common of these are private insurance plans such as Blue Cross-Blue Shield, HMOs, commercial insurance policies, or worker's compensation. Medicare is the primary payer for people over the age of 65. When these sources are not available, government programs such as Medicaid may provide coverage for those who meet eligibility requirements. A host of other Federal, State, and local programs such as CHAMPVA, Tricare, Title 5, and Indian Health Services also provide health care coverage for specific groups. Expenses not covered by private or governmental sources must be paid out-of-pocket by individuals, or absorbed as losses by health care providers.

Blincoe (1996) provided estimates of sources of payment for motor vehicle crashes that combined analysis of CODES data with previous estimates developed by the Urban Institute (Miller et al., 1991). This data was also used in the previous version of this current study

(Blincoe et al., 2002). For this current report, data from Blincoe (1996) were carried forward for insurance administration, workplace costs, legal costs, and congestion while new estimates of source of payment were developed for medical care, lost productivity, and property damage. Blincoe also estimated values for emergency services. However, in that study ambulance costs were included under emergency services, while for this current study, ambulance costs are included under medical care. Ambulance costs had been distributed across all payer categories in the same proportion as medical care. To adjust for this, the impact of ambulance costs was removed from the EMS distribution. This results in 100 percent of emergency service costs being born by States and localities (primarily localities). However, we note that over the past few years, primarily in response to budget tightening that has resulted from the economic downturn, many local fire departments and other EMS operators have begun charging insurance companies and in some cases individuals for their services. This practice has become controversial and has been banned in a number of States. Nonetheless, it is likely that some portion of emergency services are in fact now being paid through insurance, and possibly by individuals. It is not certain whether in the long run this practice will become more widespread or will decline due to legislative actions. As this is a relatively recent practice, to date we know of no studies that have examined these impacts on an aggregate basis.

We have also added an additional payer category, "unspecified government," to this study. This category represents programs that are funded primarily by government revenues, but that are lumped together in HCUP data and that therefore cannot be individually identified as belonging to either State or Federal categories. In addition, some of these programs are partially funded by participants through subsidized premium charges. Programs in these categories include Veterans' Administration, Tricare, Title 5, Indian Health Services, and State and local health care programs. These are programs that cover medical care for service personal and their families, veterans', Native Americans, and State and local employees. In previous studies, these costs were lumped under the "Other" category. We have categorized them with government programs because they are either entirely supported by, or heavily subsidized by, tax dollars, but some unknown portion of these costs are paid by individual insurance premiums.

Following are discussions of the derivation of revised source of payment estimates for medical care, lost productivity, and property damage.

Medical Costs

Miller et al. (2011) provide factors for computing the percentage of crash costs paid by State/local and Federal governments. Table 1, drawn from that paper, is built from the million-record 2007 Healthcare Cost and Utilization Project (HCUP) Nationwide Inpatient Sample (NIS). Medical costs were estimated using hospital charges, as recorded in the NIS, with hospital-specific cost-to-charge ratios supplied by HCUP. The NIS also records the expected payer for each hospital stay, which allowed estimation of the amount paid by payer. Medicaid paid an estimated 15.8 percent of hospital costs for motor vehicle crashes and Medicare paid 7.3 percent.

Zaloshnja and Miller (2012) analyzed Medicaid claims and HCUP data from 14 States. They estimated that 22 percent of adults ages 19 to 64 with hospital-admitted crash injuries covered by Medicaid (2.85% of all those admitted) became Medicaid-eligible because earnings losses and medical bills resulting from crash injury left them indigent or disabled. The crash resulted in Medicaid paying all their medical bills, not just their injury bill. Zaloshnja and Miller (2012) further estimate that 35 percent of those who convert to Medicaid to pay hospital bills stay on Medicaid indefinitely. The present value of their lifetime Medicaid health care costs averages $316,000 (computed following the article's methods but substituting fiscal year 2009 Medicaid spending of $15,840 per disabled recipient from: www.statehealthfacts.org/comparetable.jsp?ind=183&cat=4.)

Beyond crash medical costs, that adds an estimated $3,152 ($316,000×0.0285×0.35) in Medicaid costs for other care to government's crash bill per hospital-admitted non-elderly crash survivor. With roughly 5.4 percent of medically treated crash survivors admitted, added government cost due to Medicaid conversion is $170 per injured crash survivor.

The division of Medicaid costs between the Federal and State levels varies by State. On average, States paid 35.2 percent in Fiscal Year 2011 (Office of the Assistant Secretary for Planning and Evaluation, 2012).

Among the elderly, HCUP data show private insurance paid 50.0 percent of the medical cost of crash injuries but only 7.4 percent of the cost of other injuries. The 42.6 percent differential presumably is due to medical costs borne by or recovered from auto insurers. Assuming this percentage applies to injuries not requiring hospital admission is aggressive. Health insurers are less likely to pursue recovery through subrogation for smaller medical bills. Nevertheless, if we make that assumption, an estimated $14.85 billion in medical costs would be compensated by auto insurance. Since available auto insurance data do not decompose compensation for medical costs versus work losses (or even quality-of-life), we used the 42.6 percent compensation for medical care to compute the compensation level for wage and household work loss.

Productivity (Work) Losses

Estimated annual productivity losses in 2008–2010 were $92.3 billion, of which 76.8 percent was wages and fringe benefits. The remainder was lost household productivity. Above, we estimated that auto insurance compensation net of fraud was $45.516 billion ($50.017 billion* 91 percent). Subtracting the $14.850 billion in medical costs compensated ($34,860 billion* 42.6 percent), $30.666 billion in productivity losses is compensated, which equates to 33.2 percent of those losses.

From data on State income and sales tax and Federal income tax revenues, Miller et al. (2011) estimate that 4.1 percent of total productivity losses (5.27 percent of wage and fringe benefit losses) are State tax losses and 7.3 percent (9.4 percent of wage-related loss) are Federal tax losses. From old survey data, they estimate that the safety net compensates another 1.3 percent (1.65 percent of wage-related loss).

Miller et al. (2010) estimated the costs of crashes to employers. Inflating their estimates to 2010 dollars, workers' compensation covers $2.2 billion in losses (1.2 percent of total compensation), disability insurance covers $0.85 billion (0.5 percent), and sick leave $5.6 billion (2.9 percent).

Table 14-1. Primary payer for medical costs of hospital-admitted road crash injuries, by age, United States, as per 2007 HCUP-NIS

Payer*	Ages 0–18	Ages 19–64	Ages ≥65	All Ages
Medicare	0.4%	2.3%	40.5%	7.3%
Medicaid	30.6%	12.1%	2.2%	15.8%
Private	59.4%	44.4%	50.0%	56.1%
Self	4.3%	10.5%	2.8%	10.9%
Charity	0.3%	1.2%	0.0%	1.2%
Other	4.5%	7.7%	4.4%	8.5%
Unknown	0.5%	0.2%	0.2%	0.3%

* Private includes auto insurance, private health insurance, HMO/managed care, and workers' compensation. Self-pay and charity care ultimately may shift to Medicare or Medicaid. Other includes CHAMPVA, TRICARE, Indian Health Services, Title 5, and State and local government health care programs.

Table 14-2. Life insurance coverage and policy amounts, United States, 2010

	1,000s of Policies	Amount/Policy	Paid/Death
Individual	151,787	$69,067	$33,955
Group	109,462	$71,537	$25,363
Credit	23,086	$4,843	$362
All	284,335	$64,804	$59,680

Source: Computed from data in American Life Insurance Council (2012).

American Life Insurance Council (2012) provides data on life insurance policies in force by type. Table 2 starts from those data. (In this table, group policies generally are employment-related.) Dividing the amount of coverage by policies in force yields the average payment per premium. Multiplying coverage per policy times the percentage of the U.S. population with policies of the given type yields expected payout per policy. The average death in 2010 generated an estimated $59,680 in life insurance payments. With 33,012 crash deaths, life insurance payments totaled $1.97 billion (1.0 percent of productivity loss).

Together, these sources absorb an estimated 44.5 percent-45.6 percent of the productivity losses. The remaining 54.4 percent-55.5 percent are paid by people injured in crashes and their families.

Property Damage

In the Unit Cost chapter we estimated that insurers paid out $53.5 billion in property damage insurance claims for motor vehicle accidents in 2009. Private Passenger Physical Damage losses, which include both collision and comprehensive insurance, *declined* slightly in 2010 from $39.7 billion to $39.4 billion. Liability losses, which include both bodily injury and property damage, *increased* slightly in 2010 from 72.1 billion to 72.9 billion. It thus appears that overall, property damage claims for motor vehicle crashes were roughly flat from

2009 to 2010. Using this assumption, we estimate that private insurers covered 70.31 percent of all property damage in 2010 ($53.5 billion in reimbursements/$76.1 billion total property damage). The remaining 29.7 percent was paid out of pocket by crash involved drivers.

Results

Table 14-3 shows the distribution of the portion of crash related costs that are borne by private insurers, governmental sources, individual crash victims, and other sources. These distributions are quite variable depending on the nature of the cost category. Private Insurers are the primary source of payment for medical care, insurance administration, legal costs, and property damage, but tax revenues cover a significant portion of medical care, emergency services, and lost market productivity. Third parties absorb all workplace costs and congestion costs. Individual accident victims pay a modest portion of medical care, and absorb significant portions of both market and household productivity losses, as well as property damage.

In Table 14-4, total costs are distributed according to the proportions listed in Table 14-3. The results indicate that approximately $24 billion, or 9 percent of all costs are borne by public sources, with Federal revenues accounting for 5 percent and States accounting for just under 3 percent, with another 1 percent borne by a number of State and Federal programs that could not be broken out by government source. Public expenditures for economic harm caused by motor vehicle crashes are the equivalent of $205 in added taxes for every household in the United States.[60] State and local government pay almost all costs of police, fire, emergency medical, vocational rehabilitation, victim assistance, and coroner services; incident management; and roadside furniture damage. They share foregone taxes, welfare, and public medical payments. As employers, they bear their share of costs that fall on employers, including private medical care, disability compensation, property damage, auto liability insurance payments, insurance claims processing expenses, and workplace disruption/rehiring expenses.

Private insurers paid $145 billion, or 52 percent, while individual crash victims absorbed $70 billion or 25 percent. Other sources, including third parties impacted by traffic congestion from accidents, employers who pay for sick leave and workplace disruption, and health care providers and charities who absorb unpaid charges for medical care, absorbed $39 billion (14 percent) of the total cost.

To some extent it is illusory to disaggregate costs across payment categories because ultimately, it is individuals who pay for these costs through insurance premiums, taxes, direct out-of-pocket cost, or higher charges for medical care. A real distinction can be made, however, between costs borne by those directly involved in the crashes and costs that are absorbed by the rest of society. Costs paid out of Federal or State revenues are funded by taxes from the general public. Similarly, costs borne by private insurance companies are funded by insurance premiums paid by policyholders, most of whom are not involved in crashes. Even unpaid charges, which are absorbed by health care providers are ultimately translated into higher costs that are borne by a smaller segment of the general public – users of health care facilities. From this perspective, perhaps the most significant point from Table 14-4 is that society at large picks up three-quarters of all crash costs that are incurred by individual motor vehicle crash victims.

Table 14-3. Distribution of Source of Payment by Cost Category

	Federal	State	Unspecified Government	Total Government	Private Insurer	Other	Self	Total
Medical	17.54%	5.56%	8.50%	31.60%	56.10%	1.20%	11.10%	100.00%
Emergency Serv.	0.00%	100.00%	0.00%	100.00%	0.00%	0.00%	0.00%	100.00%
Market Prod.	10.44%	6.18%	0.00%	16.62%	35.95%	7.98%	39.45%	100.00%
Household Prod.	0.00%	0.00%	0.00%	0.00%	33.14%	0.00%	66.86%	100.00%
Insurance Admin.	0.89%	0.51%	0.00%	1.40%	98.60%	0.00%	0.00%	100.00%
Workplace Costs	0.00%	0.00%	0.00%	0.00%	0.00%	100.00%	0.00%	100.00%
Legal Costs	0.00%	0.00%	0.00%	0.00%	100.00%	0.00%	0.00%	100.00%
Congestion Costs	0.00%	0.00%	0.00%	0.00%	0.00%	100.00%	0.00%	100.00%
Property Damage	0.00%	0.00%	0.00%	0.00%	70.31%	0.00%	29.69%	100.00%

Table 14-4. Source of Payment by Cost Category, 2010 Motor Vehicle Crash Costs (Millions of 2010 Dollars)

	Federal	State	Unspecified Government	Total Government	Insurer	Other	Self	Total
Medical	$6,114	$1,939	$2,963	$11,016	$19,557	$418	$3,869	$34,860
Emergency Services	$0	$1,079	$0	$1,079	$0	$0	$0	$1,079
Market Productivity	$7,331	$4,337	$0	$11,668	$25,087	$5,600	$27,855	$70,210
Household Productivity	$0	$0	$0	$0	$7,549	$0	$15,380	$22,929
Insurance Admin.	$227	$130	$0	$356	$25,106	$0	$0	$25,462
Workplace Costs	$0	$0	$0	$0	$0	$4,577	$0	$4,577
Legal Costs	$0	$0	$0	$0	$13,781	$0	$0	$13,781
Congestion Costs	$0	$0	$0	$0	$0	$28,027	$0	$28,027
Property Damage	$0	$0	$0	$0	$53,500	$0	$22,596	$76,096
Total	$13,672	$7,485	$2,963	$24,120	$144,579	$38,622	$69,700	$277,020
% Total	4.94%	2.70%	1.07%	8.71%	52.19%	13.94%	25.16%	100.00%

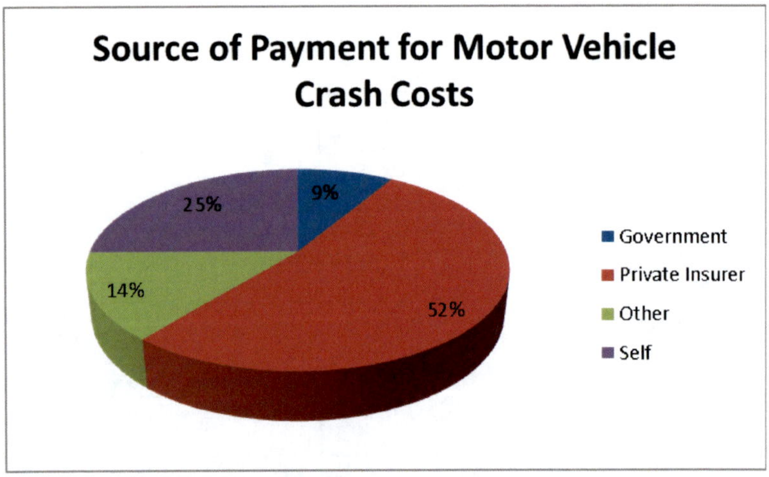

Figure 14-A. Source of Payment for Motor Vehicle Crash Costs.

APPENDIX A. SENSITIVITY ANALYSIS, VALUE OF A STATISTICAL LIFE

Previous versions of this report have focused on the economic impact of motor vehicle crashes – the societal losses that can be directly measured in economic terms. However, these costs do not represent the more intangible consequences of these events and should not, therefore, be used alone to produce cost-benefit ratios. Measurement of the dollar value of intangible consequences such as pain and suffering has been undertaken in numerous studies. These studies have estimated values based on wages for high-risk occupations and prices paid in the market place for safety products, among other measurement techniques. These "willingness to pay" based estimates of how society values risk reduction capture valuations not associated with direct monetary consequences. In this study, comprehensive costs, which include both the economic impacts of crashes and valuation of lost quality- of-life, are also examined. Comprehensive costs represent the value of the total societal harm that results from traffic crashes. The basis for these estimates is the most recent guidance issued by the U.S. Department of Transportation for valuing risk reduction. This guidance, which was issued in February 2013, establishes a new value of a statistical life (VSL) at $9.1 million in 2012 economics ($8.86 million in 2010 economics). In addition, it establishes new relative disutility factors stratified by injury severity level to estimate the lost quality-of-life for nonfatal injuries. These factors were derived in a research contract designed specifically for this current cost study. More detailed discussion of comprehensive costs is included in Chapter 4 of this report.

From Table 8 in chapter 1, the total societal harm from motor vehicle crashes in 2010 is estimated to have been $870.8 billion, roughly three times the value measured by economic impacts alone. Of this total, 68 percent represents lost quality-of-life, dwarfing the contribution of all other cost categories. This highlights the importance of accounting for all societal impacts when measuring costs and benefits from motor vehicle safety countermeasures. However, the literature on VSL estimates indicates a wide range of measured estimates of VSLs – some as low as a few million dollars, some as high as over $30 million. The U. S. DOT guidance memorandum discusses a feasible range of VSLs for sensitivity analysis in 2012 dollars from $5.2 million to $12.9 million. There is thus far more uncertainty regarding the accuracy of estimates of lost quality-of-life than there is regarding economic costs. In this appendix, comprehensive costs are estimated based on this range adjusted to the 2010 basis of an $8.86 million VSL ($5.1 million and $12.6 million, computed by applying the ratio of the 2010 VSL to the 2012 VSL to each end of the range and rounding to the nearest tenth). The results indicate a feasible range of societal harm from motor vehicle crashes of from $583 billion to $1.16 trillion in 2010, with lost quality-of-life accounting for between half and three-quarters of all societal harm respectively. The central value used in this report, $871 billion, should thus be viewed with this range in mind. Although the USDOT values were not selected statistically, they imply a central value with the equivalent of a confidence interval of approximately +-33 percent.

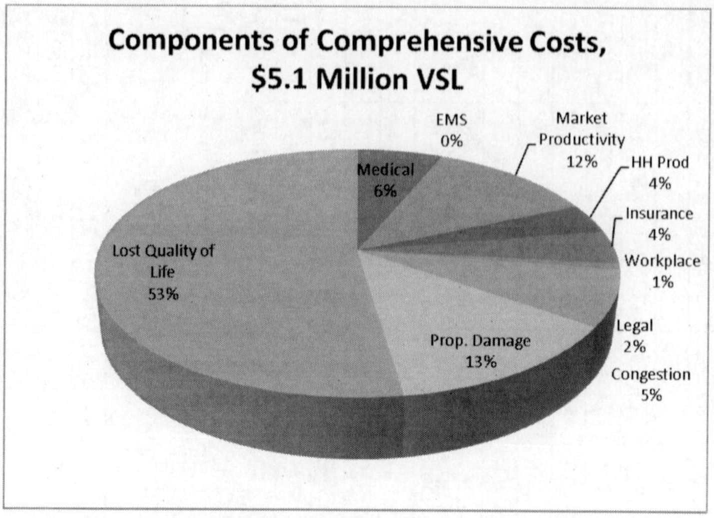

Figure A-A. Components of Comprehensive Costs, $5.1 Million VSL.

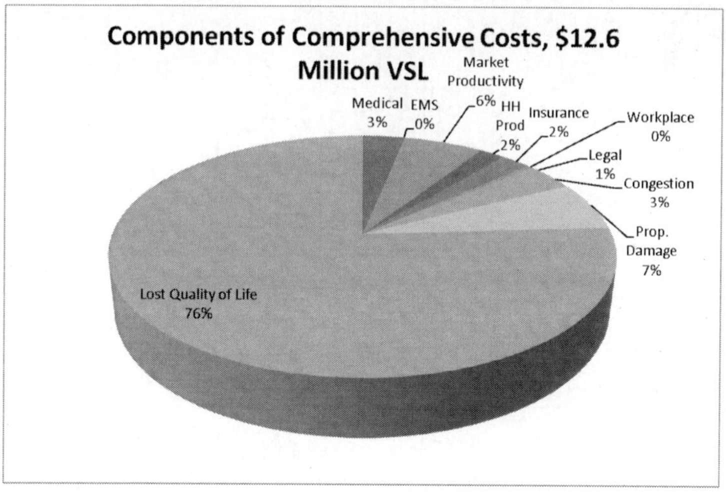

Figure A-B. Components of Comprehensive Costs, $12.6 Million VSL.

Table A-1. Total Comprehensive Costs, $5.1 Million VSL (Millions of 2010 Dollars)

	PDO	MAIS0	MAIS1	MAIS2	MAIS3	MAIS4	MAIS5	Fatal	Total	% Total
Medical	$0	$0	$13,148	$8,292	$7,143	$3,364	$2,539	$373	$34,860	6.0%
EMS	$518	$96	$308	$66	$42	$14	$5	$30	$1,079	0.2%
Market	$0	$0	$9,886	$13,426	$11,090	$2,985	$2,025	$30,797	$70,210	12.1%
Household	$1,111	$206	$3,255	$4,038	$3,375	$762	$616	$9,567	$22,929	3.9%
Insurance	$3,535	$655	$13,612	$3,174	$2,453	$639	$460	$935	$25,462	4.4%
Workplace	$1,148	$211	$1,180	$896	$582	$109	$64	$389	$4,577	0.8%
Legal Costs	$0	$0	$4,877	$2,283	$1,979	$603	$524	$3,514	$13,781	2.4%
Subtotal	**$6,311**	**$1,169**	**$46,267**	**$32,175**	**$26,664**	**$8,477**	**$6,232**	**$45,604**	**$172,898**	**29.7%**
Congestion	$19,934	$3,483	$3,836	$405	$144	$26	$9	$189	$28,027	4.8%
Prop. Damage	$45,235	$8,378	$18,694	$1,957	$1,096	$279	$87	$370	$76,096	13.1%
Subtotal	**$65,169**	**$11,861**	**$22,530**	**$2,363**	**$1,241**	**$305**	**$96**	**$559**	**$104,123**	**17.9%**
Economic Total	**$71,480**	**$13,030**	**$68,797**	**$34,537**	**$27,905**	**$8,781**	**$6,327**	**$46,163**	**$277,020**	**47.5%**
QALYs	$0	$0	$41,375	$59,424	$41,773	$17,916	$13,547	$131,570	$305,605	52.5%
Comp. Total	**$71,480**	**$13,030**	**$110,172**	**$93,961**	**$69,677**	**$26,698**	**$19,874**	**$177,733**	**$582,625**	**100.0%**
% Total	12.3%	2.2%	18.9%	16.1%	12.0%	4.6%	3.4%	30.5%	100.0%	0.0%

Table A-2. Comprehensive Unit Costs, $5.1 Million VSL (2010 Dollars)

	PDO	MAIS0	MAIS1	MAIS2	MAIS3	MAIS4	MAIS5	Fatal
Medical	$0	$0	$3,801	$24,481	$70,902	$196,911	$441,618	$11,317
EMS	$28	$21	$89	$194	$416	$838	$855	$902
Market	$0	$0	$2,858	$39,637	$110,088	$174,728	$352,178	$933,262
Household	$60	$45	$941	$11,921	$33,505	$44,593	$107,070	$289,910
Insurance	$191	$143	$3,935	$9,370	$24,348	$37,372	$79,967	$28,322
Workplace	$62	$46	$341	$2,644	$5,776	$6,361	$11,091	$11,783
Legal Costs	$0	$0	$1,410	$6,739	$19,645	$35,307	$91,197	$106,488

	PDO	MAIS0	MAIS1	MAIS2	MAIS3	MAIS4	MAIS5	Fatal
Subtotal	**$341**	**$255**	**$13,375**	**$94,986**	**$264,680**	**$496,110**	**$1,083,976**	**$1,381,984**
Congestion	$1,077	$760	$1,109	$1,197	$1,434	$1,511	$1,529	$5,720
Prop. Damage	$2,444	$1,828	$5,404	$5,778	$10,882	$16,328	$15,092	$11,212
Subtotal	**$3,521**	**$2,588**	**$6,513**	**$6,975**	**$12,316**	**$17,839**	**$16,621**	**$16,932**
Economic Total	**$3,862**	**$2,843**	**$19,888**	**$101,961**	**$276,996**	**$513,949**	**$1,100,597**	**$1,398,916**
QALYs	$0	$0	$11,961	$175,432	$414,657	$1,048,603	$2,356,365	$3,987,082
Comp. Total	**$3,862**	**$2,843**	**$31,849**	**$277,393**	**$691,653**	**$1,562,552**	**$3,456,962**	**$5,385,998**

* Note: Unit costs are expressed on a per-person basis for all injury levels. PDO costs are expressed on a per-damaged-vehicle basis.

Table A-3. Total Comprehensive Costs, $12.6 Million VSL (Millions of 2010 Dollars)

	PDO	MAIS0	MAIS1	MAIS2	MAIS3	MAIS4	MAIS5	Fatal	Total	% Total
Medical	$0	$0	$13,148	$8,292	$7,143	$3,364	$2,539	$373	$34,860	3.0%
EMS	$518	$96	$308	$66	$42	$14	$5	$30	$1,079	0.1%
Market	$0	$0	$9,886	$13,426	$11,090	$2,985	$2,025	$30,797	$70,210	6.1%
Household	$1,111	$206	$3,255	$4,038	$3,375	$762	$616	$9,567	$22,929	2.0%
Insurance	$3,535	$655	$13,612	$3,174	$2,453	$639	$460	$935	$25,462	2.2%
Workplace	$1,148	$211	$1,180	$896	$582	$109	$64	$389	$4,577	0.4%
Legal Costs	$0	$0	$4,877	$2,283	$1,979	$603	$524	$3,514	$13,781	1.2%
Subtotal	$6,311	$1,169	$46,267	$32,175	$26,664	$8,477	$6,232	$45,604	$172,898	14.9%
Congestion	$19,934	$3,483	$3,836	$405	$144	$26	$9	$189	$28,027	2.4%
Prop. Damage	$45,235	$8,378	$18,694	$1,957	$1,096	$279	$87	$370	$76,096	6.6%
Subtotal	$65,169	$11,861	$22,530	$2,363	$1,241	$305	$96	$559	$104,123	9.0%
Economic Total	$71,480	$13,030	$68,797	$34,537	$27,905	$8,781	$6,327	$46,163	$277,020	23.9%
QALYs	$0	$0	$119,207	$171,205	$120,350	$51,619	$39,029	$379,062	$880,472	76.1%
Comp. Total	$71,480	$13,030	$188,004	$205,742	$148,254	$60,400	$45,357	$425,225	$1,157,493	100.0%
% Total	6.2%	1.1%	16.2%	17.8%	12.8%	5.2%	3.9%	36.7%	100.0%	0.0%

Table A-4. Comprehensive Unit Costs, $12.6 Million VSL (2010 Dollars)

	PDO	MAIS0	MAIS1	MAIS2	MAIS3	MAIS4	MAIS5	Fatal
Medical	$0	$0	$3,801	$24,481	$70,902	$196,911	$441,618	$11,317
EMS	$28	$21	$89	$194	$416	$838	$855	$902
Market	$0	$0	$2,858	$39,637	$110,088	$174,728	$352,178	$933,262
Household	$60	$45	$941	$11,921	$33,505	$44,593	$107,070	$289,910
Insurance	$191	$143	$3,935	$9,370	$24,348	$37,372	$79,967	$28,322
Workplace	$62	$46	$341	$2,644	$5,776	$6,361	$11,091	$11,783
Legal Costs	$0	$0	$1,410	$6,739	$19,645	$35,307	$91,197	$106,488
Subtotal	$341	$255	$13,375	$94,986	$264,680	$496,110	$1,083,976	$1,381,984
Congestion	$1,077	$760	$1,109	$1,197	$1,434	$1,511	$1,529	$5,720
Prop. Damage	$2,444	$1,828	$5,404	$5,778	$10,882	$16,328	$15,092	$11,212
Subtotal	$3,521	$2,588	$6,513	$6,975	$12,316	$17,839	$16,621	$16,932
Economic Total	$3,862	$2,843	$19,888	$101,961	$276,996	$513,949	$1,100,597	$1,398,916
QALYs	$0	$0	$34,461	$505,432	$1,194,657	$3,021,103	$6,788,865	$11,487,082
Comp. Total	$3,862	$2,843	$54,349	$607,393	$1,471,653	$3,535,052	$7,889,462	$12,885,998

* Note: Unit costs are expressed on a per-person basis for all injury levels. PDO costs are expressed on a per-damaged-vehicle basis.

Appendix B. Costs by Body Region

Table B-1. Economic Costs by Body Region

MAIS	Medical	EMS	Wage	HH Prod	Ins Adm	Workplace	Legal	Congestion	Prop Dam	Total
SCI										
1	-	-	-	-	-	-	-	-	-	-
2	-	-	-	-	-	-	-	-	-	-
3	231,136	416	187,275	61,331	53,035	5,776	43,964	1,434	10,882	595,249
4	476,264	838	209,287	70,415	66,745	6,361	64,145	1,511	16,328	911,893
5	1,062,368	855	115,620	88,809	111,925	11,091	128,252	1,529	15,092	1,535,541
BRAIN										
1	3,868	89	3,500	1,198	4,228	341	1,590	1,109	5,394	21,317
2	9,145	194	23,238	6,488	5,107	2,644	3,440	1,197	5,778	57,232
3	51,455	416	107,911	38,645	22,565	5,776	18,133	1,434	10,882	257,218
4	126,021	838	173,071	59,696	32,406	6,361	30,431	1,511	16,328	446,663
5	362,210	855	393,439	117,257	77,526	11,091	88,366	1,529	15,092	1,067,364
LOWER EXTREMITY										
1	2,321	89	2,639	1,064	3,457	341	1,117	1,109	5,394	17,531
2	16,330	194	37,551	12,554	8,269	2,644	5,887	1,197	5,778	90,404
3	62,833	416	78,405	29,194	19,582	5,776	15,605	1,434	10,882	224,126
4	121,864	838	152,613	46,430	29,131	6,361	27,216	1,511	16,328	402,291
5	183,286	855	86,311	31,815	26,770	11,091	30,495	1,529	15,092	387,244
UPPER EXTREMETY										
1	1,985	89	3,838	1,455	3,837	341	1,350	1,109	5,394	19,399
2	7,854	194	18,030	6,137	4,321	2,644	2,832	1,197	5,778	48,987
3	41,372	416	69,046	27,591	16,076	5,776	12,632	1,434	10,882	185,225
4	-	-	-	-	-	-	-	-	-	-
5	-	-	-	-	-	-	-	-	-	-

Table B-1. (Continued)

MAIS	Medical	EMS	Wage	HH Prod	Ins Adm	Workplace	Legal	Congestion	Prop Dam	Total
TRUNK, ABDOMEN										
1	6,115	89	15,644	4,340	9,550	341	4,851	1,109	5,394	47,433
2	15,880	194	38,392	11,723	8,218	2,644	5,848	1,197	5,778	89,874
3	61,531	416	97,388	34,452	22,063	5,776	17,708	1,434	10,882	251,649
4	76,392	838	80,969	28,566	16,051	6,361	15,758	1,511	16,328	242,774
5	195,099	855	241,712	86,839	45,708	11,091	52,999	1,529	15,092	650,925
FACE/OTHER HEAD/NECK										
1	4,647	89	4,602	1,540	4,903	341	2,003	1,109	5,394	24,628
2	18,984	194	32,861	9,558	7,691	2,644	5,440	1,197	5,778	84,349
3	67,752	416	89,968	29,193	21,365	5,776	17,116	1,434	10,882	243,901
4	184,441	838	204,628	58,775	40,105	6,361	37,991	1,511	16,328	550,978
5	-	-	-	-	-	-	-	-	-	-
BURNS										
1	2,231	89	3,646	1,371	3,828	341	1,345	1,109	5,394	19,352
2	26,941	194	16,856	5,100	6,257	2,644	4,330	1,197	5,778	69,297
3	-	-	-	-	-	-	-	-	-	-
4	-	-	-	-	-	-	-	-	-	-
5	251,933	855	105,044	43,723	36,287	11,091	40,549	1,529	15,092	506,103
1	3,672	89	1,386	625	3,353	341	1,054	1,109	5,394	17,021

Table B-2. QALY Values by MAIS, Body Region and Injury Status (2010 Dollars)*

Maximum AIS (MAIS)	Fracture or Dislocation	Body Region	Not Hospitalized	Hospitalized	All
AIS 1	No	Head	$59,331	$489,767	$93,721
AIS 2	No	Head	$426,278	$750,522	$545,697
AIS 3	No	Head	$438,371	$1,559,620	$1,270,899
AIS 4	No	Head	$0	$3,351,274	$3,351,274
AIS 5	No	Head	$0	$5,244,208	$5,244,208
AIS 1	No	Face	$10,203	$403,604	$17,762
AIS 2	No	Face	$0	$945,900	$945,900
AIS 3	No	Face	$0	$865,784	$865,784
AIS 1	No	Neck	$0	$462,557	$462,557
AIS 2	No	Neck	$0	$1,183,981	$1,183,981
AIS 1	No	Thorax	$5,291	$125,087	$10,581
AIS 2	No	Thorax	$28,721	$440,261	$119,041
AIS 3	No	Thorax	$24,564	$196,133	$173,837
AIS 4	No	Thorax	$0	$352,209	$352,209
AIS 5	No	Thorax	$0	$572,528	$572,528
AIS 1	No	Abdomen/Pelvis	$0	$76,337	$180,639
AIS 2	No	Abdomen/Pelvis	$24,942	$193,110	$136,802
AIS 3	No	Abdomen/Pelvis	$0	$315,930	$315,930
AIS 4	No	Abdomen/Pelvis	$0	$458,023	$458,023
AIS 5	No	Abdomen/Pelvis	$0	$265,290	$265,290
AIS 3	No	Spinal Cord	$0	$2,092,468	$2,061,102
AIS 4	No	Spinal Cord	$0	$5,870,777	$5,870,777
AIS 5	No	Spinal Cord	$0	$6,837,083	$6,837,083
AIS 1	No	Upper Extremity	$9,826	$143,982	$13,605
AIS 2	No	Upper Extremity	$43,837	$468,982	$177,238
AIS 3	No	Upper Extremity	$0	$720,290	$874,475
AIS 1	No	Lower Extremity	$10,203	$169,302	$15,116
AIS 2	No	Lower Extremity	$66,512	$498,081	$82,006
AIS 3	No	Lower Extremity	$0	$741,830	$597,092
AIS 4	No	Lower Extremity	$0	$1,228,196	$1,228,196
AIS 1	No	Burns/Other	$21,919	$232,412	$23,430
AIS 2	No	Burns/Other	$0	$757,325	$757,325
AIS 2	Yes	Head	$1,104,998	$1,671,480	$1,489,329
AIS 3	Yes	Head	$0	$1,795,056	$1,795,056
AIS 4	Yes	Head	$0	$1,782,585	$1,782,585
AIS 1	Yes	Face	$622,790	$205,203	$586,889
AIS 2	Yes	Face	$18,140	$919,446	$287,587
AIS 3	Yes	Face	$0	$899,039	$899,039

Table B-2. (Continued)

Maximum AIS (MAIS)	Fracture or Dislocation	Body Region	Not Hospitalized	Hospitalized	All
AIS 1	Yes	Thorax	$0	$167,413	$167,413
AIS 2	Yes	Thorax	$0	$311,017	$311,017
AIS 3	Yes	Thorax	$0	$397,180	$397,180
AIS 4	Yes	Thorax	$0	$581,220	$581,220
AIS 1	Yes	Upper Extremity	$26,453	$298,546	$26,831
AIS 2	Yes	Upper Extremity	$53,663	$452,354	$137,180
AIS 3	Yes	Upper Extremity	$0	$911,132	$911,132
AIS 1	Yes	Lower Extremity	$9,826	$0	$9,826
AIS 2	Yes	Lower Extremity	$198,779	$535,116	$318,197
AIS 3	Yes	Lower Extremity	$828,371	$732,005	$747,121
AIS 4	Yes	Lower Extremity	$0	$1,506,713	$1,506,713

* Derived from QALY values presented in Table 5 in Spicer, Miller, Hendrie, and Blincoe (2011).

Appendix C. KABCO/MAIS Translators

Throughout this analysis translators developed from historical data records are used to translate non- fatal injury severity estimates based on police records using a KABCO scale, into the more precise Abbreviated Injury Scale measure. For nonfatal injuries and PDOs, GES data was frequently the basis for incidence. However, GES data is only recorded using the KABCO severity system, whereas this report is based on the Abbreviated Injury Scale. To translate GES data to an MAIS (Maximum Abbreviated Injury Scale) basis, we used a variety of KABCO/MAIS translators. For CDS equivalent crashes, we used a current translator derived from 2000-2008 CDS data. Since this data is relatively recent, it reflects roughly current levels of seat belt usage. For non-CDS cases, the only available data from which to develop translators were contained in the 1982-1986 NASS files. Seat belt use has increased dramatically since this time. Observed belt use during this period ranged from roughly 10 to 37 percent as public awareness of the importance of belt use and belt use laws were just beginning to take hold in 1986. Belt use has since risen dramatically, and has been between 80 and 85 percent since 2004. Belt use can influence injury reporting significantly in a number of ways. It changes the nature of injuries by preventing many more visible injuries (such as head/face contact with the windshield) but replaces them with often less visible (and also typically less serious) abdominal injuries such as bruising caused by pressure from the belt across the torso. This can influence the relationship between the KABCO reported injury severity and the corresponding MAIS injury level. For this reason, separate translators were developed from the 1982-86 NASS data for non-CDS cases where the victim was belted, unbelted, unknown belted status, and for nonoccupants/motorcyclists. These translators are presented in Tables C-1 through C-5 below.

Table C-1. All CDS Equivalent Cases

MAIS	O	C	B	A	Injured Severity Unknown	Unknown if Injured
	No Injury	Possible Injury	Non-incapacitating Injury	Incapacitating Injury		
0	0.8191	0.2188	0.0906	0.0376	0.0032	0.2429
1	0.1759	0.7014	0.7518	0.5782	0.0110	0.5961
2	0.0047	0.0674	0.1113	0.1924	0.0019	0.1039
3	0.0002	0.0101	0.0348	0.1259	0.0041	0.0406
4	0.0000	0.0021	0.0085	0.0444	0.0027	0.0047
5	0.0001	0.0001	0.0014	0.0171	0.0007	0.0117
Fatality	0.0000	0.0001	0.0015	0.0043	0.9765	0.0000
Total	1.0000	1.0000	1.0000	1.0000	1.0000	1.0000

Table C-2. Non-CDS, Unbelted

MAIS	O	C	B	A	Injured Severity Unknown	Unknown if Injured
	No Injury	Possible Injury	Non-incapacitating Injury	Incapacitating Injury		
0	0.9591	0.2655	0.0777	0.0394	0.1082	0.9196
1	0.0388	0.4360	0.6667	0.5033	0.3931	0.0681
2	0.0011	0.0389	0.0640	0.1517	0.0345	0.0004
3	0.0000	0.0045	0.0041	0.0512	0.0000	0.0013
4	0.0000	0.0000	0.0008	0.0116	0.0000	0.0000
5	0.0000	0.0000	0.0000	0.0061	0.0000	0.0000
7	0.0010	0.2545	0.1863	0.2347	0.4615	0.0106
Fatality	0.0000	0.0004	0.0005	0.0020	0.0027	0.0000
Total	1.0000	1.0000	1.0000	1.0000	1.0000	1.0000

Table C-3. Non-CDS, Belted

MAIS	O	C	B	A	Injured Severity Unknown	Unknown if Injured
	No Injury	Possible Injury	Non-incapacitating Injury	Incapacitating Injury		
0	0.9491	0.3002	0.1558	0.0962	0.5115	0.9164
1	0.0494	0.4888	0.6544	0.5418	0.1948	0.0598
2	0.0006	0.0154	0.0192	0.1966	0.0655	0.0000
3	0.0001	0.0124	0.0082	0.0505	0.0179	0.0000
4	0.0000	0.0001	0.0012	0.0073	0.0000	0.0000
5	0.0000	0.0000	0.0000	0.0000	0.0000	0.0000
7	0.0009	0.1831	0.1611	0.1076	0.2104	0.0239
Fatality	0.0000	0.0000	0.0000	0.0000	0.0000	0.0000
Total	1.0000	1.0000	1.0000	1.0000	1.0000	1.0000

Table C-4. Non-CDS, Unknown if Belted

MAIS	O	C	B	A	Injured Severity Unknown	Unknown if Injured
	No Injury	Possible Injury	Non-incapacitating Injury	Incapacitating Injury		
0	0.9960	0.1078	0.0908	0.0252	0.0413	0.9476
1	0.0001	0.1161	0.4971	0.2687	0.1866	0.0050
2	0.0000	0.0057	0.0067	0.1491	0.0067	0.0000
3	0.0000	0.0000	0.0000	0.1305	0.0000	0.0000
4	0.0000	0.0000	0.0000	0.0000	0.0000	0.0000
5	0.0000	0.0000	0.0000	0.0000	0.0000	0.0000
7	0.0038	0.7704	0.4053	0.4265	0.7654	0.0475
Fatality	0.0000	0.0000	0.0000	0.0000	0.0000	0.0000
Total	1.0000	1.0000	1.0000	1.0000	1.0000	1.0000

Table C-5. Non-CDS, Nonoccupants/Motorcyclists

MAIS	O	C	B	A	Injured Severity Unknown	Unknown if Injured
	No Injury	Possible Injury	Non-incapacitating Injury	Incapacitating Injury		
0	0.7370	0.1057	0.0221	0.0059	0.0254	0.2145
1	0.2341	0.6067	0.6439	0.2886	0.5483	0.5992
2	0.0218	0.0972	0.1455	0.2839	0.1750	0.0609
3	0.0047	0.0261	0.0516	0.2583	0.0462	0.0830
4	0.0004	0.0027	0.0024	0.0315	0.0023	0.0000
5	0.0000	0.0008	0.0005	0.0234	0.0021	0.0000
7	0.0020	0.1608	0.1334	0.0965	0.1873	0.0424
Fatality	0.0000	0.0000	0.0007	0.0119	0.0133	0.0000
Total	1.0000	1.0000	1.0000	1.0000	1.0000	1.0000

APPENDIX D. KABCO UNIT COSTS

Police reports are generally coded using a generalized injury severity estimate commonly known as the KABCO scale. Within this scale, injuries are typically coded under one of the following categories:

K = Killed
A = incapacitating injury
B = non-incapacitating injury
C= complaint of pain
O = No injury

Table D-1. Kabco Nonfatal Injury Unit Costs

	0	C	B	A
Medical	$3,512	$6,431	$7,201	$24,360
EMS	$20	$45	$56	$122
Market	$2,984	$7,467	$9,358	$28,064
Household	$971	$2,289	$2,846	$8,259
Insurance	$3,060	$5,346	$5,312	$13,514
Workplace	$7	$208	$1,459	$3,941
Legal Costs	$77	$1,648	$2,438	$9,841
Injury Subtotal	**$10,631**	**$23,433**	**$28,671**	**$88,100**
Congestion	$1,026	$1,009	$995	$1,385
Prop. Damage	$1,624	$2,407	$2,465	$3,518
Crash Subtotal	**$2,650**	**$3,416**	**$3,460**	**$4,903**
Economic Cost Total	**$13,281**	**$26,849**	**$32,131**	**$93,003**
QALYs	$31,859	$108,274	$252,268	$919,158
Comprehensive Total	**$45,140**	**$135,123**	**$284,399**	**$1,012,161**

This very general scale is used by police officers on the scene and represents their judgment regarding injury severity. While police do their best to accurately judge each case, they do not have the training or diagnostic skills or equipment to determine more precise estimates of actual injury. As noted elsewhere in this report, translators developed from data systems that collected both KABCO and MAIS injury severities indicate that KABCO ratings are not very meaningful. Large numbers of crash victims are miscoded regarding their actual injury levels, with many people coded as A – incapacitating injury actually being uninjured and many coded as uninjured actually experiencing injuries. For this reason we believe that analysis of motor vehicle injuries is more meaningful when injuries are first expressed, either directly or through a translator, using the Abbreviated Injury Scale. Nonetheless, we recognize that in many cases police-reported data will be used directly with KABCO designations. For this reason, we have supplied a KABCO based unit cost table in this appendix. Note that there is a QALY value associated with O category KABCO injuries. This reflects the fact that many O injuries are actually injured.

Appendix E. Estimating the Costs of Motor Vehicle Traffic Injuries in the United States from Healthcare Files

Data

The medical and work-loss costs of each injury death were estimated using 2008 data from the National Vital Statistics System (NVSS). The unit medical costs of non-fatal injuries were built primarily from the 2008 Nationwide Inpatient Sample and the 2003 State Emergency Department Databases (SEDD), both from the Healthcare Cost and Utilization Project. The information on initial treatment costs from the NIS and the SEDD was augmented with data from Finkelstein et al. (2006) including cost data on emergency transportation, physician fees, rehabilitation, and long-term treatment. The SEDD-based costs were merged onto another HCUP dataset, the 2007 Nationwide Emergency Department Sample (NEDS), which, unlike the SEDD, is nationally representative, and therefore more appropriate for this project. The 2008 NIS and the 2007 NEDS also served as the basis for the estimation of lifetime work loss costs, again supplemented by information from Finkelstein et al. (2006).

From the costed 2008 NVSS, 2008 NIS, and 2007 NEDS, we selected the acute injury cases that were cause-coded as unintentional motor vehicle traffic crashes. To avoid double counting, we excluded fatalities from the NIS and the NEDS and transfers from the NIS. (The NEDS already excludes ED visits that result in admission as an inpatient.) We also dropped a number of duplicate cases reported by one hospital in the NIS.

Year of Dollars, Inflator Series, and Discount Rate

All costs are reported in 2010 dollars. Individual cost elements used in developing the cost module came from datasets belonging to different time periods and were inflated to 2010 dollars. Health care costs in earlier year's dollars were inflated using the medical care component of the Consumer Price Index (CPI–Medical). Work loss costs were inflated using the index that DoT uses for inflating the value of a statistical life: CPI × ECI 0.55, where CPI is the consumer price index, ECI is the employment cost index, and 0.55 is the estimated income elasticity. Work-loss costs more than one year post-injury were discounted to present value using the 3-percent discount rate recommended by the Panel on Cost- Effectiveness in Health and Medicine (Gold, Siegel, Russell, & Weinstein, 1996) and by Haddix, Teutsch, and Corso (2003). For sensitivity analysis, we also estimated costs using alternative discount rates of 0 percent, 2 percent, 4 percent, 7 percent, and 10 percent.

Lifetime Medical Costs of Injuries

For some injuries, medical treatment and corresponding costs may persist for years or even decades after the initial injury. The medical costs presented in this study include costs associated with treatment for physical injuries only, as data required to estimate costs for mental health and psychological treatment were not available.

Fatal Injuries

Fatal medical costs were calculated using costs per case by place of death from Finkelstein et al. (2006), expanded to include deaths in hospice. Costs were computed separately for six different places of death identified in the 2008 NVSS data:

- On-scene/at home.
- Dead on arrival at a hospital (DOA).

- In a hospital emergency department (ED).
- In a hospital after inpatient admission.
- In a nursing home.
- In a hospice.

The medical costs incurred, depending on the place of death, might include charges for coroner/medical examiner, emergency medical transport, ED visit, and stays in a hospital, nursing home, or hospice. Table E-1 summarizes the components included for each place of death.

Table E-1. Data and Methods for Estimating Medical Costs of Fatal Injuries

Place of Death	Cost Categories	Description, Unit Cost (2010 U.S. $)	Source of Data
On scene/ at home	Coroner/ME (C/ME)	$43 admin, plus $1,619 if autopsy (C/ME)	Hickman, Hughes, Strom & Ropero-Miller (2007) (C/ME)
Dead on arrival (DOA) at hospital	C/ME + Transport (T)	$43[+$1,619] (C/ ME) + $315 (T)	1999 Medicare 5 percent Sample (T)
In ED	C/ME + T + ED	$43[+$1,619] (C/ME) + $315 (T) + Avg. cost for fatalities in ED by external cause and age (ED)	Estimated using fatalities in the 2007 HCUP-NEDS (ED)
In hospital after admission	C/ME + T + Fatal inpatient total (FIT)	$43[+$1,619] (C/ME) + $315 (T) + Avg. cost for fatalities in hospital by select diagnosis and mechanism group (FIT)	2008 HCUP-NIS for hospital facilities costs, 1996-97 MarketScan data for non- facility costs (FIT)
In nursing home	C/ME + 2×T + Non-fatal inpatient total (NIT) + Cost of nursing home stay ending in death (NH)	$43[+$1,619] (C/ME) + $630 (2×T) + Avg. inpatient costs for discharges to NH by diagnosis and mechanism (NIT) + Avg. days in NH by body region × $197 cost/day (NH)	2008 HCUP-NIS and 1996–97 MarketScan data (NIT), days in NH estimated from 2004 National Nursing Home Survey (NH), cost/day from Genworth 2007 Cost of Care Survey
In hospice	C/ME + 2×T + Non-fatal inpatient total (NIT) + Cost of hospice stay ending in death (HSP)	$43[+$1,619] (C/ME) + $630 (2×T) + Avg. inpatient costs for discharges to hospice by mechanism and body region (NIT) + $6,258 (HSP)	2008 HCUP-NIS and 1996– 97 MarketScan data (NIT), 2007 National Hospice and Home Care Survey (HSP)

All fatalities were assigned coroner/medical examiner costs—a $43 administrative fee, plus an additional $1,619 if an autopsy was performed, as indicated by the NVSS autopsy variable. For cases where this variable was missing, we used autopsy probabilities by external cause group (46.0 percent for motor vehicle deaths) based on Hoyert (2011). We estimated the coroner costs from Hickman, Hughes, Strom & Ropero-Miller (2007). This survey-based document provides the costs and workload of all U.S. medical examiner and coroner offices except in Louisiana. We calculated the $1,619 cost per accepted fatality under the arbitrary assumption that 5 percent of the office budget is used to determine which cases to accept, keep records about those determinations, and handle public relations and education requests unrelated to specific deaths.

DOAs and all deaths in the hospital, whether in the ED or as an inpatient, also received the cost of a one- way transport, which was based on average ambulance transport costs for injured patients in the 1999 Medicare 5 percent sample. Deaths in a nursing home or hospice were assigned the cost of two emergency transports—one from the scene to the hospital, and a second from the hospital to the facility where death eventually occurred. (This component is described in greater detail below, in the section on medical costs of hospital-admitted injuries.)

For deaths in the ED, Finkelstein et al. (2006) added average costs for injury fatalities in the ED by injury mechanism and age group, computed from 363 injury deaths in EDs in 1997 in three States. We applied this method to the fatal cases in the newer, larger 2007 HCUP-NEDS. Of the 4,336 fatalities in the NEDS, we were able to estimate costs for 3,623. AHRQ has not produced cost-to-charge ratios for the NEDS, but we were able to estimate cost-to-charge ratios for most facilities and sampling strata, as described below in the section on medical costs of ED-treated injuries. Therefore, we were able to estimate costs at the case level by multiplying hospital charges times our estimated cost-to-charge ratios. Average costs were computed by injury mechanism and, where sample size permitted, by age, intent, or body region injured.

For deaths in the hospital, costs of an inpatient admission that ended in death were added to the transport and medical examiner/coroner costs. These inpatient costs were computed by injury mechanism (fall, motor vehicle, poisoning, suffocation, other), nature of injury (fracture or dislocation, internal organ injury, other), body region (brain, spinal cord, trunk, other) and age group (0–9, 10–24, 25–49, 50–69, 70–84, 85 and older), from 16,004 injury-related cases in the 2008 HCUP-NIS in which the patient died in the hospital. For each case, the facility cost was estimated by multiplying the hospital charge times a facility-specific cost-to-charge ratio. This product, in turn, was multiplied times another factor to account for non-facility services—i.e., professional services used while in the hospital yet not included in the admissions billing (e.g., surgeon, anesthesia, physical therapy). These non-facility factors were based on Medstat's 1996 and 1997 MarketScan Commercial Claims and Encounters Database. This database contains an inpatient hospital admissions file, which summarizes each hospital admission, including total payments, facility payments, length of stay, and detailed diagnosis data. After removing non-fee-for-service claims and claims without a diagnosis of injury, a file of 19,247 inpatient injury admissions was created. Using these records, we calculated the mean ratio of total medical costs during the inpatient stay to facilities costs by body region as presented in the Barell injury-diagnosis matrix. The ratios of total costs to facilities costs ranged from 1.03 to 1.39, with an overall average of 1.26. The

HCUP-NIS cost estimate for each admission was multiplied times the ratio for the corresponding body region to yield estimated total inpatient costs for each injury admission in the HCUP-NIS. The non-facility costs of non-fatal hospital admissions were estimated using this same approach (see below). Average costs by selected mechanism, diagnosis, and age group, as described above, were computed, and these averages were applied to the corresponding cases of the 2008 NVSS data.

Deaths in a nursing home or hospice were assumed to be preceded by a stay in an acute care hospital. The method described in the previous paragraph was used to estimate hospital costs for each patient in the 2008 HCUP-NIS who was discharged to a hospice (3,336 cases) and each patient who was discharged to a nursing home following a severe (AIS≥4) injury (4,327 cases). Because of the small samples, fewer diagnosis/mechanism groups were used than for deaths in hospital, and there was no age breakdown. This cost was added to the usual coroner cost plus twice the usual emergency transport cost. Patients who died in a nursing home or hospice were assumed to have been transported by ambulance twice— first to the hospital, and then to the nursing home or hospice. The final component of medical costs for these deaths was the cost of the terminal stay in the nursing home or hospice.

For deaths in a nursing home, the cost of the nursing home stay was calculated as cost per day times the length of stay (LoS) in the nursing home. The average cost per day of nursing home care was taken from the Genworth Financial 2007 Cost of Care Survey and inflated to 2010 dollars. The average LoS in a nursing home was estimated by body region injured (head or neck, trunk, upper limb, hip, upper leg or knee, lower leg or foot, other) from 1,234 resident cases with an admitting diagnosis of injury from the 2004 National Nursing Home Survey (NNHS). Since the NNHS is based on a survey of residents rather than discharge data, it did not allow us to identify patients whose stay ended in death. Moreover, it provided only the LoS as of the survey date, not the final LoS. To estimate the average complete LoS, we assumed that each surveyed resident represented a nursing home bed that was always filled with a patient identical to the survey respondent. We further assumed that each patient was surveyed at the midpoint of the nursing home stay, unless this would have resulted in a LoS of less than 13 days, which we imposed as a minimum, based on sensitivity analysis of the nursing home data. This allowed us to account for the many residents with a short LoS who would have passed through the nursing home before and after the survey date while residents with a longer LoS remained.

The cost of a terminal hospice stay was estimated using data from the 2007 National Home Health and Hospice Care Survey. This dataset, unlike the NNHS, was based on discharge data, including both charges and payments. Only eight cases involved injury, and just five of these ended in death. We computed the average total payment for these five cases.

These costing methods were applied to the deaths in the 2008 NVSS data at the case level using the place of death variable, which specifies where the death occurred, to produce the fatal medical costs.

Hospitalized Injuries

The hospitalized injury costing methods in Finkelstein et al. (2006) were applied to 2008 acute care costs. An overview of the approach is presented in Table E-2. The details are provided in the following sections.

Table E-2. Data and Methods for Estimating Medical Costs of Non-Fatal Injuries Requiring Hospitalization

Cost Category	Description, Unit Cost (2010 U.S. $)	Source/Notes
Facilities component of inpatient stay	Inpatient facility charges for the case multiplied by inpatient cost-to-charge ratio for the facility	2008 NIS for charges; cost-to-charge ratios from AHRQ
Non-facilities component of inpatient stay	Estimated by comparing ratio of total costs to facilities costs by Barell body part	1996–97 MarketScan data
Hospital readmissions	Readmission rates by age group and Barell diagnosis group	2007 SID analysis, reported by Zaloshnja et al. (2011)
Short-to medium-term follow-up costs	Estimated as the ratio of total costs in months 1–18 (on average) to total inpatient costs by 16 diagnosis groups, excluding costs of readmission in the first 6 months	1996–99 MEPS
Follow-up costs beyond 18 months, up to 7 years	Estimated using ratios of total lifetime costs to 18-month costs for 17 diagnosis groups	1979–88 Detailed Claim Information (DCI) data from workers' compensation claims
Long-term costs beyond 7 years for SCI and TBI	SCI: All post-discharge costs were recomputed using the ratio of pre-to post-discharge costs TBI: Post-7-year costs estimated at 75 percent of SCI costs	1986 survey data reported in Berkowitz et al. (1990)
Hospital rehabilitation costs	Probabilities and average costs of rehabilitation estimated for 11 injury diagnosis groups	Probabilities from 1997 CA, MD, & PA hospital data; costs estimated using Prospective Payment System reimbursement amounts, as reported in Miller et al. (2006)
Nursing home costs	Cost/day in NH ($208) times estimated average length of stay by 7 body regions for patients discharged to NH	Cost/day from Genworth 2007 Cost of Care Survey; length of stay estimated from 2004 National Nursing Home Survey
Transport	Half of mean ambulance claims ($350) applied to each case	Mean ambulance claims for cases from the 1999 Medicare 5 percent sample with an injury E-code
Claims administration	Insurance overhead percentages by payer	Woolhandler et al. (2003)

Total inpatient costs (facility and non-facility)

The 2008 NIS included the inpatient facility charge for each admission. For each record in the NIS, this charge was multiplied times the 2008 Medicare cost-to-charge ratios provided by AHRQ. These ratios are hospital specific for 89 percent of the acute injury records in the 2008 NIS. For hospitals whose facility-specific ratio could not be calculated, a weighted group average ratio specific to the hospital's State, ownership, urban/rural location, and number of beds was used as recommended by AHRQ (Friedman, De La Mare, Andrews, & McKenzie, 2002). For Kaiser hospitals in California, which do not report charges, we computed the average facility cost by sampling stratum and diagnosis for California hospitals in the 2007 and 2008 NIS. These estimates of facility costs for each hospital admission were then multiplied times a ratio of total inpatient costs to facility costs to obtain the total cost of the admission, including non-facility costs, i.e., payments to professionals such as surgeons, anesthesiologists, and therapists who bill separately from the hospital itself. This factor is discussed in detail above, in the paragraph on medical costs of deaths in hospital.

In order to account for follow-up admissions, we used readmission rates based on HCUP's 2007 State Inpatient Databases (SID) from 13 States (AZ, CA, FL, MO, NE, NH, NV, NY, NC, SC, TN, UT, and WA), as reported by Zaloshnja et al. (2011). The SID covers all inpatient stays in participating States. In 2007 AHRQ tracked revisits for inpatients in these 13 States, providing a rare look at follow-up hospitalizations. Zaloshnja et al. computed readmission rates by Barell nature of injury and body part and age group (0–14, 15–29, 30–74, 75+). Readmission rates averaged 4.3 percent but ranged as high as 21 percent (for hip fractures, ages 75+). We assumed that, on average, follow-up admissions have the same costs as initial admissions. (We are forced to make this assumption because the NIS does not allow us to distinguish initial from follow-up admissions with any precision.) We divided the total inpatient cost of each case by (1−r), where r is the readmission rate, to factor up hospital costs for readmissions.

Short- to medium-term follow-up costs. To develop estimates of short- to medium-term costs for injuries requiring an inpatient admission, Finkelstein et al. (2006) multiplied total inpatient costs for each record in HCUP-NIS/Marketscan (as derived above) times the ratio of all costs in the first 18 months after injury, on average, (including costs for inpatient services, ED visits, ambulatory care, prescription drugs, home health care, vision aids, dental visits, and medical devices) to the total inpatient costs (including initial admissions and readmissions) for injury by diagnosis and mechanism of injury. These ratios were derived from 1996–99 MEPS data. MEPS is a nationally representative survey of the civilian non-institutionalized population that quantifies individuals' use of health services and corresponding medical expenditures for two consecutive years following enrollment. Because the MEPS analysis was limited to injuries of admitted patients with at least 12 months of follow-up and the MEPS data include costs for up to 24 months, the MEPS sample captures injuries with an average of 18 months of post-injury treatment.

Although MEPS is the best source of available data for capturing nationally representative injury costs across treatment settings (e.g., hospitals, physician's office, pharmacy), even after pooling four years of data the sample size for many injuries with low incidence rates was small. Therefore, to obtain robust direct cost estimates, injuries were collapsed into broad categories prior to quantifying average costs. Records were collapsed into ICD-9 diagnosis groupings based on the following guidelines (in priority order):

1. Groupings must be comprehensive, covering all injury diagnoses (including those for which MEPS lacks cases).
2. Groupings need to balance the goals of diagnosis-level detail and reasonable cell sizes. In some instances, cell samples as small as 5 were accepted in order to avoid combining radically dissimilar diagnoses into a single group.

3. Groupings should be similar, either in nature of injury (e.g., sprain, fracture) or in body region, if not in both.
4. Total injury costs (or the ratio of total injury costs to hospitalization costs for admitted injuries) should be similar in magnitude across diagnoses within each grouping.

Using the MEPS data grouped according to these criteria, we calculated the average ratio of 18-month costs to total inpatient costs (including inpatient facility and non-facility fees) for 15 injury-specific diagnosis groups, ranging in size from 5 to 61 unweighted cases. The ratios ranged from 1.02 to 2.12, with an overall average of 1.26 (see Supplement, Table A). The ratios were then multiplied times the corresponding inpatient cost estimates detailed in the preceding section to arrive at 18-month costs for injuries requiring an inpatient admission.

Long-term follow-up costs. While short- to medium-term costs capture the majority of costs for most injuries, some injuries continue to require treatment and costs beyond 18 months. Rice et al. (1989) estimated long-term medical costs from costs in the first six months using multipliers derived from longitudinal 1979–88 Detailed Claim Information (DCI) data on 463,174 workers' compensation claims spread across 16 States. The DCI file was unique: nothing similar in size, geographic spread, and duration has become available subsequently. Because occupational injury includes a full spectrum of external causes (e.g., motor vehicle crash, violence, fall), the DCI data by diagnosis presumably captured the medical spending pattern for an injury to a working-age adult reasonably accurately. Their applicability to childhood injuries was questionable. To address this concern, Miller, Romano, and Spicer (Miller, Romano, & Spicer, 2000b) analyzed the 30-month cost patterns (long-term costs were not available) of adult versus child injury using 1987–89 MarketScan data on private health insurance claims. They found that the ratios of 30-month costs to initial hospitalization costs for children's episodes by diagnosis did not differ significantly from the comparable ratios for adults. By diagnosis, the ratios for children ranged from 95 percent to 105 percent of the ratios for adults. Thus, it is reasonable to apply the DCI estimates to childhood injury cases.

Costs beyond 18 months were not inconsequential for some injuries. For lack of a better alternative, following Finkelstein et al. (Finkelstein, et al., 2006), we used ratios computed from the DCI expenditure patterns to adjust estimates of costs in the first 18 months to arrive at estimates of the total medical costs (including long-term) associated with injuries. This method implicitly assumed that while treatment costs varied over time, the ratio of lifetime costs to 18-month costs had remained constant between the time the DCI data was reported and 2009. The 18-month cost estimates from the previous section were multiplied times the ratio of lifetime costs to the costs in months 1–18 by Barell nature of injury (fracture, other) and body region. Although the DCI ratios varied by injury diagnosis, on average, at a 3-percent discount rate, 77 percent of the costs for admitted cases were incurred in months 1–18 (Miller et al., 2000a). The average long-term multiplier for admitted cases was 1.30.

Long-term costs of spinal cord injuries (SCI) and traumatic brain injuries (TBI). These estimates incorporate long-term SCI and TBI costs from Berkowitz et al. (1990). For several types of injuries, and especially for SCI and TBI, a substantial portion of the total medical costs occur more than seven years after the injury is sustained. For severe SCI (i.e., quadriplegia or paraplegia), the ratio of lifetime costs to costs of the initial admission (including emergency transport) was used to factor up the cost of the initial admission. Ratios were computed separately for complete quadriplegia, partial quadriplegia, complete paraplegia, and partial paraplegia, as inferred from the primary injury diagnosis. (This special procedure for severe SCI cases bypasses the medium- and long-term cost methods described in previous sections.) This ratio was generated from data collected by Berkowitz et al. (1990), who surveyed a nationally representative sample of SCI survivors and their families in 1986 and collected data on 758 SCI victims, including those residing in institutions, those living at home, and those in independent living centers. The respondents (victims, families, or guardians) provided details of care payments during the past year, including payments for medical, hospital, prescription, vocational rehabilitation, durable medical equipment, environmental modification, personal assistant, and custodial care. The long-term cost estimates for SCI rely on the assumption that the now-dated Berkowitz data on medical costs by year post-injury mirror the expected lifetime costs for recent SCI victims.

Quantifying long-term costs for TBI is more problematic. Most TBI programs do not have longitudinal data on TBI costs. However, Miller et al. (2004) estimated inpatient rehabilitation costs by diagnosis group, including SCI and TBI, finding that among patients receiving rehabilitation, the cost per case for TBI averaged 75 percent of the cost for SCI. TBI patients, however, were far less likely to receive inpatient rehabilitation (6 percent versus 31 percent). Finkelstein et al. (2006) assumed the TBI patients who received inpatient rehabilitation would follow the same cost pattern more than seven years post- injury as the SCI patients, but with costs equal to 75 percent of SCI levels. Again using the Berkowitz data, we estimated that, at a 3-percent discount rate, 46.92 percent of the medical costs of TBI are incurred in the first seven years. Therefore, we divided the seven-year costs by this percentage to arrive at lifetime medical costs of TBI. As with other long-term costs, we replicated this process for other discount rates to facilitate sensitivity analysis.

For very severe burns, amputations, and other non-SCI, non-TBI injuries requiring lifetime medical care, lack of available data will bias our lifetime cost estimates downwards.

Inpatient rehabilitation costs. Costs of inpatient rehabilitation were estimated using direct costs developed for 11 injury diagnosis groups by Miller et al. (2004). These costs came from the Health Care Financing Administration (HCFA, now the Center for Medicare and Medicaid Services, CMS) Prospective Payment System (PPS) reimbursement schedule that governs payments for all U.S. inpatient rehabilitation including professional fees. Miller et al. (2004) used PPS data on lengths of stay and cost per day to develop direct cost estimates of rehabilitative treatment. They used data from California, Maryland, and Pennsylvania hospital discharge systems to compute the probability of rehabilitation for each PPS diagnosis and mechanism group. The product of the probability of rehabilitation and the direct cost estimate of rehabilitation developed by Miller et al. (2004) were added to the HCUP- NIS/Marketscan-based cost estimates.

Transport costs. None of the data sets and analyses of non-fatal hospitalized injuries described above include transportation costs. We incorporate transportation costs from Finkelstein et al. (2006), who arbitrarily assumed that half of non-fatal injuries requiring a hospital admission also required a one-way trip via ambulance to the hospital. For each injury case, the costs include half of the one-way average emergency transport costs based on 1999 average transport costs for Medicare beneficiaries with an E code on an ambulance claim. There were 15,579 E-coded ambulance claims (including air ambulance) in the Medicare 5 percent sample, with an average cost of $350 in 2010 dollars. After our report was essentially completed, GAO (2012) reported that median cost per transport covered by Medicare was

$429 in 2010. Incorporating that estimate would raise the ambulance cost component of our medical cost estimate from $175 to $214.

The assumed 50 percent transport rate may be conservative. The National Pediatric Trauma Registry, which captures admitted serious injuries, showed that from 4/1/1994 to 11/5/2001, 58.4 percent of 48,288 pediatric patients arrived by ambulance (National Pediatric Trauma Registry, 2002). Claims administration. To estimate the claims processing expenses incurred by private insurers and government payers like Medicare and Medicaid, we drew on insurance overhead rates published by Woolhandler et al. (2003): 11.7 percent for private insurers, 3.6 percent for Medicare, and 6.8 percent for Medicaid. For each case, we applied the rate corresponding to the primary expected payer coded in the hospital record to all medical costs detailed above. When the listed payer was workers' compensation we applied the private insurance rate, and when it was another government program we applied the Medicaid rate. For cases coded as charity or self-pay, we assumed there were no claims processing expenses. And when the payer was missing we applied an average rate by sex and age group.

Injuries Treated in an Emergency Department

Table E-3 summarizes the approach for quantifying costs of non-fatal injuries treated in EDs and released without inpatient admission.

Table E-3. Data and Methods for Estimating Medical Costs of Non-Fatal, Non-Admitted Injuries Treated in Emergency Departments

Cost Category	Description, Unit Cost (2010 U.S. $)	Source/Notes
ED visit	ED facility charges times facility-specific (or hospital stratum average) cost-to-charge ratios times National/State price adjusters times ratio of 2007 NEDS charges to 2003 SEDD charges	2003 HCUP-SEDD charges, cost-to- charge ratios from AHRQ, and National/State price adjusters from ACCRA; 2007 NEDS charges
Follow-up visits and medication, months 1–18	Estimated as the ratio of all costs in the first 18 months after injury to costs of the initial ED visit by diagnosis grouping	1996–99 MEPS
Follow-up costs beyond 18 months	Estimated using ratios of total lifetime costs to 18 month costs for 17 diagnosis groups	1979–88 Detailed Claim Information (DCI) data from workers' compensation claims; adjustment factor for youth from Miller et al. (2000a)
Emergency transport	50 percent of ED visits assumed to have transport costs of $350	Mean cost estimated using 1999 Medicare ambulance claims with an injury E code
Claims administration	Insurance overhead percentages by payer	Woolhandler et al. (2003)

The primary dataset on which we constructed our cost estimates for ED-treated, non-admitted, non- fatal injuries was the 2007 Nationwide Emergency Department Sample (NEDS). However, since AHRQ has not provided cost-to-charge ratios for the NEDS, we could not estimate medical costs directly. Therefore, our estimation of direct medical costs began with the 2003 HCUP State Emergency Department Databases (SEDD). (To date, 2003 is the only year for which AHRQ has computed cost-to- charge ratios applicable to ED data.) We used data from the eight States (of fifteen) in the 2003 SEDD that provided hospital charges (CT, GA, MD, MN, NE, SC, TN, UT). AHRQ provided facility-specific cost-to-charge ratios for EDs in these States. These ratios were multiplied times the ED charges to estimate the facility cost for each case, similar to the methods described above using the HCUP-NIS to estimate hospital-admitted costs. Since these eight States were not representative of the Nation as a whole, we adjusted the costs to National prices using State-specific price adjustors based on the health care component of the ACCRA Cost of Living Index, produced by the Council for Community and Economic Research. Both average charges and average costs were computed by ICD-9-CM diagnosis. We merged these costs and charges onto the 2007 NEDS. The SEDD-based charges were much lower than NEDS charges for comparable diagnoses. The costs also were much lower than average MEPS costs for ED visits for injury. Therefore, we computed average ratios of NEDS charges to SEDD charges for three cause groups (intentional, unintentional motor-vehicle traffic accident, and other unintentional), then used these ratios to adjust the SEDD-based medical costs. This more than doubled the estimated ED medical costs. We then used these cost estimates, along with the ED charges on the NEDS, to estimate cost-to-charge ratios for each facility and sampling stratum in the NEDS. This enabled us to produce a second set of cost estimates as the product of charges and cost-to-charge ratios. We used these secondary cost estimates to fill in costs for cases that could not be assigned a SEDD-based cost by diagnosis (e.g., cases that had unusual diagnoses). We also used these cost-to-charge ratios to estimate the medical costs of the fatalities in the 2007 NEDS, which became a component in the estimation of fatal medical costs (see above).

As with costs for hospitalized injuries, the costs of the initial visit were factored up by 1996–99 MEPS- based ratios for 51 categories of non-admitted injuries treated in the ED to account for follow-up visits and medication in the first 18 months post-injury. The ratios ranged from 1.02 to 5.44, with an overall average of 1.78 (see Supplement, Table B). For follow-up costs beyond 18 months, average costs were estimated using ratios from DCI expenditure patterns and implicitly assuming that the ratio of lifetime costs to 18-month costs had remained constant between the time DCI data was reported and 2007. These long-term costs were calculated by multiplying the DCI ratios of lifetime costs to 18-month costs times the 18-month costs by diagnosis group. At a 3-percent discount rate, 88 percent of the costs for non-admitted cases occurred in months 1–18 and the average multiplier was 1.14 (Miller et al., 2000a). For age groups not represented in the DCI, the costs were adjusted using ratios from Miller et al. (2000b). As with hospital costs, half of patients were assumed to receive emergency transport, so half of the average one-way emergency transport cost was added to the medical cost of each case (see "Transport costs" section above for details). Finally, we added claims administration costs from Woolhandler et al. (2003), as described above under hospitalized medical costs.

Using the costed 2003 SEDD, we computed mean medical costs by diagnosis. We then merged these mean costs onto the 2007 NEDS, a multi-State sample of patients treated in a hospital ED and not subsequently admitted as inpatients. Unlike the SEDD, the NEDS is nationally representative. It could not be costed directly, however, because AHRQ has not made cost-to-charge ratios available for the NEDS.

Lifetime Work Losses Due to Injuries

Injuries can result in both temporary and permanent disability. When this occurs, injury victims may lose part or all of their productivity potential. Work losses due to injury may include lost earnings and accompanying fringe benefits, plus the lost ability to perform one's normal household responsibilities. For non-fatal injuries, work losses represent the value of goods and services not produced because of injury-related illness and disability. To the degree that injuries prevent or deter individuals from producing goods and services in the marketplace, the public sector, or the household, the value of these losses is a cost borne by society.

Fatal work losses represent the value of goods and services never produced because of injury-related premature death. These work loss costs were estimated by applying expected lifetime earnings by age and sex to the all deaths from injury sustained in 2008, including an imputed value for lost household services.

Consistent with the human capital approach for quantifying the burden of injuries (Rice et al., 1989), estimates of non-fatal work losses involve applying average earnings to work-years lost and the value of housekeeping services to time lost in home production. Non-fatal injuries may result in both short-term work loss and in lifetime work losses. The latter includes the value of output lost by people disabled in later years as a result of injury sustained in 2008 (or 2007 for ED-treated, non-fatal injuries).

All work loss estimates were inflated from Finkelstein et al. (2006). Non-fatal work losses were stratified into two categories: short-term losses, which represent lost earnings and accompanying fringe benefits and household services occurring in the first six months after an injury, and long-term losses, which represent the respective earnings and household loss occurring after six months from the time of the injury. The decision to use six months as the transition point between short-term and long-term work losses was driven by the availability of data on duration of work loss.

Because men earn higher earnings than women, even in the same job (Bureau of Labor Statistics, 2001) or for injuries with the same prevalence between men and women, the work loss estimates were greater for men. Finkelstein et al. (2006) view this as more of a shortcoming of the labor market than an inherent problem with the human capital approach. Regardless, this undervaluation of women's labor is reflected in the estimates.

Fatal Injuries

For someone of a given sex and age who sustained a fatal injury, Finkelstein et al. (2006) summed the sex-specific probability of surviving to each subsequent year of age times sex-specific expected earnings for someone of that age. We followed this method using updated data (Arias, 2012). We used this formula with money earnings data by sex and year of age derived from the March Supplement of the Current Population Survey, averaged across a full business cycle from 2002 through 2009. We inflated all earnings figures to 2010 dollars using the Employment Cost Index–Wages & Salaries, All Civilian. We added fringe benefits of 23.33 percent of wages based on the average ratio of wage supplements to wages for 2002–09 from the National income accounts (Economic Report of the President, various years, Table B-28). Earnings, including salary and the value of fringe benefits, at future ages were adjusted upwards to account for a historical 1 percent work growth rate (Haddix et al., 2003) and then discounted to present value using a 3-percent discount rate. (For sensitivity analysis, parallel estimates were constructed using discount rates of 0 percent, 2 percent, 4 percent, 7 percent, and 10 percent).

Parallel calculations valued lost household work. Estimates of the value of household work are also included in Haddix et al. (2003). Historically, productivity growth in household production has been negligible, so Finkelstein et al. (2006) did not adjust for it. In all cases, they assumed that the probability of surviving past the age of 102 is zero. In equation form, lifetime earnings for someone of age a and sex b (Earna,b) is computed as

$$\mathrm{Earn}_{a,b} = \sum_{k=a}^{102} \left\{ P_{a,b}(k) \times Y_{k,b} \times \left(\frac{1+g}{1+d} \right)^{k-a} \right\}$$

where Pa,b(k) = the probability that someone of age a and sex b will live until age k; Yk,b = the average value of annual earnings (including fringe benefits) or of annual household production at age k for someone of sex b; g = the productivity growth rate (0.01 for earnings, 0.00 for household production); and d is the discount rate (usually 0.03, but allowed to vary for sensitivity analysis).

These costing methods were applied to each case in the 2008 NVSS data to produce the fatal work loss costs to be used in our estimates.

Non-Fatal Injuries

For non-fatal injuries, work loss estimates included the sum of the value of wage and household work lost due to short-term disability in the acute recovery phase and of the value of wage and household work lost due to permanent or long-term disability for the subset of injuries that cause lasting impairments that restrict work choices or preclude return to work.

Short-term work losses. Finkelstein et al. (2006) quantified temporary or short-term work loss for non- fatal injuries using the approach presented in Lawrence et al. (2000). Lawrence et al. combined the probability of an injury resulting in lost workdays from 1987–96 National Health Interview Survey (NHIS) data with the mean work days lost (conditional on having missed at least one day) per injury estimated from the 1993 Annual Survey of Occupational Injury and Illness reported by the Bureau of Labor Statistics (BLS). This data was sent to BLS by employers through a mandatory reporting system. Employers reported work loss from date of occupational injury to the end of the calendar year for a sample of approximately 600,000 injury victims. All cases reported involved at least one day of work loss beyond the date of the injury. Moreover, if a worker still was out of work at the time the employer report was due to BLS, the report would undercount work days lost. On average, BLS work-loss reports cover six months post injury. Lawrence et al. (2000) used a Weibull regression model to estimate the total duration of work loss for cases still open at the end of the survey reporting period. These results were combined with those of the closed cases to estimate average work loss, conditional on having missed at least one day of work. These BLS-based estimates were then combined with the pooled 1987–96 NHIS data on probability of work loss to compute mean work loss including cases without work loss. Although BLS uses a detailed two-column coding system (body part, nature of injury), Finkelstein et al. (2006) were able to map their codes to the ICD-9-CM codes.

Although the BLS data is limited to injuries that occur on the job, Finkelstein et al.'s (2006) separate analysis of 1996–99 MEPS data (based on a much smaller sample) found that the duration of work loss did not differ significantly by whether or not the injury occurred on the job. This suggested that the BLS- NHIS work loss estimates could credibly be applied to estimate work loss associated with non-work- related injuries.

Analysis of the MEPS data revealed that work loss was roughly five times longer for hospitalized injuries than for non-hospitalized injuries with work loss. Using this ratio, Finkelstein et al. (2006) decomposed work-loss durations into separate estimates for admitted and non-admitted injuries.

To place a monetary value on temporary wage work loss, the estimated days of work lost were multiplied times average earnings per day of work, given the victim's age and sex, from the Current Population Survey, as described above in the section on fatal injuries.

Household workdays lost were estimated as 90 percent of wage workdays lost, based on findings from an unpublished nationally representative survey on household work losses following injury (S. Marquis, the Rand Corporation, personal communication, 1992). This ratio and the value of household work used in Haddix et al. (2003), were used to impute a value to household work lost. Haddix et al. (2003) valued household production lost using replacement cost. They started with national survey data on the average amount and nature of housework that people do by age group and sex, for example the hours that a woman 30–34 years old spends on cooking and on cleaning. They valued the cost of replacing these hours using BLS data on average wage rate by occupation (e.g., for cooks and maids).

Long-term work losses. Finkelstein et al. (2006) considered permanent total disability and permanent partial disability separately. For permanent total disability, the present value of age-and-sex-specific lifetime earnings and household production from the fatality analysis were multiplied times the probability of permanent total disability for each type of injury. For permanent partial disability, the earnings estimate times the probability of permanent partial disability was multiplied times an additional factor identifying the extent of disability resulting from that type of injury. The total and partial disability costs were then summed to compute the net work loss associated with permanent disability.

The probabilities of permanent total and partial disability by diagnosis and admission status came from Miller, Pindus, et al. (1995) and were based on pooled multi-State workers' compensation data from the 1979–88 Detailed Claims Information (DCI) database of the National Council on Compensation Insurance (NCCI). The disability percentage (i.e., the average extent of disability) by diagnosis came from Lawrence et al. (2000) and was based on 1992–96 DCI data. DCI records the disability status for each sampled case. Following Rice et al. (1989), Finkelstein et al. (2006) assumed that these probabilities do not vary according to whether the injury occurred on the job and that these probabilities have not changed significantly over time. This method also assumes that the probability that an injury (e.g., a skull fracture) will cause someone never to do wage or household work again is the same for children, adults, and the elderly (though the years of work lost obviously will vary with the age of onset) and that people will experience the same percentage reduction in household work ability that they experience in wage work ability.

To verify that the DCI data produce reasonable estimates, Finkelstein et al. (2006) conducted a literature review to compare their estimates to those from other sources. Because of the paucity of data on this subject, they identified only a few sources of published disability estimates, and these were generally dated and limited to specific populations. Based on the limited information available, the DCI data suggested similar probabilities of permanent disability to the other studies of long-term work loss.

Although dated and restricted to occupational injury, the DCI data have several advantages that outweigh their disadvantages. As a result of their large sample, the DCI data can be used to compute probabilities for a far wider range of specific diagnoses than all the disability studies in the literature combined. Despite its restriction to occupational injury, the DCI sample also is more representative of the mix of injuries admitted to hospitals than the few studies in the literature, notably those which are restricted to patients triaged to trauma centers. The DCI data also are virtually the only source of information about permanent disability due to injuries not admitted to the hospital. The sample includes 318,885 medically treated, non-admitted patients with valid lost-work claims in workers' compensation. Averaged across all injuries, the estimated percentage of lifetime productivity potential lost due to permanent injury-related disability was 0.26 percent per injury.

For hospital-admitted cases of traumatic brain injury (TBI), we computed modified disability probabilities using a logistic regression model developed by Selassie et al. (2008). The model took account of the severity of TBI (as per the Barell matrix, which distinguishes three types of TBI), the presence of comorbid conditions, whether the patient was transferred from the acute hospital to another medical facility, and the patient's age and sex. This new disability probability was then decomposed into separate probabilities of total and partial disability according to the total/partial ratio of the old disability probabilities. In cases where the TBI diagnosis was a secondary diagnosis, the new probability was kept only if it exceeded the old probability based on the non-TBI primary injury diagnosis.

Calculating total work loss costs. The work loss costs were computed as described for all non-fatal acute injury cases in the 2008 NIS and the 2007 NEDS. Short- and long-term costs were summed to compute total work loss costs.

Limitations of Methods for Medical and Work Loss Estimates

These cost estimates are subject to several limitations. First, the estimates focus exclusively on medical costs and work loss costs. They do not account for non-health costs (e.g., criminal justice, educational impacts, property damage etc.), pain and suffering, quality-of-life loss, or injury costs borne by family and caretakers. Also excluded are costs due to psychological treatment, e.g., for PTSD.

Second, a major limitation was the requirement to use data from a multitude of sources. Although these were the best available data at the time of the analysis, some sources are old, others are based on non- representative samples, and all are subject to reporting and measurement error. These factors may have incorporated significant bias into the cost estimates. The costing approach was designed to minimize the potential bias. More current and nationally representative data would have been preferable but were not available.

Third, combining factors from multiple data sets (sometimes with only a published mean estimates available) and unavoidable assumptions about data sets being representative make it impossible to generate standard errors around the cost estimates.

The methods for estimating work loss costs had many additional limitations. Because women, the elderly, and children have lower average earnings, the human capital approach applied undervalued injuries to these groups. The approach also placed lower values on the work of full-time homemakers than the work of people participating in the labor market, which further depressed the value placed on women's losses relative to men's losses. It also undervalued disability among those of retirement age, and did not value temporary disability among children, as they had not yet entered the labor force.

Discounting future work losses to present value meant that the loss of a lifetime of work by a 2-year-old was considered equivalent to loss of a lifetime of work by a 43-year-old. Although the child loses many more years of work, those years are far in the future and heavily discounted. The work loss cost calculations are also based on a year 2008 life table, which essentially assumes that life expectancy would have remained constant over each person's expected lifespan absent injury. Moreover, victims of serious and fatal injury may tend to be risk-takers (for example, thrill-seekers, heavy drinkers, or drug abusers) whose life expectancy may be shorter than for the average population, which would further bias the results. And, as noted above, some of the estimates are computed using fairly dated data that are based on a working population. Additionally, the estimates exclude the ability to work lost by people other than the injured person. These losses may include the time family, friends, and professionals spend caring for the injured, time spent investigating the injury, and worker retraining. All these limitations suggest that the costs and especially the available standard error information should be interpreted with caution.

SUPPLEMENT TO APPENDIX E

Short-Term Follow-Up Cost Factors

A. Multipliers for Short-Term Follow-Up Costs for Hospital-Admitted Patients

The 16 diagnosis groups and associated multipliers that were used for estimating short-term follow up costs for admitted patients were as follows:

Group No.	ICD-9-CM Diagnosis Codes	Average ratio of all costs in the first 18 months after injury to total inpatient costs
1	802, 830	1.02
2	800, 801,803, 804, 850–854	1.38
3	806, 952	2.12
4	805, 807–809, 839	1.10
5	810–819, 831–834	1.26
6	820, 835	1.35
7	821–829, 836–838	1.43
8	840–848	1.67
9	860–869	1.12
10	870–904	1.12

Group No.	ICD-9-CM Diagnosis Codes	Average ratio of all costs in the first 18 months after injury to total inpatient costs
11	910–929	1.24
12	930–939, 950–958, 990–995	1.97
13	940–949	1.13
14	959	1.16
15	960–989	1.02
16	Other	1.03
	All	1.26

B. Multipliers for Short-Term Follow-Up Costs for ED-Treated Patients

The 51 diagnosis groups and associated multipliers that were used for estimating short-term follow-up costs for injuries treated in emergency departments and released were as follows:

Group No.	ICD-9-CM Diagnosis Codes	Average ratio of all costs in the first 18 months after injury to total ED visit costs	Group No.	ICD-9-CM Diagnosis Codes	Average ratio of all costs in the first 18 months after injury to total ED visit costs
1	802, 830	2.47	27	851–854	1.38
2	800, 801, 803, 804	1.19	28	860–869	1.04
3	805–809	1.40	29	870–874	1.15
4	810–811	3.40	30	875–879	1.09
5	812	3.95	31	880–881	1.82
6	813	1.43	32	882	1.28
7	814	2.83	33	883	1.28
8	815–817	1.75	34	884–887	1.45
9	818–819	1.77	35	890–891, 894–897	1.35
10	820–822	2.01	36	892–893	1.18
11	823	2.31	37	900–904	2.73
12	824	2.19	38	910–919	1.29
13	825	1.77	39	920	1.02
14	826	1.69	40	921	1.33
15	827–829	1.38	41	922	1.32
16	831	2.44	42	923	1.28
17	832–833	3.96	43	924	1.49
18	834	1.36	44	925–929	1.53
19	835–839	1.27	45	930–934	1.11
20	840	5.44	46	935–939	1.74
21	841–842	1.22	47	940–949	1.93
22	843–844	2.25	48	950–958, 990–995	1.11
23	845	1.34	49	959	2.00
24	846–847	1.83	50	960–988	1.11
25	848	1.62	51	989	1.12
26	850	1.16		All	1.78

APPENDIX F. UNIT COSTS AND STANDARD ERRORS AT DIFFERENT DISCOUNT RATES

Table F-1. 2008 crash costs per fatal victim at different discount rates (2010 dollars)

Cost component	Discount Rate					
	3%	0%	2%	4%	7%	10%
Medical	11,317	11,317	11,317	11,317	11,317	11,317
Wage loss	933,262	1,647,638	1,107,209	799,270	543,031	403,145
Household production	289,910	544,672	348,275	246,559	167,221	125,389

Table F-2. 2007–2008 HCUP-based medical unit costs at different discount rates, AIS-90 (2010 dollars)

Body part	Fracture/ Disclocation	Mais-90	Discount rate 3%		Discount rate 0%		Discount rate 2%		Discount rate 4%		Discount rate 7%		Discount rate 10%	
			Mean	Std error	Mean	Std error	Mean	Std error	Mean	Std error	Mean	Std error	Mean	Std error
Skull	Yes	1	17,557	3,108	17,557	3,108	17,557	3,108	17,557	3,108	17,557	3,108	17,557	3,108
		2	15,131	522	15,131	522	15,131	522	15,131	522	15,131	522	15,131	522
		3	27,738	1,070	27,738	1,070	27,738	1,070	27,738	1,070	27,738	1,070	27,738	1,070
		4	185,255	5,620	194,557	5,911	187,728	5,698	183,220	5,557	179,144	5,429	176,863	5,358
		5	500,067	24,286	525,747	25,546	506,894	24,621	494,450	24,010	483,197	23,458	476,902	23,149
		6	130,200	95,046	136,888	99,990	131,978	96,361	128,737	93,965	125,806	91,799	124,167	90,586
Brain/intracarnial	No	1	3,769	30	3,769	30	3,769	30	3,769	30	3,769	30	3,769	30
		2	10,361	85	10,361	85	10,361	85	10,361	85	10,361	85	10,361	85
		3	41,167	1,674	41,167	1,674	41,167	1,674	41,167	1,674	41,167	1,674	41,167	1,674
		4	126,226	4,312	132,595	4,535	127,920	4,371	124,833	4,263	122,042	4,165	120,481	4,111
		5	290,840	18,116	305,725	19,053	294,797	18,365	287,584	17,911	281,063	17,500	277,414	17,270
		6	359,826	157,897	378,186	166,069	364,707	160,070	355,810	156,110	347,765	152,529	343,264	150,526
Ear	No	2	23,279	2,403	23,279	2,403	23,279	2,403	23,279	2,403	23,279	2,403	23,279	2,403
Eye & adnexa	No	1	3,791	118	3,791	118	3,791	118	3,791	118	3,791	118	3,791	118
		2	13,122	1,006	13,122	1,006	13,122	1,006	13,122	1,006	13,122	1,006	13,122	1,006

Body part	Fracture/ Disclocation	Mais-90	Discount rate 3%		Discount rate 0%		Discount rate 2%		Discount rate 4%		Discount rate 7%		Discount rate 10%	
			Mean	Std error	Mean	Std error	Mean	Std error	Mean	Std error	Mean	Std error	Mean	Std error
		3	14,119	1,328	14,119	1,328	14,119	1,328	14,119	1,328	14,119	1,328	14,119	1,328
		4	28,018	10,619	28,018	10,619	28,018	10,619	28,018	10,619	28,018	10,619	28,018	10,619
Nouse/ mouth /face/scalp/ neck	No	1	3,093	19	3,093	19	3,093	19	3,093	19	3,093	19	3,093	19
		2	6,152	109	6,152	109	6,152	109	6,152	109	6,152	109	6,152	109
		3	22,859	2,359	22,859	2,359	22,859	2,359	22,859	2,359	22,859	2,359	22,859	2,359
		4	28,360	10,749	28,360	10,749	28,360	10,749	28,360	10,749	28,360	10,749	28,360	10,749
		5	870,848	89,886	1,108,653	114,431	936,915	96,705	814,820	84,103	690,795	71,301	610,278	62,991
	Yes	1	6,396	161	6,396	161	6,396	161	6,396	161	6,396	161	6,396	161
		2	30,480	2,123	30,480	2,123	30,480	2,123	30,480	2,123	30,480	2,123	30,480	2,123
		3	110,600	13,946	110,600	13,946	110,600	13,946	110,600	13,946	110,600	13,946	110,600	13,946
		4	63,124	11,786	63,151	11,786	63,131	11,786	63,118	11,786	63,106	11,786	63,099	11,786
		5	590,372	65,603	717,665	81,997	624,915	70,059	561,592	61,907	499,591	54,096	460,586	49,405
Neck/internal organs/blood vessels	No	2	17,992	1,502	17,992	1,502	17,992	1,502	17,992	1,502	17,992	1,502	17,992	1,502
		3	62,784	18,237	62,784	18,237	62,784	18,237	62,784	18,237	62,784	18,237	62,784	18,237
		4	46,205	6,619	46,205	6,619	46,205	6,619	46,205	6,619	46,205	6,619	46,205	6,619
		5	62,487	5,548	62,487	5,548	62,487	5,548	62,487	5,548	62,487	5,548	62,487	5,548
Neck-spinal cord	No	3	82,125	4,752	82,125	4,575	82,125	4,702	82,125	4,794	82,125	4,899	82,125	4,984
		4	567,898	13,272	722,431	13,238	610,833	13,261	531,486	13,282	450,867	13,306	398,505	13,324
		5	974,011	56,353	1,244,924	69,353	1,049,277	60,081	910,183	53,133	768,890	45,865	677,164	41,096
		6	1,268,322	124,792	1,540,708	124,122	1,343,997	124,565	1,204,147	125,019	1,062,086	125,652	969,861	126,196
Shoulder/ clavicule/scapula/ upper arm	No	1	1,760	16	1,760	16	1,760	16	1,760	16	1,760	16	1,760	16
		2	3,557	93	3,557	93	3,557	93	3,557	93	3,557	93	3,557	93
		3	38,464	7,098	38,464	7,098	38,464	7,098	38,464	7,098	38,464	7,098	38,464	7,098
		4	32,693	2,419	32,693	2,419	32,693	2,419	32,693	2,419	32,693	2,419	32,693	2,419
	Yes	2	8,043	148	8,043	148	8,043	148	8,043	148	8,043	148	8,043	148
		3	41,269	1,799	41,269	1,799	41,269	1,799	41,269	1,799	41,269	1,799	41,269	1,799
		4	69,365	7,146	69,365	7,146	69,365	7,146	69,365	7,146	69,365	7,146	69,365	7,146
		5	161,624	25,214	161,624	25,214	161,624	25,214	161,624	25,214	161,624	25,214	161,624	25,214

Table F-2. (Continued)

Body part	Fracture/ Disclocation	Mais-90	Discount rate 3%		Discount rate 0%		Discount rate 2%		Discount rate 4%		Discount rate 7%		Discount rate 10%	
			Mean	Std error	Mean	Std error	Mean	Std error	Mean	Std error	Mean	Std error	Mean	Std error
Elbow	No	1	2,386	35	2,386	35	2,386	35	2,386	35	2,386	35	2,386	35
		2	6,120	293	6,120	293	6,120	293	6,120	293	6,120	293	6,120	293
		3	14,854	1,563	14,854	1,563	14,854	1,563	14,854	1,563	14,854	1,563	14,854	1,563
		4	71,260	11,117	71,260	11,117	71,260	11,117	71,260	11,117	71,260	11,117	71,260	11,117
	Yes	2	8,388	153	8,388	153	8,388	153	8,388	153	8,388	153	8,388	153
		3	45,606	2,012	45,606	2,012	45,606	2,012	45,606	2,012	45,606	2,012	45,606	2,012
		4	77,812	9,067	77,812	9,067	77,812	9,067	77,812	9,067	77,812	9,067	77,812	9,067
		5	248,360	41,925	248,360	41,925	248,360	41,925	248,360	41,925	248,360	41,925	248,360	41,925
Forearm	Yes	2	8,718	385	8,718	385	8,718	385	8,718	385	8,718	385	8,718	385
		3	30,985	3,610	30,985	3,610	30,985	3,610	30,985	3,610	30,985	3,610	30,985	3,610
Wrist/hand/ finger/thumb	No	1	1,835	20	1,835	20	1,835	20	1,835	20	1,835	20	1,835	20
		2	4,546	188	4,546	188	4,546	188	4,546	188	4,546	188	4,546	188
		3	14,604	2,462	14,604	2,462	14,604	2,462	14,604	2,462	14,604	2,462	14,604	2,462
		4	15,808	3,243	15,808	3,243	15,808	3,243	15,808	3,243	15,808	3,243	15,808	3,243
	Yes	1	2,774	71	2,774	71	2,774	71	2,774	71	2,774	71	2,774	71
		2	4,534	139	4,534	139	4,534	139	4,534	139	4,534	139	4,534	139
		3	41,136	4,339	41,136	4,339	41,136	4,339	41,136	4,339	41,136	4,339	41,136	4,339
		4	48,408	14,080	48,408	14,080	48,408	14,080	48,408	14,080	48,408	14,080	48,408	14,080
		5	145,076	42,197	145,076	42,197	145,076	42,197	145,076	42,197	145,076	42,197	145,076	42,197
Upper ext/mult/unsp	No	1	2,356	52	2,356	52	2,356	52	2,356	52	2,356	52	2,356	52
		2	16,060	1,487	16,060	1,487	16,060	1,487	16,060	1,487	16,060	1,487	16,060	1,487
		3	49,245	6,718	49,245	6,718	49,245	6,718	49,245	6,718	49,245	6,718	49,245	6,718
		4	60,781	6,089	60,781	6,089	60,781	6,089	60,781	6,089	60,781	6,089	60,781	6,089
		5	65,605	6,572	65,605	6,572	65,605	6,572	65,605	6,572	65,605	6,572	65,605	6,572
	Yes	2	2,392	442	2,392	442	2,392	442	2,392	442	2,392	442	2,392	442
		3	5,855	587	5,855	587	5,855	587	5,855	587	5,855	587	5,855	587
Chest/abdom	No	2	11,943	818	11,943	818	11,943	818	11,943	818	11,943	818	11,943	818
	No	1	1,863	5	1,863	5	1,863	5	1,863	5	1,863	5	1,863	5

Body part	Fracture/ Disclocation	Mais-90	Discount rate 3%		Discount rate 0%		Discount rate 2%		Discount rate 4%		Discount rate 7%		Discount rate 10%	
			Mean	Std error	Mean	Std error	Mean	Std error	Mean	Std error	Mean	Std error	Mean	Std error
		2	4,816	101	4,816	101	4,816	101	4,816	101	4,816	101	4,816	101
		3	15,511	3,459	15,511	3,459	15,511	3,459	15,511	3,459	15,511	3,459	15,511	3,459
		4	45,940	15,456	45,940	15,456	45,940	15,456	45,940	15,456	45,940	15,456	45,940	15,456
Ribs/sternum	Yes	1	5,867	634	5,867	634	5,867	634	5,867	634	5,867	634	5,867	634
		2	18,889	368	18,889	368	18,889	368	18,889	368	18,889	368	18,889	368
		3	54,028	3,828	54,028	3,828	54,028	3,828	54,028	3,828	54,028	3,828	54,028	3,828
		4	73,809	5,685	73,809	5,685	73,809	5,685	73,809	5,685	73,809	5,685	73,809	5,685
		5	100,766	13,896	100,766	13,896	100,766	13,896	100,766	13,896	100,766	13,896	100,766	13,896
		6	38,437	5,301	38,437	5,301	38,437	5,301	38,437	5,301	38,437	5,301	38,437	5,301
Back (including vertevrae)	No	2	7,280	1,014	7,280	1,014	7,280	1,014	7,280	1,014	7,280	1,014	7,280	1,014
		3	98,527	13,723	98,527	13,723	98,527	13,723	98,527	13,723	98,527	13,723	98,527	13,723
	Yes	2	17,213	488	17,213	488	17,213	488	17,213	488	17,213	488	17,213	488
		3	81,724	4,875	81,724	4,875	81,724	4,875	81,724	4,875	81,724	4,875	81,724	4,875
Trunk -spinal cord	No	2	44,626	3,489	44,626	3,489	44,626	3,489	44,626	3,489	44,626	3,489	44,626	3,489
		3	233,486	7,786	233,486	7,786	233,486	7,786	233,486	7,786	233,486	7,786	233,486	7,786
		4	396,651	31,014	396,651	31,014	396,651	31,014	396,651	31,014	396,651	31,014	396,651	31,014
		5	467,470	15,589	550,116	16,107	488,810	15,695	450,386	15,518	415,937	15,415	396,032	15,380
		6	449,466	14,989	449,466	13,160	449,466	14,432	449,466	15,486	449,466	16,658	449,466	17,455
Trunk, superf	No	2	33,962	8,767	33,962	8,767	33,962	8,767	33,962	8,767	33,962	8,767	33,962	8,767
Trunk,multiple/ unspecified	No	1	2,486	16	2,486	16	2,486	16	2,486	16	2,486	16	2,486	16
		2	6,446	375	6,446	375	6,446	375	6,446	375	6,446	375	6,446	375
		3	21,337	617	21,337	617	21,337	617	21,337	617	21,337	617	21,337	617
		4	53,426	2,962	53,426	2,962	53,426	2,962	53,426	2,962	53,426	2,962	53,426	2,962
		5	234,742	56,754	254,800	76,653	240,315	62,272	230,016	52,083	219,555	41,788	212,764	35,157
		6	120,650	42,512	120,650	42,512	120,650	42,512	120,650	42,512	120,650	42,512	120,650	42,512
	Yes	1	5,503	183	5,503	183	5,503	183	5,503	183	5,503	183	5,503	183
		2	10,412	236	10,412	236	10,412	236	10,412	236	10,412	236	10,412	236
		3	20,390	698	20,390	698	20,390	698	20,390	698	20,390	698	20,390	698
		4	29,041	1,603	29,041	1,603	29,041	1,603	29,041	1,603	29,041	1,603	29,041	1,603
		5	502,176	26,500	584,957	26,873	523,551	26,551	485,064	26,482	450,559	26,507	430,622	26,560

Table F-2. (Continued)

Body part	Fracture/ Disclocation	Mais-90	Discount rate 3%		Discount rate 0%		Discount rate 2%		Discount rate 4%		Discount rate 7%		Discount rate 10%	
			Mean	Std error	Mean	Std error	Mean	Std error	Mean	Std error	Mean	Std error	Mean	Std error
Thoracic orgs/blood vessels	No	3	32,133	3,311	32,133	3,311	32,133	3,311	32,133	3,311	32,133	3,311	32,133	3,311
		4	60,689	9,489	60,689	9,489	60,689	9,489	60,689	9,489	60,689	9,489	60,689	9,489
		5	133,644	12,811	133,644	12,811	133,644	12,811	133,644	12,811	133,644	12,811	133,644	12,811
Liver	No	1	4,861	69	4,861	69	4,861	69	4,861	69	4,861	69	4,861	69
		2	22,584	804	22,584	804	22,584	804	22,584	804	22,584	804	22,584	804
		3	45,240	3,164	45,240	3,164	45,240	3,164	45,240	3,164	45,240	3,164	45,240	3,164
		4	65,039	4,277	65,039	4,277	65,039	4,277	65,039	4,277	65,039	4,277	65,039	4,277
		5	69,500	5,589	69,500	5,589	69,500	5,589	69,500	5,589	69,500	5,589	69,500	5,589
		6	9,285	1,163	9,285	1,163	9,285	1,163	9,285	1,163	9,285	1,163	9,285	1,163
Spleen	No	4	49,795	6,237	49,795	6,237	49,795	6,237	49,795	6,237	49,795	6,237	49,795	6,237
Kidney	No	3	31,714	10,282	31,714	10,282	31,714	10,282	31,714	10,282	31,714	10,282	31,714	10,282
		4	96,447	41,104	96,447	41,104	96,447	41,104	96,447	41,104	96,447	41,104	96,447	41,104
		5	146,557	16,555	146,557	16,555	146,557	16,555	146,557	16,555	146,557	16,555	146,557	16,555
Gastrointestinal	No	3	31,202	7,745	31,202	7,745	31,202	7,745	31,202	7,745	31,202	7,745	31,202	7,745
		4	93,034	20,050	93,034	20,050	93,034	20,050	93,034	20,050	93,034	20,050	93,034	20,050
		5	139,846	28,023	139,846	28,023	139,846	28,023	139,846	28,023	139,846	28,023	139,846	28,023
Genitourinary	No	2	17,219	342	17,219	342	17,219	342	17,219	342	17,219	342	17,219	342
		3	89,941	16,104	89,941	16,104	89,941	16,104	89,941	16,104	89,941	16,104	89,941	16,104
		4	20,960	3,753	20,960	3,753	20,960	3,753	20,960	3,753	20,960	3,753	20,960	3,753
Hip/thigh	No	1	1,981	27	1,981	27	1,981	27	1,981	27	1,981	27	1,981	27
		2	10,224	472	10,224	472	10,224	472	10,224	472	10,224	472	10,224	472
		3	47,429	7,327	47,429	7,327	47,429	7,327	47,429	7,327	47,429	7,327	47,429	7,327
		4	42,372	8,387	42,372	8,387	42,372	8,387	42,372	8,387	42,372	8,387	42,372	8,387
	Yes	2	30,508	1,203	30,508	1,203	30,508	1,203	30,508	1,203	30,508	1,203	30,508	1,203
		3	57,240	995	57,240	995	57,240	995	57,240	995	57,240	995	57,240	995
		4	116,872	5,916	116,872	5,916	116,872	5,916	116,872	5,916	116,872	5,916	116,872	5,916
		5	183,024	21,611	183,024	21,611	183,024	21,611	183,024	21,611	183,024	21,611	183,024	21,611

Body part	Fracture/ Dislocation	Mais-90	Discount rate 3%		Discount rate 0%		Discount rate 2%		Discount rate 4%		Discount rate 7%		Discount rate 10%	
			Mean	Std error	Mean	Std error	Mean	Std error	Mean	Std error	Mean	Std error	Mean	Std error
L. extremity superfic	No	1	3,261	77	3,261	77	3,261	77	3,261	77	3,261	77	3,261	77
		2	2,721	89	2,721	89	2,721	89	2,721	89	2,721	89	2,721	89
Knee	No	1	1,723	21	1,723	21	1,723	21	1,723	21	1,723	21	1,723	21
		2	8,328	656	8,328	656	8,328	656	8,328	656	8,328	656	8,328	656
		3	27,520	3,936	27,520	3,936	27,520	3,936	27,520	3,936	27,520	3,936	27,520	3,936
		4	138,409	28,931	138,409	28,931	138,409	28,931	138,409	28,931	138,409	28,931	138,409	28,931
	Yes	2	17,682	670	17,682	670	17,682	670	17,682	670	17,682	670	17,682	670
		3	47,232	2,741	47,232	2,741	47,232	2,741	47,232	2,741	47,232	2,741	47,232	2,741
		4	62,279	21,499	62,279	21,499	62,279	21,499	62,279	21,499	62,279	21,499	62,279	21,499
Lower leg	No	1	1,895	28	1,895	28	1,895	28	1,895	28	1,895	28	1,895	28
		2	5,267	499	5,267	499	5,267	499	5,267	499	5,267	499	5,267	499
		3	125,735	12,922	125,735	12,922	125,735	12,922	125,735	12,922	125,735	12,922	125,735	12,922
	Yes	2	25,602	324	25,602	324	25,602	324	25,602	324	25,602	324	25,602	324
		3	78,487	2,200	78,487	2,200	78,487	2,200	78,487	2,200	78,487	2,200	78,487	2,200
Ankle/foot/toes	No	1	1,951	55	1,951	55	1,951	55	1,951	55	1,951	55	1,951	55
		2	8,722	1,020	8,722	1,020	8,722	1,020	8,722	1,020	8,722	1,020	8,722	1,020
		3	60,320	12,200	60,320	12,200	60,320	12,200	60,320	12,200	60,320	12,200	60,320	12,200
	Yes	1	2,823	118	2,823	118	2,823	118	2,823	118	2,823	118	2,823	118
		2	11,513	273	11,513	273	11,513	273	11,513	273	11,513	273	11,513	273
		3	72,590	5,507	72,590	5,507	72,590	5,507	72,590	5,507	72,590	5,507	72,590	5,507
Burns	No	1	2,443	410	2,443	410	2,443	410	2,443	410	2,443	410	2,443	410
		2	30,112	4,834	30,112	4,834	30,112	4,834	30,112	4,834	30,112	4,834	30,112	4,834
		3	113,565	17,726	113,565	17,726	113,565	17,726	113,565	17,726	113,565	17,726	113,565	17,726
		4	132,183	13,668	132,183	13,668	132,183	13,668	132,183	13,668	132,183	13,668	132,183	13,668
		5	183,076	28,197	183,076	28,197	183,076	28,197	183,076	28,197	183,076	28,197	183,076	28,197
Min. extern.	No	1	3,896	35	3,896	35	3,896	35	3,896	35	3,896	35	3,896	35

Table F-3. 2007–2008 HCUP-based unit earnings loss at different discount rates, AIS-90 (2010 dollars)

Body part	Fracture/ Disclocation	Mais-90	Discount rate 3%		Discount rate 0%		Discount rate 2%		Discount rate 4%		Discount rate 7%		Discount rate 10%	
			Mean	Std error	Mean	Std error	Mean	Std error	Mean	Std error	Mean	Std error	Mean	Std error
Skull	Yes	1	97,345	38,645	200,339	96,195	121,282	51,375	79,628	29,643	48,146	15,105	32,836	9,289
		2	17,786	1,021	31,954	2,233	21,089	1,283	15,316	837	10,813	532	8,506	394
		3	53,509	2,284	90,843	4,438	62,466	2,756	46,676	1,942	33,779	1,340	26,838	1,034
		4	209,400	4,293	375,076	8,500	249,070	5,224	179,223	3,623	122,673	2,461	92,630	1,887
		5	325,245	12,132	575,467	25,039	386,324	15,003	278,209	10,077	188,637	6,613	140,328	5,008
		6	57,840	31,855	71,673	41,648	61,700	34,605	54,540	29,495	47,066	24,130	41,992	20,488
Brain/ intracarnial	No	1	3,188	76	5,236	146	3,672	92	2,822	65	2,145	45	1,792	35
		2	13,473	262	22,638	472	15,703	311	11,756	225	8,479	157	6,694	121
		3	97,773	3,231	163,556	5,966	113,989	3,868	85,157	2,756	60,605	1,884	46,855	1,427
		4	177,952	4,269	306,474	8,139	209,410	5,161	153,629	3,610	106,835	2,424	81,067	1,820
		5	338,368	12,401	616,596	25,158	405,448	15,267	287,141	10,333	190,738	6,779	139,420	5,073
		6	341,664	119,750	677,432	233,879	421,426	147,219	281,661	98,911	172,230	60,483	117,124	40,924
Ear	No	2	116,088	9,484	159,239	13,643	127,915	10,520	106,140	8,667	84,298	7,007	70,158	5,970
Eye & adnexa	No	1	4,300	232	7,942	452	5,155	281	3,659	196	2,488	134	1,887	102
		2	22,239	3,232	41,810	5,935	26,809	3,903	18,792	2,747	12,433	1,892	9,129	1,451
		3	43,817	9,687	67,389	17,790	49,912	11,669	38,910	8,169	28,853	5,356	22,861	3,939
		4	215,722	48,952	273,833	71,013	232,395	55,197	201,139	43,590	166,847	31,597	142,376	23,897
Nouse/ mouth /face/scalp/ neck	No	1	2,433	24	4,272	47	2,861	29	2,114	20	1,536	14	1,243	12
		2	7,705	251	13,550	510	9,087	301	6,662	218	4,736	170	3,732	149
		3	56,339	4,603	92,712	7,943	65,299	5,370	49,366	4,031	35,779	2,974	28,143	2,395
		4	142,854	32,417	261,812	67,896	171,573	40,751	120,961	26,214	80,058	15,161	58,652	9,844
		5	34,763	2,840	43,279	3,708	37,108	3,052	32,774	2,676	28,293	2,352	25,225	2,146

Body part	Fracture/ Disclocation	Mais-90	Discount rate 3%		Discount rate 0%		Discount rate 2%		Discount rate 4%		Discount rate 7%		Discount rate 10%	
			Mean	Std error	Mean	Std error	Mean	Std error	Mean	Std error	Mean	Std error	Mean	Std error
	Yes	1	4,724	110	8,027	190	5,518	127	4,119	98	2,989	76	2,394	65
		2	28,610	972	49,796	1,696	33,670	1,113	24,773	878	17,643	724	13,916	643
		3	127,260	8,138	207,609	14,185	147,414	9,576	111,403	7,048	80,092	5,003	62,326	3,898
		4	98,612	9,017	156,641	16,395	112,969	10,682	87,423	7,817	65,626	5,765	53,434	4,780
		5	574,870	84,753	1,016,362	135,417	683,277	97,764	491,051	74,319	330,469	53,070	243,227	40,550
Neck/internal organs/blood vessels	No	2	52,182	30,241	92,316	64,095	61,696	38,173	45,039	24,347	32,056	13,848	25,551	8,772
		3	195,295	4,227	331,753	5,190	229,578	4,533	168,366	3,943	115,635	3,225	86,368	2,683
		4	46,830	8,362	59,913	7,531	50,140	8,222	44,198	8,405	38,869	8,211	35,696	7,837
		5	93,461	20,077	120,722	24,839	100,934	21,488	87,141	18,812	73,041	15,741	63,604	13,501
Neck-spinal cord	No	3	189,020	18,862	317,732	34,075	221,625	22,559	163,264	16,040	112,413	10,788	83,924	8,063
		4	243,933	11,712	402,160	23,024	282,764	14,238	213,438	9,891	153,323	6,793	119,187	5,333
		5	21,946	2,190	23,807	2,553	22,532	2,294	21,391	2,102	19,889	1,909	18,597	1,787
		6	607,621	42,314	889,183	104,084	683,014	56,247	545,339	32,793	412,578	20,588	329,727	18,610
Shoulder/ clavicule/ scapula/ upper arm	No	1	2,442	18	3,138	24	2,612	19	2,310	17	2,058	16	1,920	15
		2	7,058	101	9,544	130	7,669	106	6,585	99	5,676	96	5,178	94
		3	23,730	2,907	41,484	8,315	27,691	4,068	20,715	2,083	15,095	875	12,161	759
		4	58,944	4,716	78,852	6,641	63,993	5,201	54,937	4,339	46,892	3,622	42,205	3,250
	Yes	2	13,834	142	20,782	223	15,516	157	12,543	131	10,089	116	8,759	108
		3	58,579	1,545	89,472	2,626	66,032	1,763	52,855	1,402	41,959	1,180	36,046	1,080
		4	95,756	7,705	143,957	13,867	107,703	9,131	86,428	6,660	68,195	4,852	57,948	4,010
		5	103,825	35,525	195,490	55,775	125,257	40,897	87,858	31,185	59,046	22,447	44,588	17,514
Elbow	No	1	1,458	24	2,146	39	1,623	27	1,333	22	1,097	18	973	17
		2	4,351	546	6,561	867	4,892	625	3,933	484	3,130	367	2,692	303
		3	21,910	1,815	33,565	2,462	24,657	1,942	19,842	1,729	16,018	1,589	14,009	1,515
		4	49,322	16,876	69,984	19,967	54,639	17,840	45,041	15,987	36,209	13,765	30,864	12,123

Table F-3. (Continued)

Body part	Fracture/ Disclocation	Mais-90	Discount rate 3%		Discount rate 0%		Discount rate 2%		Discount rate 4%		Discount rate 7%		Discount rate 10%	
			Mean	Std error	Mean	Std error	Mean	Std error	Mean	Std error	Mean	Std error	Mean	Std error
	Yes	2	18,998	205	29,503	352	21,542	235	17,046	185	13,337	154	11,332	141
		3	78,022	1,434	116,196	2,444	87,389	1,634	70,746	1,305	56,650	1,121	48,831	1,043
		4	114,724	7,327	176,567	14,320	129,949	8,842	102,901	6,284	80,001	4,676	67,276	4,024
		5	237,473	60,484	407,510	96,852	279,393	69,942	205,008	52,857	142,836	37,274	109,295	28,128
Forearm	Yes	2	22,142	407	26,749	563	23,362	437	21,141	390	19,013	376	17,669	377
		3	43,140	2,755	46,295	3,755	44,067	2,998	42,310	2,584	40,283	2,355	38,755	2,318
Wrist/hand/ finger/thumb	No	1	898	17	1,218	29	975	19	840	15	730	11	673	10
		2	10,890	677	19,115	1,381	12,833	833	9,426	565	6,721	374	5,309	283
		3	29,668	3,622	51,520	6,765	34,909	4,365	25,684	3,066	18,247	2,065	14,340	1,570
		4	111,126	44,025	246,648	84,584	141,120	53,924	89,644	36,432	53,265	22,159	36,585	14,690
	Yes	1	6,533	126	10,817	227	7,557	147	5,754	112	4,299	90	3,529	79
		2	11,128	166	17,617	289	12,707	192	9,915	148	7,611	120	6,370	108
		3	61,199	4,663	90,599	6,976	68,437	5,133	55,579	4,344	44,695	3,814	38,655	3,541
		4	168,932	16,488	273,870	31,220	195,102	19,716	148,478	14,200	108,666	10,343	86,648	8,492
		5	157,151	15,338	323,780	36,910	196,347	19,842	127,930	12,235	75,566	7,192	49,947	4,895
Upper ext/mult/unsp	No	1	1,019	40	1,444	73	1,120	47	941	35	796	27	720	24
		2	17,401	1,348	29,448	2,620	20,329	1,628	15,153	1,152	10,897	834	8,618	695
		3	93,576	11,850	153,891	22,813	108,478	14,402	82,000	9,966	59,664	6,655	47,423	5,079
		4	193,324	6,531	344,165	7,248	230,506	6,857	164,540	6,188	109,485	5,235	79,862	4,500
		5	221,656	7,489	295,399	6,221	241,982	7,198	204,475	7,690	166,447	7,958	141,548	7,977
	Yes	2	7,728	225	13,054	376	9,006	250	6,757	211	4,952	198	4,012	192
		3	10,921	369	18,971	400	12,886	383	9,412	354	6,574	314	5,086	287

Body part	Fracture/ Disclocation	Mais-90	Discount rate 3%		Discount rate 0%		Discount rate 2%		Discount rate 4%		Discount rate 7%		Discount rate 10%	
			Mean	Std error	Mean	Std error	Mean	Std error	Mean	Std error	Mean	Std error	Mean	Std error
Chest/abdom	No	2	26,140	7,879	42,665	13,626	30,172	9,302	23,043	6,781	17,194	4,720	14,097	3,668
	No	1	3,090	6	3,251	6	3,129	6	3,059	6	2,999	6	2,967	6
		2	8,003	173	9,809	242	8,460	188	7,639	163	6,908	145	6,481	135
		3	32,408	9,408	43,019	13,473	35,216	10,500	30,105	8,502	25,239	6,557	22,225	5,336
		4	53,978	2,325	99,726	9,979	63,632	3,539	47,274	1,910	36,363	2,257	31,499	2,603
Ribs/sternum	Yes	1	5,850	645	8,120	1,009	6,429	725	5,389	589	4,457	496	3,911	451
		2	28,681	410	41,805	655	31,935	465	26,132	371	21,129	302	18,303	268
		3	79,504	4,227	127,170	7,422	90,963	4,957	70,631	3,691	53,562	2,748	44,210	2,286
		4	67,634	4,371	98,992	7,412	75,526	5,073	61,400	3,852	48,976	2,917	41,804	2,431
		5	142,177	22,127	210,618	18,946	159,685	21,835	128,176	22,015	99,719	20,558	82,864	18,669
		6	116,643	18,153	172,428	15,511	130,819	17,888	105,404	18,104	83,036	17,119	70,320	15,843
Back (including vertevrae)	No	2	73,821	3,815	101,155	5,595	81,121	4,264	67,805	3,461	55,050	2,757	47,169	2,357
		3	67,039	3,465	106,389	5,884	77,083	4,052	59,039	3,013	42,955	2,151	33,648	1,682
	Yes	2	31,519	529	46,526	875	35,244	605	28,604	475	22,888	385	19,665	342
		3	111,125	4,144	182,127	7,628	128,866	4,951	97,217	3,543	69,960	2,448	54,702	1,878
Trunk -spinal cord	No	2	162,123	10,663	291,258	24,876	194,028	13,817	137,409	8,448	90,174	4,941	64,862	3,541
		3	227,262	6,329	521,933	17,879	292,169	8,715	181,082	4,731	104,074	2,414	69,760	1,682
		4	251,902	16,568	430,923	36,804	296,027	21,081	217,652	13,381	151,498	8,301	115,029	6,279
		5	432,892	12,056	877,666	30,065	537,820	16,043	354,424	9,260	212,872	4,938	142,815	3,444
		6	321,557	8,955	756,238	25,905	419,725	12,520	250,649	6,549	130,288	3,022	76,128	1,836
Trunk, superf	No	2	82,662	13,641	122,215	11,691	92,713	12,962	74,666	14,067	58,556	14,220	49,137	13,593
Trunk,multiple/ unspecified	No	1	1,649	11	1,963	14	1,725	12	1,591	10	1,479	10	1,419	9
		2	4,589	332	6,215	526	4,990	377	4,278	298	3,672	235	3,333	202
		3	21,291	631	30,318	1,095	23,477	737	19,602	553	16,357	413	14,575	342
		4	42,648	2,419	61,994	4,332	47,403	2,864	38,915	2,086	31,546	1,474	27,341	1,157

Table F-3. (Continued)

Body part	Fracture/ Disclocation	Mais-90	Discount rate 3%		Discount rate 0%		Discount rate 2%		Discount rate 4%		Discount rate 7%		Discount rate 10%	
			Mean	Std error	Mean	Std error	Mean	Std error	Mean	Std error	Mean	Std error	Mean	Std error
		5	86,074	28,105	152,712	58,571	102,165	35,372	73,816	22,631	50,976	12,670	39,086	7,716
		6	46,495	5,151	76,477	5,898	53,695	5,360	41,019	4,965	30,786	4,502	25,379	4,142
	Yes	1	4,991	113	6,443	149	5,361	121	4,694	106	4,088	94	3,726	88
		2	10,896	297	13,585	407	11,587	324	10,335	276	9,164	235	8,444	212
		3	23,602	576	29,731	817	25,143	632	22,358	533	19,783	453	18,214	408
		4	34,353	1,485	43,495	2,329	36,693	1,681	32,457	1,340	28,517	1,075	26,108	935
		5	431,420	14,020	863,587	31,685	533,760	17,792	354,659	11,473	215,465	7,717	146,019	6,272
Thoracic orgs/blood vessels	No	3	16,845	2,364	23,026	3,490	18,379	2,649	15,645	2,139	13,298	1,694	11,975	1,442
		4	54,319	10,431	78,675	18,213	60,461	12,355	49,469	8,937	39,836	6,072	34,325	4,533
		5	81,042	16,133	128,197	29,869	92,842	19,393	71,759	13,685	53,385	9,236	42,883	6,987
Liver	No	1	1,661	36	2,483	65	1,852	42	1,518	33	1,259	27	1,129	25
		2	17,965	593	26,844	1,190	20,076	718	16,366	507	13,394	372	11,834	315
		3	29,013	1,784	46,186	3,446	33,096	2,162	25,915	1,508	20,147	1,028	17,117	801
		4	45,549	3,299	73,972	6,413	52,433	4,021	40,277	2,767	30,316	1,832	24,985	1,384
		5	48,765	4,890	78,853	9,258	56,055	5,894	43,175	4,149	32,579	2,821	26,879	2,151
		6	13,014	2,325	17,546	4,213	14,181	2,795	12,082	1,962	10,204	1,292	9,125	983
Spleen	No	4	79,660	14,230	132,689	31,857	92,581	18,249	69,725	11,321	50,853	6,441	40,692	4,384
Kidney	No	3	29,935	16,139	48,726	31,041	34,534	19,747	26,394	13,392	19,679	8,310	16,097	5,726
		4	49,312	1,963	66,338	3,266	53,850	2,283	45,565	1,716	37,552	1,247	32,503	998
		5	102,948	9,065	169,007	17,479	119,320	11,097	90,171	7,519	65,229	4,663	51,241	3,207
Gastrointestinal	No	3	17,251	4,054	26,500	6,692	19,469	4,624	15,562	3,652	12,392	2,963	10,697	2,605
		4	89,605	14,671	152,451	28,722	104,937	17,897	77,815	12,307	55,457	8,149	43,510	6,122
		5	103,629	14,200	180,963	32,001	122,390	18,441	89,292	11,017	62,458	5,282	48,424	2,500

Body part	Fracture/ Disclocation	Mais-90	Discount rate 3%		Discount rate 0%		Discount rate 2%		Discount rate 4%		Discount rate 7%		Discount rate 10%	
			Mean	Std error	Mean	Std error	Mean	Std error	Mean	Std error	Mean	Std error	Mean	Std error
Genitourinary	No	2	32,710	9,555	53,125	18,341	37,674	11,701	28,909	7,909	21,766	4,814	18,016	3,195
		3	63,142	20,261	106,086	37,503	73,701	24,338	54,967	17,208	39,260	11,667	30,680	8,854
		4	15,297	4,909	17,628	6,232	15,978	5,276	14,693	4,600	13,233	3,932	12,151	3,507
Hip/thigh	No	1	1,920	24	2,239	31	1,997	25	1,861	23	1,748	22	1,688	21
		2	13,424	702	20,733	1,451	15,170	858	12,101	599	9,652	461	8,383	419
		3	50,585	7,665	75,009	13,185	56,687	8,995	45,803	6,655	36,442	4,788	31,207	3,835
		4	64,894	9,685	97,113	12,770	72,801	10,573	58,730	8,905	46,599	7,085	39,619	5,841
	Yes	2	54,009	1,136	90,138	1,932	62,511	1,307	47,605	1,014	35,774	797	29,590	685
		3	48,929	568	77,327	984	55,756	659	43,691	503	33,755	394	28,403	341
		4	88,675	3,725	143,095	6,646	101,923	4,387	78,476	3,242	58,994	2,387	48,382	1,956
		5	73,227	5,564	122,647	10,995	85,297	6,730	63,935	4,767	46,276	3,554	36,802	3,072
L. extremity superfic	No	1	1,743	26	2,565	43	1,939	30	1,594	24	1,318	21	1,173	19
		2	3,610	72	4,410	120	3,803	81	3,462	65	3,183	56	3,032	52
Knee	No	1	619	11	681	15	634	12	606	10	582	8	568	8
		2	8,760	350	11,620	456	9,472	372	8,202	335	7,109	308	6,496	293
		3	28,451	1,997	38,297	3,024	30,970	2,263	26,439	1,784	22,361	1,347	19,968	1,091
		4	218,075	18,159	284,303	22,237	235,213	19,192	204,259	17,323	175,860	15,541	158,861	14,385
	Yes	2	33,402	918	49,579	1,565	37,409	1,061	30,263	817	24,103	649	20,634	572
		3	71,422	2,136	111,038	3,658	81,089	2,464	63,933	1,908	49,518	1,538	41,607	1,368
		4	153,068	34,991	244,550	53,681	175,793	39,869	135,329	31,033	100,773	22,817	81,546	17,856
Lower leg	No	1	809	11	966	17	846	12	781	10	729	9	702	8
		2	7,445	431	10,655	717	8,235	493	6,829	388	5,631	312	4,958	273
		3	156,504	10,450	211,573	15,005	170,339	11,329	145,593	9,899	123,915	9,147	111,459	8,862
	Yes	2	37,441	264	54,358	424	41,591	295	34,224	244	27,996	214	24,533	200
		3	84,777	1,265	124,856	2,152	94,582	1,446	77,183	1,144	62,534	958	54,442	876

Table F-3. (Continued)

Body part	Fracture/ Disclocation	Mais-90	Discount rate 3%		Discount rate 0%		Discount rate 2%		Discount rate 4%		Discount rate 7%		Discount rate 10%	
			Mean	Std error	Mean	Std error	Mean	Std error	Mean	Std error	Mean	Std error	Mean	Std error
Ankle/foot/toes	No	1	1,941	38	2,981	66	2,188	44	1,753	34	1,405	27	1,223	25
		2	11,620	643	19,075	1,392	13,364	777	10,313	563	7,916	453	6,667	407
		3	68,551	3,725	104,443	10,992	77,463	4,975	61,594	3,266	48,024	3,642	40,444	4,096
	Yes	1	3,344	84	4,699	138	3,672	94	3,092	78	2,617	71	2,361	68
		2	20,469	267	29,616	421	22,726	298	18,715	247	15,316	216	13,433	202
		3	76,643	4,479	111,477	8,032	85,328	5,264	69,843	3,932	56,520	3,071	49,033	2,717
Burns	No		3,731	129	6,335	255	4,355	157	3,257	109	2,373	75	1,909	59
			19,238	1,637	32,605	3,119	22,450	1,975	16,790	1,389	12,198	960	9,764	758
			44,219	11,341	72,833	21,586	51,249	13,818	38,780	9,457	28,352	5,977	22,683	4,211
			56,755	15,978	104,900	34,466	68,220	20,276	48,117	12,806	32,334	7,246	24,373	4,667
			64,495	32,346	103,945	48,833	74,254	36,521	56,877	29,011	41,966	22,181	33,552	18,048
Min. extern.	No	1	1,221	17	1,658	27	1,326	19	1,140	16	989	13	908	12

Table F-4. 2007–2008 HCUP-based unit household production loss at different discount rates, AIS-90 (2010 dollars)

Body part	Fracture/ Disclocation	Mais-90	Discount rate 3%		Discount rate 0%		Discount rate 2%		Discount rate 4%		Discount rate 7%		Discount rate 10%	
			Mean	Std error	Mean	Std error	Mean	Std error	Mean	Std error	Mean	Std error	Mean	Std error
Skull	Yes	1	34,072	7,272	66,788	18,386	41,280	9,558	28,862	5,709	19,699	3,215	15,074	2,131
		2	5,490	265	10,014	571	6,492	328	4,764	221	3,480	148	2,832	114
		3	16,256	604	28,449	1,195	19,029	729	14,206	515	10,475	360	8,518	282

Body part	Fracture/ Disclocation	Mais-90	Discount rate 3%		Discount rate 0%		Discount rate 2%		Discount rate 4%		Discount rate 7%		Discount rate 10%	
			Mean	Std error	Mean	Std error	Mean	Std error	Mean	Std error	Mean	Std error	Mean	Std error
		4	59,667	1,061	113,060	2,252	71,675	1,307	50,865	890	35,084	603	26,984	463
		5	92,362	2,906	177,033	6,554	111,536	3,653	78,271	2,395	52,987	1,564	40,040	1,181
		6	18,393	2,321	24,871	5,685	20,079	3,148	17,028	1,686	14,199	509	12,479	77
Brain/ intracarnial	No	1	1,167	26	1,904	54	1,332	32	1,047	22	832	15	723	12
		2	4,785	82	8,386	156	5,612	98	4,170	70	3,040	48	2,440	37
		3	36,487	919	63,531	1,795	42,834	1,110	31,687	780	22,649	533	17,704	406
		4	60,579	1,021	107,809	2,130	71,558	1,257	52,333	854	36,947	568	28,609	427
		5	100,722	3,141	196,551	6,950	122,321	3,937	84,884	2,586	56,518	1,658	41,998	1,217
		6	113,006	18,125	231,442	40,629	139,167	23,090	94,131	14,564	61,205	8,485	44,909	5,638
Ear	No	2	24,340	1,722	39,151	2,805	28,016	1,984	21,476	1,519	15,925	1,124	12,859	902
Eye & adnexa	No	1	1,445	75	2,738	150	1,729	90	1,239	64	878	46	698	36
		2	6,462	553	12,829	1,578	7,857	733	5,455	447	3,689	318	2,802	273
		3	17,537	1,539	29,244	4,411	20,354	2,108	15,375	1,184	11,216	761	8,886	658
		4	64,603	11,356	97,528	19,741	73,307	13,429	57,487	9,756	42,581	6,712	33,495	5,066
Nouse/ mouth /face/scalp/ neck	No	1	846	7	1,479	14	985	8	746	6	571	4	485	4
		2	2,514	74	4,597	147	2,977	88	2,177	64	1,581	48	1,280	41
		3	20,299	1,436	35,576	2,548	23,863	1,690	17,620	1,246	12,637	892	9,965	699
		4	51,950	9,132	99,998	20,241	62,749	11,495	44,050	7,476	29,947	4,720	22,751	3,441
		5	73,780	5,220	86,702	6,211	77,572	5,494	70,384	4,978	62,003	4,375	55,567	3,896
	Yes	1	1,537	23	2,687	45	1,794	27	1,350	20	1,018	16	851	14
		2	8,834	323	15,834	525	10,399	362	7,693	295	5,668	246	4,642	219
		3	44,418	2,119	78,204	3,866	52,310	2,512	38,490	1,829	27,493	1,300	21,617	1,019
		4	34,231	3,195	57,812	5,991	39,687	3,811	30,157	2,748	22,665	1,960	18,710	1,562
		5	136,109	15,193	258,677	27,858	164,159	18,125	115,353	13,005	77,735	8,998	58,253	6,901
Neck/internal organs/blood vessels	No	2	17,971	4,148	27,494	10,997	20,109	5,610	16,406	3,127	13,605	1,468	12,162	789
		3	107,474	3,726	199,668	6,109	128,597	4,324	91,866	3,258	63,700	2,342	49,178	1,830
		4	12,547	2,741	16,772	3,457	13,576	2,962	11,749	2,543	10,192	2,079	9,309	1,765
		5	26,648	4,787	38,778	7,359	29,753	5,463	24,166	4,239	19,132	3,113	16,172	2,450

Table F-4. (Continued)

Body part	Fracture/ Disclocation	Mais-90	Discount rate 3%		Discount rate 0%		Discount rate 2%		Discount rate 4%		Discount rate 7%		Discount rate 10%	
			Mean	Std error	Mean	Std error	Mean	Std error	Mean	Std error	Mean	Std error	Mean	Std error
Neck-spinal cord	No	3	116,012	8,566	217,236	17,343	139,100	10,441	98,994	7,239	68,362	4,965	52,591	3,843
		4	77,444	2,756	135,703	5,794	91,122	3,380	67,121	2,327	47,798	1,617	37,348	1,273
		5	80,519	5,945	92,955	7,421	84,288	6,327	77,069	5,636	68,309	4,961	61,397	4,487
		6	191,069	15,907	294,462	29,379	217,119	18,869	170,415	13,797	128,750	10,184	104,009	8,376
Shoulder/ clavicule/ scapula/upper arm	No	1	1,124	6	1,440	8	1,196	7	1,071	6	974	6	924	6
		2	2,263	26	3,055	36	2,445	28	2,128	25	1,883	23	1,755	23
		3	8,417	515	14,021	1,685	9,676	753	7,495	361	5,850	198	5,009	215
		4	24,245	2,209	32,679	1,648	26,319	2,141	22,623	2,218	19,408	2,089	17,530	1,901
	Yes	2	5,060	36	7,713	59	5,665	39	4,613	33	3,800	31	3,374	29
		3	20,323	403	31,383	692	22,854	458	18,446	367	15,024	311	13,231	285
		4	34,237	1,817	51,828	3,068	38,396	2,077	31,081	1,636	25,118	1,330	21,851	1,178
		5	35,495	7,214	61,474	11,867	41,349	8,334	31,200	6,363	23,484	4,767	19,502	3,918
Elbow	No	1	599	8	870	13	660	9	555	7	476	6	435	6
		2	1,622	139	2,463	225	1,813	158	1,482	126	1,229	104	1,100	93
		3	7,193	629	11,016	857	8,060	681	6,553	589	5,392	513	4,788	472
		4	18,417	3,743	27,669	5,341	20,668	4,166	16,678	3,401	13,312	2,702	11,422	2,295
	Yes	2	7,491	46	11,629	81	8,439	52	6,789	43	5,507	39	4,834	37
		3	28,722	391	43,216	633	32,102	433	26,186	364	21,476	322	18,954	301
		4	40,555	2,453	64,339	4,892	46,079	2,939	36,412	2,127	28,718	1,610	24,593	1,373
		5	47,086	11,232	86,844	20,901	56,078	13,512	40,510	9,519	28,872	6,367	23,062	4,724
Forearm	Yes	2	15,279	208	19,602	287	16,406	221	14,364	200	12,462	187	11,305	179
		3	34,916	2,112	43,681	3,322	37,246	2,375	33,003	1,928	28,969	1,624	26,489	1,479
Wrist/hand/ finger/thumb	No	1	386	4	501	8	412	5	367	4	333	3	316	3
		2	3,389	173	6,084	369	3,990	215	2,951	144	2,173	96	1,778	73
		3	11,288	899	19,858	1,733	13,217	1,074	9,874	778	7,335	573	6,028	474
		4	31,745	7,216	82,550	16,279	41,772	9,239	25,036	5,752	14,621	3,199	10,233	1,947

Body part	Fracture/ Disclocation	Mais-90	Discount rate 3%		Discount rate 0%		Discount rate 2%		Discount rate 4%		Discount rate 7%		Discount rate 10%	
			Mean	Std error	Mean	Std error	Mean	Std error	Mean	Std error	Mean	Std error	Mean	Std error
	Yes	1	2,390	37	4,076	75	2,767	44	2,115	32	1,626	25	1,377	22
		2	3,918	36	6,189	59	4,437	40	3,534	34	2,837	32	2,475	30
		3	22,685	787	34,938	1,803	25,504	957	20,583	693	16,714	591	14,659	564
		4	45,417	2,804	75,179	5,262	52,374	3,253	40,199	2,526	30,555	2,128	25,445	1,962
		5	94,551	5,838	205,729	14,400	118,445	7,356	77,617	4,877	48,825	3,401	35,018	2,700
Upper ext/mult/unsp	No	1	426	11	576	19	460	13	402	10	357	8	335	7
		2	5,849	440	9,987	646	6,805	489	5,139	400	3,841	317	3,160	266
		3	26,338	2,940	45,334	5,640	30,700	3,527	23,103	2,521	17,228	1,804	14,176	1,457
		4	45,014	9,514	86,388	16,316	54,383	11,101	38,155	8,327	25,985	6,146	19,877	4,995
		5	47,389	10,016	73,681	13,916	53,956	11,014	42,252	9,221	32,227	7,622	26,647	6,697
	Yes	2	3,031	75	5,126	173	3,498	91	2,693	65	2,096	57	1,797	55
		3	2,237	473	4,040	763	2,640	539	1,945	425	1,436	340	1,186	298
Chest/abdom	No	2	5,650	1,397	9,505	2,715	6,519	1,693	5,016	1,182	3,892	808	3,327	625
	No	1	1,604	2	1,685	2	1,622	2	1,590	2	1,565	2	1,552	2
		2	3,535	50	4,426	74	3,746	55	3,375	46	3,074	40	2,910	38
		3	9,650	1,719	13,489	2,918	10,584	2,013	8,931	1,494	7,554	1,071	6,800	850
		4	15,604	352	27,259	1,438	18,056	463	13,881	345	10,967	398	9,566	414
Ribs/sternum	Yes	1	2,298	95	3,268	139	2,531	98	2,119	94	1,777	96	1,588	96
		2	11,955	109	17,400	188	13,233	125	10,990	99	9,182	81	8,201	73
		3	30,157	1,196	48,384	2,278	34,341	1,426	27,050	1,035	21,372	767	18,391	644
		4	25,896	1,338	38,477	2,347	28,914	1,552	23,579	1,186	19,112	920	16,596	781
		5	48,660	1,202	80,280	6,257	56,168	2,116	42,955	654	32,188	1	26,343	134
		6	24,297	600	38,428	2,995	27,579	1,039	21,854	333	17,417	1	15,135	77
Back (including vertevrae)	No	2	15,163	484	23,383	831	17,155	556	13,638	433	10,754	342	9,204	294
		3	26,462	845	46,962	1,669	31,229	1,013	22,900	727	16,360	520	12,929	412
	Yes	2	13,103	116	19,693	221	14,642	135	11,945	104	9,787	87	8,625	81
		3	50,118	1,152	88,891	2,314	59,052	1,408	43,482	967	31,381	642	25,051	480

Table F-4. (Continued)

Body part	Fracture/ Disclocation	Mais-90	Discount rate 3%		Discount rate 0%		Discount rate 2%		Discount rate 4%		Discount rate 7%		Discount rate 10%	
			Mean	Std error	Mean	Std error	Mean	Std error	Mean	Std error	Mean	Std error	Mean	Std error
Trunk -spinal cord	No	2	30,140	1,937	59,851	4,879	36,845	2,527	25,250	1,540	16,624	923	12,334	659
		3	85,755	2,558	212,580	7,179	111,312	3,393	68,380	2,031	40,583	1,261	28,271	943
		4	92,171	5,923	172,726	14,080	110,569	7,585	78,555	4,792	53,828	2,988	40,961	2,190
		5	96,040	2,865	203,851	6,884	119,617	3,646	79,174	2,352	50,207	1,561	36,227	1,209
		6	61,814	1,844	148,153	5,003	79,895	2,435	49,279	1,464	28,808	895	19,599	654
Trunk, superf	No	2	21,334	3,432	35,487	3,884	24,631	3,638	18,862	3,220	14,296	2,662	11,882	2,270
Trunk,multiple/ unspecified	No	1	867	4	1,010	5	900	4	842	4	798	3	774	3
		2	2,035	97	2,754	189	2,201	116	1,911	83	1,682	61	1,560	50
		3	8,127	193	11,520	347	8,911	227	7,542	168	6,462	124	5,885	103
		4	14,355	658	21,522	1,237	16,051	787	13,066	564	10,624	395	9,282	307
		5	45,448	7,940	80,209	18,511	53,328	10,113	39,624	6,482	28,967	4,292	23,260	3,432
		6	19,149	5,706	29,544	8,640	21,619	6,492	17,265	5,057	13,670	3,673	11,676	2,813
	Yes	1	2,278	19	2,966	30	2,445	21	2,150	18	1,901	17	1,760	16
		2	5,209	70	6,529	111	5,536	79	4,953	63	4,443	52	4,146	47
		3	10,594	139	13,502	216	11,311	156	10,032	127	8,919	106	8,273	94
		4	13,684	483	18,217	844	14,791	562	12,825	425	11,143	321	10,184	268
		5	94,320	1,192	197,873	3,383	117,062	1,459	77,997	1,102	49,807	1,051	36,090	991
Thoracic orgs/blood vessels	No	3	8,332	2,204	11,587	3,044	9,131	2,434	7,707	2,011	6,467	1,587	5,743	1,312
		4	16,133	2,011	25,738	3,757	18,361	2,409	14,465	1,718	11,387	1,192	9,753	924
		5	30,530	5,139	52,728	10,448	35,661	6,305	26,700	4,299	19,641	2,838	15,886	2,116
Liver	No	1	807	12	1,162	19	886	13	751	12	651	11	601	10
		2	6,713	178	9,963	326	7,442	209	6,180	156	5,224	118	4,734	100
		3	10,302	508	16,333	952	11,659	605	9,308	438	7,525	316	6,610	256
		4	17,061	1,018	27,978	1,962	19,565	1,225	15,200	869	11,783	608	9,971	475
		5	17,371	1,621	29,231	3,275	20,057	1,977	15,392	1,366	11,814	925	9,955	707
		6	9,410	1,371	13,122	2,109	10,266	1,539	8,774	1,246	7,622	1,017	7,024	897

Body part	Fracture/ Disclocation	Mais-90	Discount rate 3%		Discount rate 0%		Discount rate 2%		Discount rate 4%		Discount rate 7%		Discount rate 10%	
			Mean	Std error	Mean	Std error	Mean	Std error	Mean	Std error	Mean	Std error	Mean	Std error
Spleen	No	4	33,898	4,937	57,481	9,240	39,345	5,898	29,828	4,235	22,309	2,977	18,287	2,335
Kidney	No	3	17,227	8,795	30,393	18,901	20,193	10,998	15,052	7,220	11,146	4,512	9,130	3,193
		4	21,524	231	33,633	44	24,465	171	19,265	274	14,964	345	12,626	376
		5	38,842	5,399	69,768	11,359	45,894	6,704	33,620	4,463	24,099	2,841	19,085	2,045
Gastrointestinal	No	3	7,955	2,645	12,080	4,068	8,927	3,011	7,219	2,351	5,833	1,754	5,081	1,406
		4	31,432	4,742	57,950	9,625	37,393	5,803	27,076	3,982	19,332	2,661	15,411	2,008
		5	57,306	6,836	111,553	18,379	69,312	9,269	48,612	5,153	33,296	2,444	25,565	1,279
Genitourinary	No	2	8,168	851	13,707	2,683	9,397	1,254	7,279	564	5,725	76	4,956	153
		3	19,782	6,429	37,357	13,406	23,724	7,954	16,903	5,335	11,776	3,435	9,174	2,499
		4	12,540	4,075	17,759	6,373	13,908	4,663	11,430	3,607	9,131	2,664	7,751	2,111
Hip/thigh	No	1	933	9	1,071	10	964	9	909	9	867	9	844	8
		2	4,636	180	7,017	350	5,168	212	4,249	159	3,565	128	3,221	117
		3	13,943	1,521	21,514	2,824	15,695	1,809	12,637	1,314	10,245	956	8,988	785
		4	25,510	3,473	42,010	6,287	29,286	4,124	22,702	2,984	17,545	2,082	14,817	1,609
	Yes	2	18,121	316	30,736	553	20,932	363	16,073	284	12,428	230	10,572	203
		3	19,053	171	29,915	317	21,527	201	17,224	151	13,896	118	12,154	103
		4	31,459	1,152	52,751	2,218	36,316	1,379	27,864	992	21,317	719	17,888	586
		5	30,839	3,684	51,987	7,142	35,656	4,402	27,280	3,186	20,825	2,357	17,457	1,956
L. extremity superfic	No	1	708	8	1,018	13	777	9	657	7	565	7	518	6
		2	1,492	21	1,750	31	1,551	23	1,449	20	1,370	18	1,329	17
Knee	No	1	362	3	396	5	370	3	355	3	344	2	337	2
		2	3,270	99	4,429	150	3,538	109	3,069	92	2,700	80	2,505	74
		3	10,817	805	15,851	1,127	11,995	881	9,931	745	8,287	623	7,411	549
		4	61,817	2,632	86,165	3,401	67,671	2,793	57,341	2,519	48,851	2,334	44,230	2,250
	Yes	2	13,056	200	19,239	348	14,511	228	11,958	182	9,899	155	8,784	143
		3	29,155	657	46,559	1,269	33,144	779	26,195	574	20,791	445	17,952	388
		4	41,954	5,837	68,159	10,144	48,026	6,810	37,427	5,137	29,145	3,955	24,817	3,421

Table F-4. (Continued)

Body part	Fracture/ Disclocation	Mais-90	Discount rate 3%		Discount rate 0%		Discount rate 2%		Discount rate 4%		Discount rate 7%		Discount rate 10%	
			Mean	Std error	Mean	Std error	Mean	Std error	Mean	Std error	Mean	Std error	Mean	Std error
Lower leg	No	1	392	3	456	5	406	4	381	3	362	3	352	3
		2	2,788	96	4,081	152	3,089	106	2,562	89	2,142	76	1,916	69
		3	45,435	2,343	65,186	3,000	50,107	2,455	41,896	2,269	35,254	2,142	31,672	2,071
	Yes	2	13,860	68	20,549	110	15,412	75	12,698	63	10,548	57	9,400	54
		3	29,071	337	43,834	591	32,481	385	26,529	306	21,859	257	19,393	236
Ankle/foot/toes	No	1	792	12	1,207	22	885	14	724	11	603	9	542	8
		2	3,874	223	6,593	498	4,476	276	3,438	188	2,666	133	2,276	109
		3	18,667	2,002	30,510	4,172	21,395	2,434	16,641	1,715	12,944	1,276	11,017	1,087
	Yes	1	1,360	24	1,897	30	1,482	25	1,270	24	1,108	23	1,024	23
		2	8,126	74	11,938	121	9,006	83	7,470	69	6,268	62	5,635	59
		3	28,338	1,518	42,986	2,991	31,735	1,824	25,798	1,308	21,114	972	18,629	825
Burns	No	1	1,385	40	2,437	82	1,620	49	1,214	34	912	24	759	19
		2	5,798	306	10,556	648	6,850	374	5,036	261	3,701	189	3,036	157
		3	15,622	7,054	29,118	14,723	18,630	8,740	13,434	5,840	9,563	3,730	7,610	2,687
		4	18,679	7,065	34,751	14,692	22,233	8,742	16,111	5,859	11,621	3,768	9,384	2,741
		5	26,909	11,399	48,260	22,720	31,741	13,908	23,347	9,581	16,894	6,373	13,520	4,759
Min. extern.	No	1	532	6	710	10	572	7	502	6	449	5	422	5

Table F-5. 2007–2008 HCUP-based medical unit costs at different discount rates, AIS-85 (2010 dollars)

Body part	Fracture/ Disclocation	Mais-90	Discount rate 3%		Discount rate 0%		Discount rate 2%		Discount rate 4%		Discount rate 7%		Discount rate 10%	
			Mean	Std error	Mean	Std error	Mean	Std error	Mean	Std error	Mean	Std error	Mean	Std error
Skull	Yes	1	7,605	1,316	7,605	1,316	7,605	1,316	7,605	1,316	7,605	1,316	7,605	1,316
		2	13,503	407	13,503	407	13,503	407	13,503	407	13,503	407	13,503	407
		3	103,193	3,244	107,713	3,414	104,394	3,289	102,204	3,207	100,223	3,133	99,115	3,091
		4	205,821	14,757	215,686	15,534	208,444	14,963	203,664	14,587	199,341	14,246	196,923	14,056
		5	506,014	24,888	531,822	26,183	512,875	25,232	500,369	24,604	489,061	24,037	482,734	23,719
Brain/ intracarnial	No	1	4,166	124	4,166	124	4,166	124	4,166	124	4,166	124	4,166	124
		2	7,050	55	7,050	55	7,050	55	7,050	55	7,050	55	7,050	55
		3	68,727	2,302	70,970	2,408	69,323	2,330	68,237	2,278	67,254	2,232	66,704	2,206
		4	122,964	5,972	129,037	6,281	124,579	6,054	121,636	5,905	118,975	5,770	117,486	5,694
		5	401,223	24,885	421,770	26,175	406,685	25,228	396,728	24,603	387,725	24,038	382,688	23,721
Ear	No	2	24,249	3,440	24,249	3,440	24,249	3,440	24,249	3,440	24,249	3,440	24,249	3,440
Eye & adnexa	No	1	3,346	107	3,346	107	3,346	107	3,346	107	3,346	107	3,346	107
		2	10,013	530	10,013	530	10,013	530	10,013	530	10,013	530	10,013	530
		3	15,825	2,245	15,825	2,245	15,825	2,245	15,825	2,245	15,825	2,245	15,825	2,245
		4	38,486	3,416	38,486	3,416	38,486	3,416	38,486	3,416	38,486	3,416	38,486	3,416
Nouse/ mouth /face/scalp/ neck	No	1	3,086	20	3,086	20	3,086	20	3,086	20	3,086	20	3,086	20
		2	5,293	69	5,293	69	5,293	69	5,293	69	5,293	69	5,293	69
		3	16,742	1,486	16,742	1,486	16,742	1,486	16,742	1,486	16,742	1,486	16,742	1,486
		4	39,413	10,561	39,413	10,561	39,413	10,561	39,413	10,561	39,413	10,561	39,413	10,561
	Yes	1	5,999	135	5,999	135	5,999	135	5,999	135	5,999	135	5,999	135
		2	18,210	1,112	18,210	1,112	18,210	1,112	18,210	1,112	18,210	1,112	18,210	1,112
		3	77,981	8,062	78,571	8,251	78,145	8,111	77,842	8,024	77,535	7,945	77,335	7,900
		4	136,312	34,426	136,312	34,426	136,312	34,426	136,312	34,426	136,312	34,426	136,312	34,426
		5	603,497	62,882	733,494	78,435	638,742	67,084	574,152	59,418	511,004	52,181	471,330	47,921

Table F-5. (Continued)

Body part	Fracture/ Disclocation	Mais-90	Discount rate 3%		Discount rate 0%		Discount rate 2%		Discount rate 4%		Discount rate 7%		Discount rate 10%	
			Mean	Std error	Mean	Std error	Mean	Std error	Mean	Std error	Mean	Std error	Mean	Std error
Neck/internal organs/blood vessels	No	1	2,521	202	2,521	202	2,521	202	2,521	202	2,521	202	2,521	202
		2	13,895	1,114	13,895	1,114	13,895	1,114	13,895	1,114	13,895	1,114	13,895	1,114
		3	29,313	5,969	29,313	5,969	29,313	5,969	29,313	5,969	29,313	5,969	29,313	5,969
		4	218,034	44,400	218,034	44,400	218,034	44,400	218,034	44,400	218,034	44,400	218,034	44,400
Neck-spinal cord	No	3	279,201	240,638	333,263	294,699	294,222	255,658	266,463	227,900	238,259	199,696	219,941	181,378
		4	465,098	14,822	617,562	15,934	507,458	15,109	429,173	14,592	349,633	14,135	297,972	13,877
		5	1,021,847	36,986	1,293,000	37,225	1,097,179	37,052	957,962	36,930	816,545	36,805	724,737	36,724
Shoulder/ clavicule/ scapula/ upper arm	No	1	1,772	16	1,772	16	1,772	16	1,772	16	1,772	16	1,772	16
		2	3,688	84	3,688	84	3,688	84	3,688	84	3,688	84	3,688	84
		3	28,778	8,545	28,778	8,545	28,778	8,545	28,778	8,545	28,778	8,545	28,778	8,545
	Yes	2	8,067	125	8,067	125	8,067	125	8,067	125	8,067	125	8,067	125
		3	42,055	1,726	42,055	1,726	42,055	1,726	42,055	1,726	42,055	1,726	42,055	1,726
		4	109,034	13,676	109,034	13,676	109,034	13,676	109,034	13,676	109,034	13,676	109,034	13,676
		5	54,319	6,813	54,319	6,813	54,319	6,813	54,319	6,813	54,319	6,813	54,319	6,813
Elbow	No	1	2,122	33	2,122	33	2,122	33	2,122	33	2,122	33	2,122	33
		2	5,894	176	5,894	176	5,894	176	5,894	176	5,894	176	5,894	176
		3	18,371	1,239	18,371	1,239	18,371	1,239	18,371	1,239	18,371	1,239	18,371	1,239
		4	60,476	16,687	60,476	16,687	60,476	16,687	60,476	16,687	60,476	16,687	60,476	16,687
	Yes	2	8,208	150	8,208	150	8,208	150	8,208	150	8,208	150	8,208	150
		3	43,648	1,686	43,648	1,686	43,648	1,686	43,648	1,686	43,648	1,686	43,648	1,686
		4	92,475	19,699	92,475	19,699	92,475	19,699	92,475	19,699	92,475	19,699	92,475	19,699
Forearm	Yes	2	10,688	2,277	10,688	2,277	10,688	2,277	10,688	2,277	10,688	2,277	10,688	2,277
		3	52,870	6,092	52,870	6,092	52,870	6,092	52,870	6,092	52,870	6,092	52,870	6,092
Wrist/hand/ finger/thumb	No	1	1,802	20	1,802	20	1,802	20	1,802	20	1,802	20	1,802	20
		2	4,243	146	4,243	146	4,243	146	4,243	146	4,243	146	4,243	146
		3	12,837	1,479	12,837	1,479	12,837	1,479	12,837	1,479	12,837	1,479	12,837	1,479

Body part	Fracture/ Dislocation	Mais-90	Discount rate 3%		Discount rate 0%		Discount rate 2%		Discount rate 4%		Discount rate 7%		Discount rate 10%	
			Mean	Std error	Mean	Std error	Mean	Std error	Mean	Std error	Mean	Std error	Mean	Std error
	Yes	1	2,552	65	2,552	65	2,552	65	2,552	65	2,552	65	2,552	65
		2	4,642	123	4,642	123	4,642	123	4,642	123	4,642	123	4,642	123
		3	34,303	3,244	34,303	3,244	34,303	3,244	34,303	3,244	34,303	3,244	34,303	3,244
Upper ext/mult/unsp	No	1	2,268	53	2,268	53	2,268	53	2,268	53	2,268	53	2,268	53
		2	8,569	832	8,569	832	8,569	832	8,569	832	8,569	832	8,569	832
		3	40,837	3,916	40,837	3,916	40,837	3,916	40,837	3,916	40,837	3,916	40,837	3,916
		4	31,677	12,627	31,677	12,627	31,677	12,627	31,677	12,627	31,677	12,627	31,677	12,627
	Yes	2	2,491	460	2,491	460	2,491	460	2,491	460	2,491	460	2,491	460
		3	5,855	79	5,855	79	5,855	79	5,855	79	5,855	79	5,855	79
Chest/abdom	No	2	8,929	713	8,929	713	8,929	713	8,929	713	8,929	713	8,929	713
Ribs/sternum	No	1	1,896	5	1,896	5	1,896	5	1,896	5	1,896	5	1,896	5
		2	4,104	55	4,104	55	4,104	55	4,104	55	4,104	55	4,104	55
		3	8,375	993	8,375	993	8,375	993	8,375	993	8,375	993	8,375	993
		4	38,942	7,569	38,942	7,569	38,942	7,569	38,942	7,569	38,942	7,569	38,942	7,569
	Yes	2	14,304	375	14,304	375	14,304	375	14,304	375	14,304	375	14,304	375
		3	23,146	583	23,146	583	23,146	583	23,146	583	23,146	583	23,146	583
		4	81,361	7,686	81,361	7,686	81,361	7,686	81,361	7,686	81,361	7,686	81,361	7,686
		5	165,271	15,603	165,271	15,603	165,271	15,603	165,271	15,603	165,271	15,603	165,271	15,603
Back (including vertevrae)	No	2	5,668	49	5,668	49	5,668	49	5,668	49	5,668	49	5,668	49
		3	88,311	8,337	88,311	8,337	88,311	8,337	88,311	8,337	88,311	8,337	88,311	8,337
		4	151,302	1,304	151,302	1,304	151,302	1,304	151,302	1,304	151,302	1,304	151,302	1,304
	Yes	2	12,232	892	12,232	892	12,232	892	12,232	892	12,232	892	12,232	892
		3	28,493	1,173	28,493	1,173	28,493	1,173	28,493	1,173	28,493	1,173	28,493	1,173
		4	63,564	14,773	63,564	14,773	63,564	14,773	63,564	14,773	63,564	14,773	63,564	14,773
Trunk -spinal cord	No	3	446,860	20,975	446,860	20,975	446,860	20,975	446,860	20,975	446,860	20,975	446,860	20,975
		4	386,106	37,787	386,106	37,787	386,106	37,787	386,106	37,787	386,106	37,787	386,106	37,787
		5	464,190	15,395	548,062	16,171	485,847	15,579	446,853	15,257	411,892	15,004	391,692	14,873

Table F-5. (Continued)

Body part	Fracture/ Disclocation	Mais-90	Discount rate 3%		Discount rate 0%		Discount rate 2%		Discount rate 4%		Discount rate 7%		Discount rate 10%	
			Mean	Std error	Mean	Std error	Mean	Std error	Mean	Std error	Mean	Std error	Mean	Std error
Trunk, superf	No	2	21,331	1,844	21,331	1,844	21,331	1,844	21,331	1,844	21,331	1,844	21,331	1,844
Trunk,multiple/ unspecified	No	1	2,532	16	2,532	16	2,532	16	2,532	16	2,532	16	2,532	16
		2	5,649	245	5,649	245	5,649	245	5,649	245	5,649	245	5,649	245
		3	23,283	711	23,283	711	23,283	711	23,283	711	23,283	711	23,283	711
		4	60,095	3,941	60,095	3,941	60,095	3,941	60,095	3,941	60,095	3,941	60,095	3,941
		5	222,924	50,028	239,248	64,992	227,459	54,106	219,078	46,635	210,564	39,423	205,037	35,055
	Yes	1	5,557	196	5,557	196	5,557	196	5,557	196	5,557	196	5,557	196
		2	10,681	262	10,681	262	10,681	262	10,681	262	10,681	262	10,681	262
		3	20,903	584	20,903	584	20,903	584	20,903	584	20,903	584	20,903	584
		4	62,951	16,085	62,951	16,085	62,951	16,085	62,951	16,085	62,951	16,085	62,951	16,085
		5	496,569	26,390	581,841	27,332	518,587	26,611	478,942	26,224	443,398	25,922	422,861	25,767
Thoracic orgs/blood vessels	No	3	53,841	6,263	53,841	6,263	53,841	6,263	53,841	6,263	53,841	6,263	53,841	6,263
		4	67,642	15,084	67,642	15,084	67,642	15,084	67,642	15,084	67,642	15,084	67,642	15,084
		5	164,472	18,793	164,472	18,793	164,472	18,793	164,472	18,793	164,472	18,793	164,472	18,793
Liver	No	1	4,873	66	4,873	66	4,873	66	4,873	66	4,873	66	4,873	66
		2	19,439	612	19,439	612	19,439	612	19,439	612	19,439	612	19,439	612
		3	44,191	2,190	44,191	2,190	44,191	2,190	44,191	2,190	44,191	2,190	44,191	2,190
		4	69,550	4,750	69,550	4,750	69,550	4,750	69,550	4,750	69,550	4,750	69,550	4,750
		5	110,418	12,568	110,418	12,568	110,418	12,568	110,418	12,568	110,418	12,568	110,418	12,568
Spleen	No	3	45,923	3,487	45,923	3,487	45,923	3,487	45,923	3,487	45,923	3,487	45,923	3,487
		4	54,820	16,628	54,820	16,628	54,820	16,628	54,820	16,628	54,820	16,628	54,820	16,628
Kidney	No	3	41,831	10,531	41,831	10,531	41,831	10,531	41,831	10,531	41,831	10,531	41,831	10,531
		4	78,227	20,649	78,227	20,649	78,227	20,649	78,227	20,649	78,227	20,649	78,227	20,649
		5	209,445	20,910	209,445	20,910	209,445	20,910	209,445	20,910	209,445	20,910	209,445	20,910
Gastrointestinal	No	2	78,681	53,493	78,681	53,493	78,681	53,493	78,681	53,493	78,681	53,493	78,681	53,493
		3	60,723	11,300	60,723	11,300	60,723	11,300	60,723	11,300	60,723	11,300	60,723	11,300

Body part	Fracture/ Disclocation	Mais-90	Discount rate 3%		Discount rate 0%		Discount rate 2%		Discount rate 4%		Discount rate 7%		Discount rate 10%	
			Mean	Std error	Mean	Std error	Mean	Std error	Mean	Std error	Mean	Std error	Mean	Std error
		4	73,340	15,948	73,340	15,948	73,340	15,948	73,340	15,948	73,340	15,948	73,340	15,948
		5	200,047	52,238	200,047	52,238	200,047	52,238	200,047	52,238	200,047	52,238	200,047	52,238
Genitourinary	No	2	22,371	2,150	22,371	2,150	22,371	2,150	22,371	2,150	22,371	2,150	22,371	2,150
		3	93,809	15,103	93,809	15,103	93,809	15,103	93,809	15,103	93,809	15,103	93,809	15,103
		4	20,960	3,374	20,960	3,374	20,960	3,374	20,960	3,374	20,960	3,374	20,960	3,374
Hip/thigh	No	1	2,007	28	2,007	28	2,007	28	2,007	28	2,007	28	2,007	28
		2	6,273	363	6,273	363	6,273	363	6,273	363	6,273	363	6,273	363
		3	23,778	2,271	23,778	2,271	23,778	2,271	23,778	2,271	23,778	2,271	23,778	2,271
		4	37,292	6,608	37,292	6,608	37,292	6,608	37,292	6,608	37,292	6,608	37,292	6,608
	Yes	2	26,324	910	26,324	910	26,324	910	26,324	910	26,324	910	26,324	910
		3	63,923	1,078	63,923	1,078	63,923	1,078	63,923	1,078	63,923	1,078	63,923	1,078
		4	145,908	9,947	145,908	9,947	145,908	9,947	145,908	9,947	145,908	9,947	145,908	9,947
		5	186,347	49,407	186,347	49,407	186,347	49,407	186,347	49,407	186,347	49,407	186,347	49,407
L. extremity superfic	No	1	2,537	44	2,537	44	2,537	44	2,537	44	2,537	44	2,537	44
		2	3,380	92	3,380	92	3,380	92	3,380	92	3,380	92	3,380	92
Knee	No	1	1,767	22	1,767	22	1,767	22	1,767	22	1,767	22	1,767	22
		2	4,424	398	4,424	398	4,424	398	4,424	398	4,424	398	4,424	398
		3	15,474	1,624	15,474	1,624	15,474	1,624	15,474	1,624	15,474	1,624	15,474	1,624
		4	127,358	26,425	127,358	26,425	127,358	26,425	127,358	26,425	127,358	26,425	127,358	26,425
	Yes	2	18,015	606	18,015	606	18,015	606	18,015	606	18,015	606	18,015	606
		3	56,782	3,314	56,782	3,314	56,782	3,314	56,782	3,314	56,782	3,314	56,782	3,314
		4	46,803	13,831	46,803	13,831	46,803	13,831	46,803	13,831	46,803	13,831	46,803	13,831
Lower leg	No	1	1,942	28	1,942	28	1,942	28	1,942	28	1,942	28	1,942	28
		2	3,120	207	3,120	207	3,120	207	3,120	207	3,120	207	3,120	207
		3	35,598	3,970	35,598	3,970	35,598	3,970	35,598	3,970	35,598	3,970	35,598	3,970
	Yes	2	22,370	309	22,370	309	22,370	309	22,370	309	22,370	309	22,370	309
		3	66,178	1,414	66,178	1,414	66,178	1,414	66,178	1,414	66,178	1,414	66,178	1,414

Table F-5. (Continued)

Body part	Fracture/ Disclocation	Mais-90	Discount rate 3%		Discount rate 0%		Discount rate 2%		Discount rate 4%		Discount rate 7%		Discount rate 10%	
			Mean	Std error	Mean	Std error	Mean	Std error	Mean	Std error	Mean	Std error	Mean	Std error
Ankle/foot/toes	No	1	1,752	41	1,752	41	1,752	41	1,752	41	1,752	41	1,752	41
		2	7,913	627	7,913	627	7,913	627	7,913	627	7,913	627	7,913	627
		3	9,675	1,360	9,675	1,360	9,675	1,360	9,675	1,360	9,675	1,360	9,675	1,360
	Yes	1	2,722	120	2,722	120	2,722	120	2,722	120	2,722	120	2,722	120
		2	11,113	279	11,113	279	11,113	279	11,113	279	11,113	279	11,113	279
		3	56,333	3,130	56,333	3,130	56,333	3,130	56,333	3,130	56,333	3,130	56,333	3,130
		4	84,727	31,426	84,727	31,426	84,727	31,426	84,727	31,426	84,727	31,426	84,727	31,426
Burns	No	1	1,889	116	1,889	116	1,889	116	1,889	116	1,889	116	1,889	116
		2	17,878	3,041	17,878	3,041	17,878	3,041	17,878	3,041	17,878	3,041	17,878	3,041
		3	119,526	15,041	119,526	15,041	119,526	15,041	119,526	15,041	119,526	15,041	119,526	15,041
		4	110,391	21,056	110,391	21,056	110,391	21,056	110,391	21,056	110,391	21,056	110,391	21,056
		5	251,671	32,167	251,671	32,167	251,671	32,167	251,671	32,167	251,671	32,167	251,671	32,167
Min. extern.	No	1	3,313	29	3,313	29	3,313	29	3,313	29	3,313	29	3,313	29

Table F-6. 2007–2008 HCUP-based unit earnings loss at different discount rates, AIS-85 (2010 dollars)

Body part	Fracture/ Disclocation	Mais-90	Discount rate 3%		Discount rate 0%		Discount rate 2%		Discount rate 4%		Discount rate 7%		Discount rate 10%	
			Mean	Std error	Mean	Std error	Mean	Std error	Mean	Std error	Mean	Std error	Mean	Std error
Skull	Yes	1	14,176	84	21,301	356	15,994	151	12,725	33	9,790	64	8,073	110
		2	14,442	713	25,167	1,471	16,962	876	12,547	598	9,057	409	7,243	321
		3	131,419	2,789	233,887	5,503	155,920	3,394	112,797	2,351	77,939	1,586	59,437	1,207
		4	196,499	10,638	346,443	20,073	232,656	12,749	168,836	9,095	116,487	6,325	88,305	4,883

Body part	Fracture/ Disclocation	Mais-90	Discount rate 3%		Discount rate 0%		Discount rate 2%		Discount rate 4%		Discount rate 7%		Discount rate 10%	
			Mean	Std error	Mean	Std error	Mean	Std error	Mean	Std error	Mean	Std error	Mean	Std error
		5	322,173	12,201	570,799	25,248	382,818	15,114	275,497	10,113	186,684	6,588	138,832	4,963
Brain/ intracarnial	No	1	3,714	269	6,666	530	4,417	330	3,182	224	2,197	141	1,683	100
		2	7,798	129	13,017	235	9,058	153	6,833	110	5,008	76	4,026	59
		3	132,791	2,962	229,106	5,617	156,345	3,577	114,594	2,506	79,642	1,680	60,453	1,257
		4	145,272	5,252	239,876	9,737	168,711	6,290	126,960	4,480	91,059	3,075	70,718	2,343
		5	365,430	13,127	665,037	27,705	437,517	16,331	310,434	10,864	207,008	7,132	151,902	5,432
Ear	No	2	116,088	14,908	159,239	25,169	127,915	17,479	106,140	12,937	84,298	9,357	70,158	7,641
Eye & adnexa	No	1	2,717	132	4,946	273	3,238	163	2,329	110	1,623	72	1,264	55
		2	19,439	1,554	37,114	2,901	23,599	1,852	16,318	1,337	10,603	942	7,669	730
		3	44,754	5,747	73,434	11,607	51,575	7,048	39,367	4,798	28,648	3,180	22,483	2,449
		4	310,429	23,453	395,936	32,680	334,942	25,886	289,005	21,485	238,691	17,421	202,855	14,887
Nouse/ mouth /face/scalp/ neck	No	1	2,211	23	3,869	46	2,596	28	1,925	19	1,409	14	1,148	11
		2	6,336	130	11,347	270	7,517	158	5,449	112	3,818	85	2,976	73
		3	42,804	3,234	71,514	5,903	49,857	3,853	37,338	2,776	26,786	1,955	20,957	1,538
		4	123,391	18,978	202,648	39,273	143,062	23,632	108,013	15,571	77,874	9,629	60,856	6,794
	Yes	1	4,099	78	7,001	155	4,793	94	3,572	68	2,592	53	2,080	46
		2	16,525	625	28,956	1,037	19,491	711	14,279	565	10,115	456	7,945	395
		3	97,492	5,799	161,736	10,674	113,410	6,928	85,085	4,963	60,959	3,460	47,545	2,693
		4	149,050	15,271	231,051	25,419	170,004	17,783	132,325	13,313	98,508	9,502	78,704	7,370
		5	596,779	83,084	1,049,788	135,789	708,229	96,544	510,475	72,345	344,715	50,693	254,345	38,143
Neck/internal organs/blood vessels	No	1	1,599	315	3,613	575	2,042	376	1,284	269	763	185	537	144
		2	50,805	10,013	100,031	15,915	62,424	11,502	42,100	8,821	26,312	6,388	18,406	4,946
		3	34,953	13,498	49,883	16,833	38,928	14,511	31,672	12,585	24,656	10,368	20,224	8,767
		4	237,565	91,742	378,254	127,642	273,758	102,047	208,590	82,884	149,959	63,059	115,847	50,218
Neck-spinal cord	No	3	74,250	60,023	102,610	83,922	82,069	66,609	67,641	54,456	53,009	42,139	43,424	34,079
		4	177,009	13,990	251,322	24,479	196,527	16,442	161,037	12,147	127,186	8,710	105,890	6,846
		5	117,139	13,113	164,461	15,537	129,884	13,830	106,551	12,470	83,715	10,894	69,167	9,705

Table F-6. (Continued)

Body part	Fracture/ Disclocation	Mais-90	Discount rate 3%		Discount rate 0%		Discount rate 2%		Discount rate 4%		Discount rate 7%		Discount rate 10%	
			Mean	Std error	Mean	Std error	Mean	Std error	Mean	Std error	Mean	Std error	Mean	Std error
Shoulder/ clavicule/ scapula/ upper arm	No	1	2,416	18	3,089	24	2,580	20	2,288	18	2,043	16	1,910	15
		2	6,373	117	8,706	192	6,945	132	5,931	108	5,084	97	4,623	93
		3	27,569	2,384	41,683	4,293	30,875	2,841	25,013	2,095	20,109	1,731	17,437	1,635
	Yes	2	13,728	143	20,681	232	15,410	161	12,437	132	9,984	115	8,655	107
		3	58,904	1,537	88,740	2,652	66,161	1,771	53,317	1,378	42,632	1,124	36,794	1,009
		4	91,027	9,084	143,486	17,013	103,883	10,829	81,083	7,870	61,947	5,971	51,415	5,201
		5	164,379	16,404	282,028	33,440	192,707	20,088	142,742	13,854	101,893	9,822	79,916	8,084
Elbow	No	1	1,202	21	1,754	34	1,334	24	1,102	18	914	15	816	13
		2	4,112	198	6,251	311	4,630	224	3,715	179	2,964	143	2,561	125
		3	20,957	1,564	31,263	2,147	23,420	1,683	19,082	1,483	15,553	1,349	13,657	1,280
		4	60,990	26,474	84,809	28,872	67,189	27,531	55,948	25,362	45,367	22,240	38,807	19,750
	Yes	2	18,126	201	28,177	346	20,557	231	16,263	181	12,726	151	10,817	138
		3	76,895	1,314	115,212	2,273	86,308	1,508	69,582	1,187	55,420	1,000	47,568	920
		4	119,821	13,681	184,836	24,915	136,061	16,302	107,099	11,751	82,185	8,387	68,235	6,805
Forearm	Yes	2	26,763	3,056	30,624	4,128	27,813	3,332	25,879	2,839	23,920	2,441	22,606	2,255
		3	52,356	3,814	56,130	4,813	53,465	4,078	51,367	3,606	48,951	3,218	47,135	3,042
Wrist/hand/ finger/thumb	No	1	763	14	1,010	26	822	16	719	12	636	9	594	8
		2	7,847	498	13,933	1,019	9,282	614	6,767	413	4,775	267	3,738	195
		3	30,396	2,215	49,387	4,234	34,968	2,667	26,906	1,889	20,336	1,337	16,833	1,086
	Yes	1	5,981	119	9,941	217	6,924	140	5,266	105	3,935	82	3,234	71
		2	11,082	161	17,544	282	12,655	187	9,873	144	7,574	117	6,335	105
		3	61,076	4,886	95,004	8,730	69,421	5,769	54,611	4,242	42,170	3,131	35,357	2,613
Upper ext/ mult/unsp	No	1	754	21	997	38	813	24	710	18	626	14	582	13
		2	8,138	489	12,900	887	9,292	567	7,252	437	5,572	355	4,668	316
		3	81,393	7,733	138,664	14,053	95,435	9,185	70,551	6,667	49,832	4,792	38,626	3,869
		4	64,612	13,545	109,261	15,031	75,836	14,220	55,798	12,832	38,562	10,856	29,027	9,333
	Yes	2	7,728	225	13,054	376	9,006	250	6,757	211	4,952	198	4,012	192

Body part	Fracture/ Disclocation	Mais-90	Discount rate 3%		Discount rate 0%		Discount rate 2%		Discount rate 4%		Discount rate 7%		Discount rate 10%	
			Mean	Std error	Mean	Std error	Mean	Std error	Mean	Std error	Mean	Std error	Mean	Std error
		3	10,921	177	18,971	346	12,886	215	9,412	150	6,574	102	5,086	77
Chest/abdom	No	2	16,374	489	24,606	2,848	18,379	1,064	14,834	100	11,915	789	10,353	1,196
Ribs/sternum	No	1	3,091	6	3,252	6	3,130	6	3,060	6	3,001	6	2,968	6
		2	5,149	84	5,984	109	5,362	89	4,979	79	4,635	72	4,433	67
		3	15,670	1,995	21,702	2,942	17,139	2,235	14,532	1,801	12,323	1,397	11,079	1,151
		4	73,323	4,527	109,284	6,379	81,280	4,887	67,551	4,301	57,312	3,984	52,048	3,845
	Yes	2	24,596	565	35,468	909	27,307	643	22,465	509	18,258	411	15,865	363
		3	36,937	676	55,223	1,143	41,445	781	33,423	600	26,577	469	22,751	406
		4	55,978	4,988	78,260	7,720	61,572	5,643	51,487	4,481	42,292	3,494	36,786	2,929
		5	193,658	11,553	245,159	16,675	208,315	12,908	180,927	10,446	151,369	8,149	130,667	6,785
Back (including vertevrae)	No	2	29,305	1,935	45,412	5,530	33,506	2,777	25,909	1,319	18,945	284	14,835	142
		3	62,807	3,747	99,612	6,775	72,201	4,474	55,326	3,194	40,289	2,169	31,594	1,641
		4	183,680	12,129	284,491	34,645	209,582	17,370	162,940	8,296	120,841	1,809	96,087	918
	Yes	2	25,096	1,207	39,410	2,144	28,573	1,391	22,424	1,085	17,334	898	14,572	811
		3	48,638	992	75,429	1,795	55,312	1,177	43,405	855	33,137	611	27,363	489
		4	47,283	12,295	70,973	22,677	53,228	14,783	42,602	10,413	33,340	6,951	28,060	5,193
Trunk -spinal cord	No	3	274,178	14,085	480,233	44,495	324,386	20,034	235,528	10,666	161,815	8,041	121,801	7,995
		4	254,472	20,631	437,989	44,068	299,690	25,968	219,399	16,770	151,788	10,265	114,674	7,398
		5	429,965	11,884	873,649	30,281	534,556	15,964	351,796	9,018	210,928	4,592	141,314	3,121
Trunk, superf	No	2	39,120	1,167	56,504	6,981	43,393	2,591	35,808	77	29,422	1,940	25,909	2,945
Trunk,multiple/ unspecified	No	1	1,641	11	1,942	14	1,714	12	1,584	11	1,477	10	1,418	9
		2	3,248	125	4,532	178	3,558	136	3,011	116	2,563	101	2,320	92
		3	23,929	697	33,787	1,197	26,344	814	22,050	610	18,401	449	16,367	366
		4	35,500	2,899	54,369	5,385	39,988	3,451	32,023	2,494	25,286	1,772	21,526	1,403
		5	93,482	23,274	170,186	48,174	111,989	29,182	79,392	18,844	53,176	10,851	39,569	6,939
	Yes	1	4,877	112	6,302	150	5,240	121	4,586	106	3,991	94	3,637	87
		2	10,393	288	12,925	391	11,041	314	9,867	269	8,767	230	8,091	208
		3	25,448	539	32,135	793	27,150	598	24,073	494	21,223	412	19,487	368
		4	38,203	14,032	56,780	26,645	42,449	16,959	34,880	11,773	28,388	7,494	24,760	5,245

Table F-6. (Continued)

Body part	Fracture/ Dislocation	Mais-90	Discount rate 3%		Discount rate 0%		Discount rate 2%		Discount rate 4%		Discount rate 7%		Discount rate 10%	
			Mean	Std error	Mean	Std error	Mean	Std error	Mean	Std error	Mean	Std error	Mean	Std error
		5	425,439	13,315	855,016	31,847	527,017	17,271	349,336	10,667	211,601	6,949	143,072	5,708
Thoracic orgs/blood vessels	No	3	37,719	7,367	53,372	12,690	41,654	8,684	34,618	6,344	28,484	4,378	24,996	3,320
		4	27,432	6,482	36,931	8,248	29,806	6,964	25,553	6,070	21,765	5,134	19,512	4,486
		5	144,889	15,417	234,264	28,626	167,288	18,538	127,276	13,079	92,521	8,840	72,798	6,711
Liver	No	1	1,432	23	2,122	44	1,592	26	1,314	21	1,100	19	993	18
		2	15,385	487	23,023	1,024	17,196	597	14,018	413	11,493	302	10,180	259
		3	33,278	1,626	53,100	3,175	38,014	1,980	29,671	1,367	22,916	918	19,345	707
		4	40,157	3,016	63,133	5,414	45,778	3,583	35,816	2,590	27,490	1,805	22,939	1,399
		5	87,322	9,684	162,025	21,871	104,915	12,409	74,133	7,725	50,062	4,406	37,783	2,904
Spleen	No	3	80,584	25,481	144,188	57,086	95,799	32,747	69,066	20,163	47,793	10,951	36,842	6,728
		4	78,460	484	117,762	295	88,405	270	70,581	664	54,825	1,057	45,689	1,304
Kidney	No	3	84,992	30,932	149,174	68,383	100,368	39,594	73,330	24,562	51,682	13,430	40,415	8,237
		4	17,734	463	26,749	1,238	19,893	616	16,083	365	12,933	277	11,193	313
		5	137,340	9,677	212,991	19,702	156,667	12,082	121,914	7,858	90,704	4,522	72,389	2,840
Gastrointestinal	No	2	27,988	11,098	36,293	13,379	30,084	11,780	26,314	10,481	22,849	8,944	20,678	7,762
		3	46,497	4,574	78,524	10,389	54,220	5,871	40,614	3,662	29,645	2,262	23,929	1,797
		4	103,496	33,412	160,400	59,047	117,833	39,642	92,189	28,629	69,799	19,551	57,078	14,664
		5	129,766	28,415	250,388	66,362	158,503	37,047	108,109	22,155	68,528	11,498	48,552	6,741
Genitourinary	No	2	29,714	8,556	48,641	16,241	34,306	10,429	26,202	7,122	19,623	4,445	16,182	3,071
		3	109,946	31,564	183,643	57,363	128,131	37,643	95,832	27,019	68,611	18,765	53,672	14,531
		4	15,297	4,392	17,628	5,506	15,978	4,694	14,693	4,143	13,233	3,619	12,151	3,290
Hip/thigh	No	1	1,892	23	2,184	28	1,963	24	1,837	22	1,732	21	1,675	21
		2	5,249	535	7,982	1,190	5,882	675	4,781	441	3,943	310	3,528	272
		3	35,348	2,479	53,802	4,056	39,885	2,855	31,836	2,195	25,089	1,667	21,409	1,392
		4	51,389	6,800	78,498	10,654	57,856	7,718	46,457	6,089	37,083	4,678	31,914	3,842
	Yes	2	47,471	949	75,233	1,689	54,139	1,108	42,346	837	32,578	650	27,272	561
		3	53,172	645	85,592	1,126	60,923	752	47,259	569	36,142	439	30,220	374

Body part	Fracture/ Disclocation	Mais-90	Discount rate 3%		Discount rate 0%		Discount rate 2%		Discount rate 4%		Discount rate 7%		Discount rate 10%	
			Mean	Std error	Mean	Std error	Mean	Std error	Mean	Std error	Mean	Std error	Mean	Std error
		4	74,680	4,493	115,729	7,651	84,741	5,210	66,887	3,965	51,831	3,015	43,484	2,516
		5	67,469	3,799	106,158	6,816	77,240	4,177	59,749	3,650	44,433	3,527	35,729	3,420
L. extremity superfic	No	1	1,307	26	1,894	44	1,446	30	1,203	24	1,010	21	911	20
		2	3,041	47	3,954	76	3,261	53	2,874	44	2,559	38	2,391	36
Knee	No	1	622	11	685	15	638	12	609	10	585	8	571	8
		2	3,565	229	4,511	296	3,803	244	3,376	218	3,000	197	2,785	186
		3	18,701	1,129	25,175	1,379	20,322	1,190	17,427	1,081	14,910	979	13,481	911
		4	209,495	15,030	274,733	18,419	226,295	15,827	196,006	14,415	168,470	13,183	152,150	12,420
	Yes	2	34,762	903	52,521	1,546	39,126	1,045	31,358	802	24,730	638	21,038	563
		3	77,750	2,548	119,172	4,494	87,974	2,973	69,803	2,250	54,416	1,768	45,903	1,553
		4	87,050	23,016	117,481	33,811	95,103	25,892	80,442	20,647	66,452	15,618	57,753	12,500
Lower leg	No	1	817	10	965	16	853	11	790	10	740	9	714	8
		2	1,231	74	1,589	183	1,314	95	1,169	62	1,058	49	1,001	47
		3	50,556	2,876	71,329	4,219	55,718	3,154	46,517	2,688	38,582	2,389	34,086	2,252
	Yes	2	33,844	264	48,863	418	37,524	294	30,992	245	25,474	217	22,406	204
		3	78,500	915	115,625	1,531	87,630	1,044	71,407	827	57,652	686	50,001	619
Ankle/foot/toes	No	1	1,688	35	2,624	65	1,911	42	1,520	31	1,209	23	1,047	20
		2	8,087	547	13,050	1,437	9,230	700	7,241	463	5,710	375	4,926	348
		3	14,820	581	23,339	1,064	16,868	670	13,256	527	10,305	458	8,719	432
	Yes	1	3,161	84	4,398	131	3,462	93	2,930	79	2,490	72	2,252	70
		2	19,566	263	28,358	414	21,733	293	17,883	243	14,626	213	12,824	199
		3	69,193	3,008	101,238	5,552	77,155	3,565	62,975	2,622	50,838	2,027	44,046	1,790
		4	113,633	10,296	162,538	11,158	126,057	10,485	103,768	10,128	84,013	9,652	72,610	9,205
Burns	No	1	3,525	147	5,990	280	4,116	177	3,075	125	2,234	86	1,790	66
		2	10,343	622	17,358	1,227	12,024	757	9,064	526	6,672	367	5,408	297
		3	45,738	8,804	79,836	19,648	53,947	11,288	39,489	6,995	27,842	3,910	21,764	2,556
		4	49,040	12,552	87,715	23,490	58,347	15,218	41,967	10,511	28,841	6,707	22,063	4,766
		5	105,044	32,695	166,285	41,419	120,524	35,323	92,788	30,349	68,316	24,774	54,200	20,871
Min. extern.	No	1	1,618	17	2,026	26	1,717	19	1,543	16	1,402	14	1,326	13

Table F-7. 2007–2008 HCUP-based unit household production loss at different discount rates, AIS-85 (2010 dollars)

Body part	Fracture/ Disclocation	Mais-90	Discount rate 3%		Discount rate 0%		Discount rate 2%		Discount rate 4%		Discount rate 7%		Discount rate 10%	
			Mean	Std error	Mean	Std error	Mean	Std error	Mean	Std error	Mean	Std error	Mean	Std error
Skull	Yes	1	4,374	32	7,527	175	5,099	61	3,836	12	2,858	19	2,352	31
		2	4,528	177	8,135	414	5,326	223	3,948	146	2,925	97	2,408	76
		3	38,161	717	71,146	1,484	45,593	878	32,705	603	22,897	410	17,842	314
		4	53,041	2,476	100,672	5,269	63,757	3,056	45,189	2,071	31,129	1,389	23,933	1,059
		5	92,592	2,970	177,624	6,704	111,837	3,736	78,453	2,445	53,089	1,592	40,104	1,199
Brain/ intracarnial	No	1	1,155	67	2,110	132	1,369	81	998	56	719	38	577	29
		2	2,828	40	4,856	77	3,291	48	2,486	34	1,862	24	1,533	18
		3	44,940	769	80,889	1,569	53,259	941	38,715	646	27,179	432	20,992	324
		4	54,669	1,268	92,379	2,544	63,636	1,544	47,821	1,071	34,705	727	27,357	553
		5	109,361	3,543	214,688	8,217	133,082	4,499	91,976	2,887	60,863	1,820	44,965	1,327
Ear	No	2	24,340	2,988	39,151	5,396	28,016		21,476	2,580	15,925	1,838	12,859	1,453
Eye & adnexa	No	1	914	32	1,690	83	1,084	42	792	26	580	15	475	11
		2	5,778	326	11,772	713	7,086	399	4,838	279	3,204	203	2,396	166
		3	19,247	2,363	32,029	4,415	22,311	2,819	16,898	2,030	12,384	1,429	9,853	1,113
		4	89,721	6,881	136,801	11,425	102,130		79,602	5,994	58,497	4,231	45,711	3,210
Nouse/ mouth /face/scalp/ neck	No	1	785	7	1,361	13	911	8	694	6	536	4	458	4
		2	2,025	38	3,740	78	2,405	46	1,750	33	1,264	25	1,020	21
		3	14,293	1,096	25,625	2,140	16,897	1,325	12,357	931	8,810	637	6,941	487
		4	48,299	3,663	83,479	7,782	56,642	4,527	41,968	3,058	30,061	2,046	23,624	1,568
	Yes	1	1,366	16	2,385	34	1,593	18	1,201	14	909	13	761	12
		2	5,275	182	9,448	295	6,207	205	4,597	166	3,395	137	2,786	121
		3	36,008	1,896	64,035	3,653	42,483	2,280	31,182	1,617	22,325	1,124	17,654	871
		4	43,272	3,010	73,053	5,500	50,336	3,571	37,911	2,597	27,816	1,849	22,339	1,456
		5	136,791	14,031	257,004	25,488	164,466	16,709	116,234	12,029	78,778	8,369	59,262	6,462

Body part	Fracture/ Disclocation	Mais-90	Discount rate 3%		Discount rate 0%		Discount rate 2%		Discount rate 4%		Discount rate 7%		Discount rate 10%	
			Mean	Std error	Mean	Std error	Mean	Std error	Mean	Std error	Mean	Std error	Mean	Std error
Neck/internal organs/blood vessels	No	1	893	111	2,142	240	1,148		719	91	437	59	312	44
		2	12,113	1,500	24,470	2,737	14,830	1,780	10,162	1,293	6,786	916	5,141	720
		3	13,287	4,557	21,783	7,362	15,377	5,267	11,666	3,997	8,538	2,891	6,811	2,268
		4	46,057	15,795	83,374	28,177	54,831	0	39,468	13,521	27,348	9,261	21,020	6,999
Neck-spinal cord	No	3	18,287	10,702	28,629	18,998	20,884	12,761	16,246	9,101	12,221	6,017	9,938	4,342
		4	59,604	2,186	91,808	4,402	67,720	2,662	53,166	1,849	40,198	1,272	32,570	989
		5	98,487	5,709	125,707	8,460	105,878	6,467	92,241	5,069	78,133	3,665	68,323	2,764
Shoulder/ clavicule/ scapula/ upper arm	No	1	1,118	6	1,425	8	1,188	6	1,066	6	971	6	922	6
		2	2,100	28	2,868	41	2,276	30	1,970	26	1,734	24	1,612	23
		3	9,017	551	13,529	1,031	10,056	636	8,244	501	6,827	434	6,078	408
	Yes	2	5,032	37	7,672	61	5,634	41	4,587	34	3,778	31	3,355	30
		3	20,428	391	31,367	674	22,941	446	18,559	354	15,134	295	13,327	267
		4	34,421	2,495	53,466	3,704	38,883	2,737	31,054	2,322	24,740	1,996	21,308	1,803
		5	35,606	2,580	64,968	4,500	42,303		30,654	2,292	21,681	1,750	17,040	1,442
Elbow	No	1	528	7	758	12	580	8	491	6	424	5	390	5
		2	1,384	55	2,113	85	1,549	62	1,262	51	1,042	43	929	39
		3	7,236	488	10,732	667	8,038	528	6,640	457	5,549	400	4,974	369
		4	21,600	5,926	33,572	8,058	24,524	6,523	19,338	5,428	14,961	4,381	12,516	3,755
	Yes	2	7,220	45	11,218	80	8,135	51	6,542	42	5,305	38	4,657	36
		3	27,847	373	42,186	636	31,187	421	25,341	341	20,695	292	18,212	268
		4	39,993	3,486	62,567	5,104	45,284	3,862	36,008	3,198	28,581	2,630	24,587	2,294
Forearm	Yes	2	20,502	1,787	26,025	2,123	21,959		19,311	1,715	16,812	1,547	15,279	1,425
		3	40,878	2,178	50,938	3,544	43,545		38,693	1,995	34,100	1,727	31,288	1,626
Wrist/hand/ finger/thumb	No	1	341	3	427	5	360	3	327	2	302	2	289	2
		2	2,500	131	4,549	277	2,955	162	2,168	109	1,582	71	1,285	53
		3	10,705	570	17,900	1,245	12,324	695	9,519	491	7,392	374	6,298	327
	Yes	1	2,222	36	3,778	73	2,569	43	1,969	32	1,518	24	1,289	21
		2	3,880	35	6,152	57	4,398	38	3,496	33	2,801	31	2,439	29
		3	20,720	794	32,773	1,662	23,489	957	18,661	692	14,896	555	12,914	507

Table F-7. (Continued)

Body part	Fracture/ Disclocation	Mais-90	Discount rate 3%		Discount rate 0%		Discount rate 2%		Discount rate 4%		Discount rate 7%		Discount rate 10%	
			Mean	Std error	Mean	Std error	Mean	Std error	Mean	Std error	Mean	Std error	Mean	Std error
Upper ext/ mult/unsp	No	1	346	7	439	11	367	8	331	6	304	5	289	5
		2	2,602	135	4,203	219	2,967	152	2,333	122	1,846	99	1,594	88
		3	23,591	1,935	41,495	3,506	27,706	2,276	20,541	1,689	15,001	1,253	12,118	1,027
		4	29,272	19,730	53,058	33,836	34,726	23,021	25,242	17,269	17,979	12,745	14,246	10,359
	Yes	2	3,031	75	5,126	173	3,498	91	2,693	65	2,096	57	1,797	55
		3	2,237	21	4,040	49	2,640		1,945	18	1,436	13	1,186	10
Chest/abdom	No	2	3,795	50	5,811	377	4,254	59	3,457	99	2,849	179	2,537	208
Ribs/sternum	No	1	1,603	2	1,684	2	1,622	2	1,590	2	1,565	2	1,552	2
		2	2,426	23	2,855	34	2,527	25	2,349	22	2,204	19	2,125	18
		3	5,925	412	8,093	665	6,433	472	5,541	366	4,823	282	4,435	238
		4	24,897	1,017	36,693	1,362	27,548	1,088	22,946	965	19,406	865	17,556	804
	Yes	2	9,614	137	13,877	246	10,618	160	8,856	122	7,432	98	6,660	87
		3	15,402	189	22,977	339	17,170	220	14,074	168	11,599	134	10,266	118
		4	24,205	1,591	34,587	2,595	26,747	1,812	22,224	1,429	18,308	1,128	16,032	960
		5	77,905	2,504	118,301	4,060	88,340		69,530	2,234	52,513	1,702	42,552	1,387
Back (including vertevrae)	No	2	5,658	386	9,989	1,262	6,682	0	4,887	275	3,464	114	2,721	51
		3	24,405	785	43,405	1,490	28,816	933	21,114	678	15,091	489	11,945	389
		4	75,969	5,188	130,708	16,518	88,916	7,413	66,156	3,729	47,689	1,570	37,641	712
	Yes	2	10,555	335	16,782	654	11,981	396	9,499	295	7,575	236	6,567	209
		3	21,028	269	34,378	540	24,122	328	18,717	227	14,463	156	12,209	123
		4	20,379	5,318	31,836	9,922	23,078	6,378	18,339	4,533	14,513	3,111	12,437	2,379
Trunk -spinal cord	No	3	113,046	8,228	214,669	20,160	136,150	10,507	95,991	6,740	65,123	4,482	49,129	3,512
		4	84,603	5,397	158,719	12,778	101,519	6,897	72,097	4,380	49,429	2,768	37,664	2,058
		5	95,745	2,876	203,401	6,953	119,277	3,667	78,915	2,357	50,024	1,562	36,086	1,210
Trunk, superf	No	2	13,046	97	21,834	913	15,006	111	11,618	236	9,089	440	7,814	513

Body part	Fracture/ Disclocation	Mais-90	Discount rate 3%		Discount rate 0%		Discount rate 2%		Discount rate 4%		Discount rate 7%		Discount rate 10%	
			Mean	Std error	Mean	Std error	Mean	Std error	Mean	Std error	Mean	Std error	Mean	Std error
Trunk,multiple/ unspecified	No	1	865	4	1,004	5	897	4	841	4	798	3	775	3
		2	1,446	41	1,960	64	1,563	46	1,358	38	1,198	31	1,113	28
		3	8,657	203	12,382	367	9,523	239	8,007	176	6,799	129	6,149	105
		4	13,968	815	20,944	1,559	15,607	978	12,727	697	10,384	486	9,099	378
		5	43,618	4,189	77,427	11,975	51,256	5,747	37,992	3,200	27,780	2,029	22,384	1,835
	Yes	1	2,246	20	2,916	30	2,409	21	2,121	18	1,878	17	1,740	17
		2	4,977	55	6,212	84	5,284	61	4,736	51	4,256	45	3,975	41
		3	11,024	152	14,248	252	11,816	174	10,405	136	9,186	107	8,484	93
		4	14,914	2,556	20,910	5,174	16,318	3,139	13,853	2,136	11,857	1,415	10,761	1,073
		5	93,669	862	196,782	3,177	116,296	1,104	77,438	847	49,423	945	35,801	939
Thoracic orgs/blood vessels	No	3	14,211	3,081	21,670	5,708	15,969	3,664	12,879	2,654	10,359	1,879	8,974	1,466
		4	9,309	1,545	12,876	2,334	10,158	1,750	8,660	1,381	7,422	1,047	6,737	857
		5	51,473	7,134	91,352	14,754	60,692	8,816	44,593	5,919	31,930	3,793	25,211	2,736
Liver	No	1	729	9	1,038	13	797	10	679	9	593	9	549	9
		2	5,914	150	8,716	276	6,542	176	5,455	132	4,634	102	4,213	88
		3	11,602	470	18,572	898	13,177	563	10,442	404	8,347	287	7,258	228
		4	15,044	999	24,339	1,860	17,183	1,191	13,452	858	10,527	605	8,976	475
		5	35,430	5,650	67,628	13,302	42,514	7,241	30,313	4,548	21,335	2,736	16,835	1,905
Spleen	No	3	34,887	7,314	63,773	14,762	41,354	8,938	30,170	6,156	21,781	4,180	17,509	3,239
		4	32,613	3,648	49,314	6,769	36,737	4,411	29,385	3,064	22,995	1,983	19,296	1,441
Kidney	No	3	34,397	9,847	61,957	21,745	40,670	12,412	29,767	8,029	21,386	4,951	17,035	3,487
		4	7,668	83	11,356	452	8,500	130	7,057	97	5,953	173	5,382	214
		5	54,847	6,375	97,694	13,777	64,699	7,988	47,513	5,221	34,030	3,233	26,865	2,263
Gastrointestinal	No	2	19,702	10,850	25,517	14,410	21,258	11,876	18,403	9,953	15,550	7,852	13,665	6,372
		3	18,579	3,762	34,198	7,709	22,034	4,622	16,083	3,146	11,717	2,082	9,550	1,564
		4	33,835	7,568	57,268	13,968	39,329	9,047	29,700	6,462	22,012	4,421	17,906	3,336
		5	62,112	9,704	125,010	21,342	75,946	12,197	52,158	7,944	34,855	4,971	26,323	3,565

Table F-7. (Continued)

Body part	Fracture/ Disclocation	Mais-90	Discount rate 3%		Discount rate 0%		Discount rate 2%		Discount rate 4%		Discount rate 7%		Discount rate 10%	
			Mean	Std error	Mean	Std error	Mean	Std error	Mean	Std error	Mean	Std error	Mean	Std error
Genitourinary	No	2	7,925	767	13,440	2,414	9,143	1,127	7,046	510	5,516	80	4,762	135
		3	33,934	5,428	63,635	10,578	40,626	6,572	29,031	4,594	20,266	3,108	15,795	2,349
		4	12,540	2,006	17,759	2,952	13,908		11,430	1,809	9,131	1,400	7,751	1,152
Hip/thigh	No	1	927	9	1,060	10	958	9	905	9	864	9	843	8
		2	1,761	104	2,473	228	1,917	128	1,648	89	1,451	69	1,354	62
		3	11,443	681	17,900	1,213	12,916	802	10,357	593	8,399	437	7,391	360
		4	20,174	1,554	32,982	3,653	23,048	1,980	18,061	1,263	14,227	797	12,216	600
	Yes	2	20,438	245	32,298	483	23,149	291	18,426	216	14,745	173	12,801	155
		3	18,940	186	30,472	344	21,546	219	17,023	163	13,567	126	11,779	108
		4	29,735	1,663	48,713	3,336	34,075	2,014	26,514	1,418	20,619	1,007	17,507	812
		5	46,410	5,326	76,764	10,624	53,451	6,457	41,143	4,525	31,412	3,160	26,219	2,499
L. extremity superfic	No	1	533	8	758	13	583	9	496	7	432	7	399	6
		2	1,263	15	1,587	21	1,337	16	1,209	14	1,111	12	1,060	12
Knee	No	1	363	3	397	6	371	4	356	3	344	2	338	2
		2	1,527	57	1,974	80	1,631	62	1,448	54	1,304	48	1,227	45
		3	6,500	310	9,113	455	7,108	343	6,045	286	5,205	240	4,760	216
		4	60,451	2,647	84,295	3,516	66,162	2,841	56,096	2,503	47,864	2,230	43,400	2,077
	Yes	2	14,397	203	21,597	369	16,077	234	13,134	183	10,788	154	9,527	142
		3	29,044	829	46,288	1,599	33,007	986	26,097	722	20,702	549	17,861	471
		4	30,035	5,190	43,224	9,163	33,240	6,134	27,567	4,477	22,840	3,160	20,237	2,476
Lower leg	No	1	398	3	460	5	412	4	387	3	369	3	359	3
		2	475	18	573	30	497	19	459	18	431	17	417	17
		3	16,393	673	24,286	965	18,244	730	14,999	634	12,397	568	10,998	535
	Yes	2	12,533	65	18,450	103	13,906	71	11,504	61	9,599	56	8,581	54
		3	27,504	241	41,634	414	30,781	274	25,053	219	20,531	184	18,129	167

Body part	Fracture/ Disclocation	Mais-90	Discount rate 3%		Discount rate 0%		Discount rate 2%		Discount rate 4%		Discount rate 7%		Discount rate 10%	
			Mean	Std error	Mean	Std error	Mean	Std error	Mean	Std error	Mean	Std error	Mean	Std error
Ankle/foot/toes	No	1	717	12	1,102	22	803	14	655	11	544	9	488	8
		2	2,895	245	4,750	553	3,304	305	2,599	206	2,075	144	1,811	116
		3	4,499	197	7,428	389	5,158	235	4,016	171	3,152	132	2,710	114
	Yes	1	1,294	24	1,780	24	1,405	24	1,212	24	1,063	24	987	24
		2	7,767	74	11,426	121	8,611	83	7,138	69	5,986	62	5,379	58
		3	26,279	974	39,505	1,767	29,358	1,139	23,971	860	19,700	674	17,426	587
		4	28,393	2,088	44,662	2,961	32,168	2,247	25,577	1,977	20,422	1,773	17,731	1,654
Burns	No	1	1,350	45	2,379	91	1,580	55	1,182	38	886	26	736	20
		2	3,192	120	5,667	267	3,741	149	2,795	101	2,093	73	1,742	61
		3	11,257	1,520	20,312	3,680	13,272	1,972	9,793	1,208	7,212	710	5,914	503
		4	24,740	7,870	47,591	16,079	29,789	9,689	21,094	6,554	14,721	4,251	11,552	3,106
		5	43,723	6,712	79,263	8,386	51,815	7,317	37,737	6,156	26,831	4,872	21,092	4,056
Min. extern.	No	1	757	7	933	11	796	8	728	6	675	6	648	5

APPENDIX G. DEFINITIONS

Comprehensive Costs: Comprehensive costs are a measure of total societal harm that results from traffic crashes. They present the value of lost quality-of-life as measured by society's willingness to pay to avoid risk, together with the economic impacts that result from death or injury in traffic crashes.

Congestion Costs: The value of travel delay, added fuel usage, greenhouse gas and criteria pollutants that result from congestion that results from motor vehicle crashes.

Economic Costs: The monetary impact of traffic crashes resulting from goods and services expended to respond to the crash, treat injuries, repair or replace damaged property, litigate restitution, administer insurance programs, and retrain or replace injured employees. Economic costs also include the health and environmental impacts that result from congestion, the value of workplace and household productivity that is lost due to death and injury, and the value of productivity and added travel time that is incurred by uninvolved motorists due to congestion from traffic crashes.

Emergency Services: Police and fire department response costs.

Household Productivity: The present value of lost productive household activity, valued at the market price for hiring a person to accomplish the same tasks.

Insurance Administration: The administrative costs associated with processing insurance claims resulting from motor vehicle crashes and defense attorney costs.

Legal Costs: The legal fees and court costs associated with civil litigation resulting from traffic crashes. Market Productivity: The present discounted value (using a 3-percent discount rate for 2010 dollars) of the lost wages and benefits over the victim's remaining life span.

Medical Care: The cost of all medical treatment associated with motor vehicle injuries including that given during ambulance transport. Medical costs include emergency room and inpatient costs, follow-up visits, physical therapy, rehabilitation, prescriptions, prosthetic devices, and home modifications.

Property Damage: The value of vehicles, cargo, roadways and other items damaged in traffic crashes.

Travel delay: The value of travel time delay for people who are not involved in traffic crashes, but who are delayed in the resulting traffic congestion from these crashes.

Vocational Rehabilitation: The cost of job or career retraining required as a result of disability caused by motor vehicle injuries.

Workplace Costs: The costs of workplace disruption that is due to the loss or absence of an employee. This includes the cost of retraining new employees, overtime required to accomplish work of the injured employee, and the administrative costs of processing personnel changes.

BIBLIOGRAPHY

Association for the Advancement of Automotive Medicine. (1985). *The Abbreviated Injury Scale 1985*. Des Plaines IL: Author.

AAAM. (1990). *The Abbreviated Injury Scale 1990.* Des Plaines IL: Association for the Advancement of Automotive Medicine.

Academic ED SBIRT Research Collaborative. (2010). The impact of screening, brief intervention and referral for treatment in emergency department patients' alcohol use: A 3-, 6- and 12-month follow-up. *Alcohol and Alcoholism*, *45*, 514–519.

AIRAC. (1988). *Attorney Involvement in Auto Injury Claims*. Oak Brook, IL: All-Industry Research Advisory Council.

Aldy, J. E. & Viscusi, W. K. (2007, April). *Age Differences in the Value of Statistical Life, Revealed Preference Evidence.* (Discussion Paper RFF DP 07-05). Washington, DC: Resources for the Future. Available at www.rff.org/documents/rff-dp-07-05.pdf.

American Council of Life Insurers. (2012). *Life Insurers Fact Book 2012*, Washington DC.

Annest, J. L., Mercy, J. A., Gibson, D. R. & Ryan, G. W. (1995). National estimates of nonfatal firearm- related injuries: beyond the tip of the iceberg. *Journal of the American Medical Association*, *273*, 22, 1749-1754.

Arias, E. (2012). United States Life Tables, 2008. *National Vital Statistics Reports*, *61*(3). Hyattsville, MD: National Center for Health Statistics.

Assistant Secretary for Planning and Evaluation. (2012). *Federal percentages and Federal medical assistance percentages, FY 1961 – FY 2011*. Washington, DC: Department of Health and Human Services. Available at http://aspe.hhs.gov/health/fmapearly.htm, accessed June 15, 2012.

Association for the Advancement of Automotive Medicine. (1985). T*he abbreviated injury scale*, *1985*. Des Plaines IL.

Association for the Advancement of Automotive Medicine. (1990). T*he abbreviated injury scale*, *1990*. Des Plaines, IL.

Bahar, P. E. & Miller, T. R. (2010, December). *Estimating the Costs to State Governments Due to Highway Related Debilitating and Fatal Crashes. (Working Paper 4*, Project 20-24 [068]). Washington, DC: National Academies of Sciences, National Cooperative Highway Research Program.

Barton, K. A., Blanchard, E. B. & Hickling, E. J. (1996). Antecedents and consequences of acute stress disorder among motor vehicle accident victims. *Behav Res Ther*, *34*, 805-13.

Berkowitz, M., Harvey, C., Greene, C. & Wilson, S. (1990). *The economic consequences of spinal cord injury*. Washington, DC: Paralysis Society of America of the Paralyzed Veterans of America.

Berkowitz, M.; O'Leary, P. K., Kruse, D. L. & Harvey, C. (1998). Spinal Cord Injury: *An Analysis of Medical and Societal Costs*. New York, NY: Demos Medical Publishing, Inc.

Blanchard, E. B., Hickling, E. J., Taylor, A. E., Loos, W. R., Forneris, C. A. & Jaccard, J. (1996). Who develops PTSD from motor vehicle accidents. *Behav Res Ther*, *34*(1), 1-10.

Blanchard, E. B., Hickling, E. J., Forneris, C. A., Taylor, A. E., Buckley, T. C., Loos, W. R. & Jaccard, J. (1997). Prediction of remission of acute posttraumatic stress disorder in motor vehicle accident victims. *J Trauma Stress*, *10*(2), 215-34.

Blanchard, E. B., Buckley, T. C., Hickling, E.J. & Taylor, A.E. (1998). Postraumatic stress disorder and comorbid major depression: is the correlation an illusion? *J Anxiety Disord*, 1998 Jan- Feb, *12*(1), 21-37.

Blaszczynski , A., Gordon, K., Silove, D., Sloane, D., Hillman, K. & Panasetis, P. (1998). Psychiatric morbidity following motor vehicle accidents: a review of methodological issues. *Compr Psychiatry*, *39*(3), 111-21.

Blincoe, L. & Shankar, U. (2007, January). *The Impact of Safety Standards and Behavioral Trends on Motor Vehicle Fatality Rates*. (Report No. DOT HS 810 777). Washington, DC: National Highway Traffic Safety Administration.

Blincoe, L., Seay, A., Zaloshnja, E., Miller, T., Romano, E., Luchter, S. & Spicer, R. (2002, May). *The Economic Impact of Motor Vehicle Crashes, 2000*. (Report No. DOT HS 809 446, May, 2002.

Blincoe, L. J. (1988, March). *A Model for Estimating the Economic Savings from Increased Motorcycle Helmet Use* ((Report No. DOT HS 807 251). Washington, DC: National Highway Traffic Safety Administration.

Blincoe, L. J. (1994). *Estimating the Benefits From Increased Safety belt Use*. (Report No. DOT HS 808 134). Washington, DC: National Highway Traffic Safety Administration.

Blincoe, L. J. (1996). T*he economic costs of motor vehicle crashes, 1994*. (Report No. DOT HS 808 425). Washington, DC: National Highway Traffic Safety Administration.

Blincoe, L. J. & Faigin, B. M. (1992). *The Economic Cost of Motor Vehicle Crashes, 1990*. (Report No. DOT HS 807 876). Washington DC: National Highway Traffic Safety Administration.

Blincoe, L. J. & Luchter, S. (1983). *The Economic Cost to Society of Motor Vehicle Accidents*. (Report No. DOT HS 806 342). Washington, DC: National Highway Traffic Safety Administration.

Blincoe, L. J. (1994). *Estimating the Benefits From Increased Safety Belt Use*, (Report No. DOT HS 808 133). Washington, DC: National Highway Traffic Safety Administration.

Blincoe, L. J., Seay, A., Zaloshnja, E., Miller, T., Romano, E., Luchter, S. & Spicer, R. (2002, May). *The Economic Impact of Motor Vehicle Crashes, 2000.* (Report No. DOT HS 809 446). Washington DC: National Highway Traffic Safety Administration.

Blomberg, R., Peck, R. C., Moskowitz, H., Burns, M. & Fiorentino, D. (2005, September). *Crash Risk of Alcohol Involved Driving: A Case-Control Study, Final Report*. Stamford, CT: Dunlap and Associates.

Blumentritt, C. W., Ross, D. W., Glazer, J., Pinnell, C. &McCasland, W. R. (1981, May). Guidelines for selection of ramp control systems. (NCHRP Report No. 232). Washington, DC: Transportation Research Board.

Bondy, N. (in review). How Well Do the National Estimates of the General Estimates System Compare to the State's Crash Numbers. Washington DC: National Highway Traffic Safety Administration.

Brach, R. M, & Brach, M. R. (2005). *Vehicle Analysis and Reconstruction Methods*. Warrendale, PA: SAE International.

Bureau of Labor Statistics. (2001). *Highlights of Women's Earnings in 2000*. Washington, DC: Bureau of Labor Statistics.

Bureau of the Census. (1994). Statistical Abstract of the United States 2010. Suitland, MD.

Bureau of the Census. (1999). 1997 Economic Census – Vehicle Inventory and Use Survey. Suitland, MD. Center for Substance Abuse Treatment. (1995). *Alcohol and other drug screening of hospitalized trauma patients. Treatment Improvement Protocol (TIP)*. (Series No. 16). Rockville, MD: Substance Abuse and Mental Health Services Administration.

Chang, J. L. & Hu, P. S., *Delays and Congestion Costs Caused by Truck Incidents – An Exploratory Analysis*. Proceedings: of the Second International Truck and Bus Safety

Symposium, October 6-8, 1999, University of Tennessee Transportation Center, Knoxville, TN.

Chaterjee, K. & McDonald, M. (2004). Effectiveness of Using Variable Message Signs to Disseminate DynamicTraffic Information: Evidence from Field Trials in European Cities. *Transport Reviews*, *Vol 24*, No. 5.

Chezem, L. (2005). Legal Barriers to Alcohol Screening in Emergency Departments and Trauma Centers. *RCHE Publications*, Paper 18. Regenstrief Center for Healthcare Engineering, Purdue University: Bloomington, IN, from http://docs.lib.purdue.edu/rche_rp/18

Chin, S. M, Franzese, O., Greene, D. L. & Hwang, H. L. (2002). *Temporary Losses of Capacity Study*. Oak Ridge TN: National Transportation Research Center, Oak Ridge National Laboratory.

Chin, S. M., Franzese, O., Greene, D. L., Hwang, H. L. & Gibson, R. C. (2004, November). *Temporary Losses of Highway Capacity and Impacts on Performance: Phase 2*. Oak Ridge, TN: Oak Ridge National Laboratory.

Chin, S. M., Franzese, O., Greene, D. L., Hwang, H. L. & Gibson, R. C. (2004, November). *Temporary Losses of Highway Capacity and Impacts on Performance: Phase 2* (Report No. ORNL/TM-2004/209. Oak Ridge, TN: Oak Ridge National Laboratory.

Consumer Product Safety Commission. (Last updated Aug 2009). *NEISS (National Electronic Injury Surveillance System)*. Retrieved August 13, 2009, from www.cpsc.gov/LIBRARY/neiss.html.

Council of Economic Advisers (2000). *Economic Report of the President*. Washington, DC: U.S. Government Printing Office.

Davidson, J. R.T. (2000). Trauma: the impact of post-traumatic stress disorder. *J Psychopharmacol*, *14*(2) Suppl I, S5-12.

Deutermann, W. (2004, March). Motorcycle Helmet Effectiveness Revisited. (Technical Report No. DOT HS 809 715). Washington, DC: National Highway Traffic Safety Administration.

Dischinger, P.C. & Cowley, R. A. (1989, October). Alcohol Use Among Victims of Vehicular Crashes Admitted to a Level I Trauma Center. 33rd Proceedings of the Association for the Advancement of Automotive Medicine, October 2-4, 1989, Baltimore, MD.

Environmental Protection Agency. (2009). Draft Motor Vehicle Emissions Simulator (MOVES) 2009: Software Design and Reference Manual. Washington, DC: Author. Available at www.epa.gov/otaq/models/moves/420b09007.pdf

Environmental Protection Agency. (2010). *Light Duty Automotive Technology, Carbon Dioxide Emissions, and Fuel Economy Trends: 1975 Through 2010*. Available at www.epa.gov/otaq/fetrends.htm, Updated November 17, 2010

Expectancy Data. (1999). *The Dollar Value of a Day – 1997 Dollar Valuation*. Shawnee, KS: Expectancy Data.

Fahrenthold, D. (2013, June 9). What does rural mean? Feds have 15 answers. Washington, DC: *The Washington Post*, p. A-1.

Festin, S. M. (1996, May). *Summary of National and Regional Travel Trends: 1970-1995*. Washington, D.C.: Federal Highway Administration.

Finkelstein, E. A., Corso, P. C., Miller, T. R., Fiebelkorn, I. A., Zaloshnja, E. & Lawrence, B. A. (2006). *Incidence and Economic Burden of Injuries in the United States, 2000*. New York: Oxford University Press.

Friedman, B., De La Mare, J., Andrews, R. & McKenzie, D. H. (2002). Practical options for estimating cost of hospital inpatient stays. *Journal of Health Care Finance*, *29*(1), 1–13.

Gaines, L., Levinson, T. & McConnell, S. (2010). *To Idle or Not to Idle: That is the Question*, Argonne, IL: Argonne National Laboratory. Available at www.transportation.anl.gov/pdfs/EE/642.PDF

Garthe, E. A., Ferguson, S. A. & Early, N. (1996). A Method for Converting Injury Severity in NASS93 (AIS90) to NASS88 (AIS 85). 40th Annual Proceedings, Association for the Advancement of Automotive Medicine, October 7-9, 1996; Vancouver, BC, 477-494.

Glassbrenner, D. & Starnes, M. (2009, December). *Lives Saved Calculations for Seat Belts and Frontal Air Bags* (Report No. DOT HS 811 206). Washington, DC: National Highway Traffic Safety Administration.

Glenn, D. P. (2010). U.S. property/casualty loss ratio improved in 2009. A.M. Best & Company, *Best's Review*, 91–94.

Gold, M. R., Siegel, J. E., Russell, L. B. & Weinstein, M. C. (Eds.). (1996). *Cost-effectiveness in Health and Medicine: Report of the Panel on Cost-Effectiveness in Health and Medicine*. New York, NY: Oxford University Press.

Government Accountability Office. (2012). *Ambulance providers: Costs and Medicare margins varied widely; Transports of beneficiries have increased*. (Report No. GAO-13-6). Washington DC.

Greenblatt, J., Merrin, M. B. & Morganstein, D. (1981, November). National Accident Sampling System Nonreported Accident Survey. (Report No. DOT HS 806 198).Washington, DC: National Highway Traffic Safety Administration.

Guo, H., Eskridge, K. M., Christensen, D., Qu, M. & Safranek, T. (2007). Statistical adjustment for misclassification of seat belt and alcohol use in the analysis of motor vehicle accident data. *Accident Analysis and Prevention*, *39*, 117–124.

Haddix, A. C., Teutsch, S. M. & Corso, P. S. (2003). *Prevention Effectiveness, A Guide to Decision Analysis and Economic Evaluation*. Second ed. New York: Oxford University Press.

Hagemann, G., Hymel, K., Klauber, A., Lee, D. B., Noel, G., Pace, D. & Taylor, C. (2013, March). *Delay and Environmental Costs of Truck Crashes*. (Report No. FMCSA-RRA-13-043). Washington, DC: Federal Motor Carrier Safety Administration. Available at www.fmcsa.dot.gov/facts-research/research-technology/report/Crash-Costs-Final-Report-508.pdf

Harder, K. A., Bloomfield, J. & Chihak, B. J. (2004). *The Effectiveness and Safety and Non-Traffic Related Messages Presented on Changeable Message Signs (CMS)*. (Technical Report 2004-27). Mineapolis, MN: Minnesota Department of Transportation.

Hartwick, J. M. (2008, November 17). *The Discount Rate and the Value of Remaining Years of Life*, Kingston, Ontario: Queen's University.

Hickman, M. J., Hughes, K. A., Strom, K. J. & Ropero-Miller, J. D. (2007). *Medical Examiners and Coroners Offices, 2004, Special Report*. Washington, DC: Bureau of Justice Statistics, Department of Justice.

Hirsch, A., Eppinger, R., Shame, T., Nguyen, T., Levine, R., Mackenzie, J., Marks, M. & Ommaya, A. (1983). *Impairment scaling from the abbreviated injury scale*. Washington D.C: National Highway Traffic Safety Administration.

Horowitz, A. J., Weisser, I. & Notbohm, T. (2003). Diversion From a Rural Work Zone With Traffic Responsive Variable Message Sineage System. *Journal of the Transportation Research Board*, No. 1824. Washington, DC: Transportation Research Board of the National Academies.

Hoyert, D. L. (2011). Unpublished supplemental table accompanying *The Changing Profile of Autopsied Deaths in the United States, 1972–2007*. (NCHS Data Brief No. 67). Hyattsville, MD: National Center for Health Statistics.

Insurance Research Council. (208, November 24). Fraud and buildup add 13 to 18 percent in excess payments to auto injury claims. (Press Release). Retrieved from www.insurance-research.org/sites/default/files/downloads/IRC_Fraud_NR.pdf

Insurance Research Council. (2011, April 11). Recession marked by bump in uninsured motorists. (Press Release). Retrieved from www.insurance- research.org/sites/default/files/downloads/IRCUM2011_042111.pdf

Interagency Working Group on Social Cost of Carbon. (2010). *Social cost of Carbon for Regulatory Impact Analysis Under Executive Order 12866*. Washington, DC: U. S. Government.

Interagency Working Group on the Social Cost of Carbon. (2013, May). *Technical Update of the Social Cost of Carbon for Regulatory Impact Analysis – Under Executive Order 12866*. Washington, DC: U. S. Government. Available at www.whitehouse.gov/sites/default/files/omb/assets/inforeg/technical-update-social-cost-of- carbon-for-regulator-impact-analysis.pdf

Kahane, Charles J. (2000). *Fatality Reduction by Safety Belts for Front Seat Occupants of Cars and LightTrucks, Updated and Expanded Estimates Based on 1986 – 99 FARS Data*. (Report No. DOT HS 809 199). Washington, DC: National Highway Traffic Safety Administration.

Kakalik, J. S. & Pace, N. M. (1986). *Costs and Compensation Paid In Tort Litigation*. (Report R-3391-ICJ). Santa Monica, CA: RAND Corp.

Knoop, V. L., Hoogendoorn, S. P. & van Zuylen, H. J. (2008). Capacity Reduction at Incidents, Empirical Data Collected from a Helicopter. (Journal of the Transport Research Board, No. 2071) Washington, DC: Transportation Research Board of the National Acadamies.

Kochi, I., Hubbell, T. & Kramer, R. (2006, July). An Empirical Bayes Approach to Combining and Comparing Estimates of the Value of a Statistical Life for Environmental Policy Analysis. *Environmental and Resource Economics, 34.3*.

Lacey, J., Kelley-Baker, T., Furr-Holden, D., Voas, R., Romano, E., Torres, P. & Berning, A. (2009). *2007 National Roadside Survey of Alcohol and Drug Use by Drivers: Alcohol Results*. (Report No. DOT HS 811 248) Washington, DC: National Highway Traffic Safety Administration.

Lawrence, B., Miller, T. R., Jensen, A., Fisher, D. & Zamula, W. (1999). Estimating the Costs of Nonfatal Consumer Product Injuries in the United States. *Proceedings of the 7th*

International Conference on Product Safety Research, September 30-October 1, 1999, Washington, DC.

Lawrence, B. A., Miller, T. R., Jensen, A. F., Fisher, D. A. & Zamula, W. W. (2000). Estimating the costs of non-fatal consumer product injuries in the United States. *Injury Control & Safety Promotion*, *7*(2), 97–113.

Lu, S. (2006, January). Vehicle Survivability and Travel Mileage Schedules (NHTSA Technical Report No. DOT HS 809 952). Washington, DC: National Highway Traffic Safety Administration. Available at www-nrd.nhtsa.dot.gov/Pubs/809952.pdf.

MacKenzie, E. J. & Sacco, W. J. (1997). ICDMAP-90 Software: User's Guide. Baltimore: Johns Hopkins University and Tri-Analytics, Inc.

Malt, U. (1988). The long-term psychiatric consequences of accidental injury. A longitudinal study of 107 adults. *Br J Psychiatry. 153*:810-8.

Masinick, J. P. & Teng, H. (2004, August). An Analysis on the Impact of Rubbernecking on Urban Freeway Traffic. (Research Report No. UVACTS-15-0-62). Charlottesville, VA: Center for Transportation Studies, University of Virginia. Available at cts.virginia.edu/docs/UVACTS-15-0-62.pdf

Maull, K. I., Kunning, L. S. & Hickman, J. K. (1984, October 12) Culpability and Accountability of Hospitalized Injured Alcohol-Impaired Drivers. *Journal of the American Medical Association*, *252*, 14.

Miller, T. R., Whiting, B., Kragh, B. & Zegeer, C. (1987). Sensitivity of a Highway Safety Resource Allocation Model to Variations in Benefit Computation Parameters. *Transportation Research Record*, *1124*, 58-65.

Miller, T. R.; Cohen, M. & Wiersema, B. (1996). *Victim Costs and Consequences: A New Look*. Rockville, MD: National Institute of Justice.

Miller, T. R. (2000). Valuing non-fatal quality-of-life losses with quality-adjusted life years: The health economist's meow. *J. Forensic Econ*, *13*(2), 145-168.

Miller, T. R., Viner, J., Rossman, S., Pindus, N., Gellert, W., Dillingham, A. & Blomquist, G. (1991). *The Costs of Highway Crashes*. Washington, DC: The Urban Institute.

Miller, T. R., Pindus, N. & Douglass, J. (1993). Motor vehicle injury costs by body region and severity, *J. Trauma*, *34*, 270-275.

Miller, T. R. (1990). The Plausible Range for the Value of Life – Red Herrings among the Mackerel. *Journal of Forensic Economics*, *3*, 1912-46.

Miller, T. R. (1993). Costs and Functional Consequences of US Roadway Crashes. *Accident Analysis and Prevention.*, *25*, 593-607.

Miller, T. R. (1997). Societal Costs of Transportation Crashes. In D. Greene, D. Jones, & M. Delucchi (Eds.), *The Full Social Costs and Benefits of Transportation*, Heidelberger, Germany: Springer- Verlag, 281-314.

Miller, T. R. (2000). Variations between Countries in Value of Statistical Life. *Journal of Transport Economics and Policy*, *34*, 169-188.

Miller, T. R. & Blincoe, L. J. (1994). Incidence and cost of alcohol-involved crashes in the United States. *Accident Analysis and Prevention*, *26*(5), 583–592.

Miller, T. R. & Lawrence, B. A. (2003). *Motor vehicle insurance in the United States: A 1998-1999 snapshot with emphasis on motorcycle coverage*. (Report No. DOT HS 809 494). Washington, DC: National Highway Traffic Safety Administration.

Miller, T. R., Bhattacharya, S., Zaloshnja, E., Taylor, D. M., Bahar, G. & David, I. (2011). Costs of crashes to government, United States. *Annals of Advances in Automotive Medicine*, *55*, 347–356.

Miller, T. R., Brigham P. A., Cohen M. A., Douglas J. B., Galbraith M. S., Lestina D. C., Nelkin V. S., Pindus N. M. & Smith-Regojo P. (1993). *Societal Cost of Cigarette Fires.* Washington, DC: Consumer Product Safety Commission.

Miller, T. R., Galbraith, M., Lestina, D., Schlax, T., Mabery, P., Deering, R., Massie, D. & Campbell, K. (1995). *Understanding the harm from U.S. motor vehicle crashes.* 39th Annual Proceedings, Association for the Advancement of Automotive Medicine, October 16-18, 1995, Chicago, IL, 327–342.

Miller, T. R., Langston, E. A., Lawrence, B. A., Becker, L. R., Snowden, C. B., Granger, C. V., Russell, C., Kreutzer, J. S. & Marwitz, J. (2006). *Rehabilitation costs and long-term consequences of motor vehicle injury.* (Report No. DOT HS 810 581).Washington, DC: National Highway Traffic Safety Administration. Available at www.nhtsa.gov/people/injury/research/RehabCosts/images/Final%20ReportLo.pdf.

Miller, T. R., Lawrence, B. A., Jensen, A. F., Waehrer, G. M., Spicer, R. S., Lestina, D. C. & Cohen, M. A. (2000a). *The Consumer Product Safety Commission's Revised Injury Cost Model.* Landover MD: Public Services Research Institute.

Miller, T. R., Lestina, D. C. & Spicer, R. S. (1998). Highway Crash Costs in the United States by Driver Age, Blood Alcohol Level, Victim Age and Restraint Use. *Accident Analysis and Prevention*, *30*(2) 137-150.

Miller, T. R., Levy, D. T., Spicer, R. S., Lestina, D. C. (1998). Allocating the Costs of Motor Vehicle Crashes Between Vehicle Types, *Transportation Research Record*, *1635*, 81-87.

Miller, T. R., Pindus, N., Douglass, J. & Rossman, S. (1995). *Databook on Nonfatal Injury – Incidence, Costs, and Consequences.* Washington, DC: The Urban Institute Press.

Miller, T. R., Romano, E. & Spicer, R. S. (2000). The Cost of Unintentional Childhood Injuries and the Value of Prevention. *The Future of Children*, *10*(1) 137-163.

Miller, T. R., Spicer R. S. & Levy, D. T. (1999). How Intoxicated are Drivers in the United States? Estimating the Extent, Risks, and Cost per Kilometer of Driving by Blood Alcohol Level, *Accident Analysis and Prevention*, *V.31* No. 5, 1999.

Miller, T. R., Spicer R. S., D Lestina, D. C. & Levy, D. T. (1999). Is It Safest to Travel By Bicycle, Car or Big Truck? *Journal of Crash Prevention and Injury Control*, *1*(1) 25-34.

Miller, T. R., Taylor, D. M. & Zaloshnja, E. (2010). *The economic burden of motor vehicle crashes on employers - 2010.* Washington, DC: National Highway Traffic Safety Administration.

Miller, T. R., Lawrence, B., Jensen, A., Waehrer, G., Spicer, R., Lestina, D. & Cohen, M. (1998). *Estimating the Cost to Society of Consumer Product Injuries: The Revised Injury Cost Model.* Bethesda, MD: Consumer Product Safety Commission.

Miller T., Gibson, R. Zaloshnja, E., Blincoe, L., Kindelberger, J., Strashny, A., Thomas, A., Ho, S., Bauer, M., Sperry, S., Peng, J., Singleton, M., Smith, T. & Zhang, Y. (2012). Underreporting of driver alcohol involvement in United States police and hospital records: capture-recapture estimates. *Annals of Advances in Automotive Medicine*, *56*, 87-96.

Mirza, K. A. H., Bhadrinath, B. R., Goodyer, I. M. & Gilmour, C. (1998). Post-traumatic stress disorder in children and adolescents following road traffic accidents. *Br J Psychiatry 172*, 443-7

Morgan, C. (1999, June). *Effectiveness of Lap/Shoulder Belts in the Back Outboard Seating Positions* (Report No. DOT HS 808 945). Washington, DC: National Highway Traffic Safety Administration. Available at *www*-nrd.nhtsa.dot.gov/Pubs/808945.PDF

Mrozek, J. R. & Taylor, L. O. (2002). Who Determines the Value of Life? A Meta-Analysis. *Journal of Policy Analysis and Management*, *21*.2 253-270.

National Highway Traffic Safety Administration. (1984, July). *Final Regulatory Impact Analysis Amendment to FMVSS No 208. Passenger Car Front Seat Occupant Protection*, page IV-2. National Highway Traffic Safety Administration, Plans and Programs Office of Planning and Analysis.

NHTSA. (1988, March). A Model for Estimating the Economic Savings from Increased Motorcycle Helmet Use. National Highway Traffic Safety Administration, Report No. DOT HS 807 251. NHTSA. (2011a). *National Accident Sampling System 2010*. Washington, DC.

NHTSA. (1990, November). *Final Regulatory Impact Analysis Extension of the Automatic Restraint Requirements of FMVSS 208 to Trucks, Buses, and Multi-Purpose Passenger Vehicles with a Gross Weight Rating of 8,500 Pounds or Less and an Unloaded Vehicle Weight of 5,500 Pounds or Less*, page 23. National Highway Traffic Safety Administration, Plans and Policy, Office of Regulatory Analysis.

NHTSA. (1995). *National Automotive Sampling System Crashworthiness Data System 2010*. Washington, DC.

NHTSA. (2011b, March) *Determining Estimates of Lives and Costs Saved by Motorcycle Helmets*. (Traffic Safety Facts Research Note, Report No. DOT HS 811 433). Washington, DC.

NHTSA. *Final Regulatory Impact Analysis, Corporate Average Fuel Economy for MY 2017-MY 2025 Passenger Cars and Light Trucks*, Office of Regulatory Analysis and Evaluation, National Center for Statistics and Analysis, August 2012.

NHTSA. (2008, July). *National Motor Vehicle Crash Causation Survey, Report to Congress*. (Report No. DOT HS 811 059). Washington, DC.

NHTSA.(2011c). *National Telephone Survey of Reported and Unreported Motor Vehicle Crashes*.(Project No. DTNH22-08-C-0065, Findings Report). M. Davis and Company.

NHTSA. (2012, September). *Distracted Driving 2010*. (Traffic Safety Facts Research Note, Report No. DOT HS 811 650). Washington, DC: Author. Available at www-nrd.nhtsa.dot.gov/Pubs/811650.pdf

NHTSA. (2012, November). *Seat Belt Use in 2012- Overall Results*, NHTSA, (Traffic Safety Facts Research Note, Report No. DOT HS 811 691) Washington, DC.

National Pediatric Trauma Registry. (2002). Biannual report, December 2001. Boston, MA: Tufts University.

National Safety Council. (1990). *Manual on Classification of Motor Vehicle Traffic Accidents*. Fifth Edition. (ANSI D-16.1-1989). Itasca, IL.

National Vital Statistics Reports. (1997). Vol 47, No. 28 United States Life Tables.

O'Day, J. (Ed.). (1993). Accident data quality: A synthesis of highway practice. *National Cooperative Highway Research Program Synthesis 192*. Washington, DC: Transportation Research Board.

Office of the Assistant Secretary for Planning and Evaluation. Federal Percentages and Federal Medical Assistance Percentages, FY 1961 - FY 2011, US Department of Health and Human Services, http://aspe.hhs.gov/health/fmapearly.htm, accessed June 15, 2012.

Partyka, S. C. & Womble, K. B. (1989, June). *Projected Lives Savings From Greater Belt Use*. (Research Note). Washington, DC: National Highway Traffic Safety Administration.

Piccinelli, M., Patterson M., Braithwaite I, Boot D. & Wilkinson G. (1999). Anxiety and depression disorders 5 years after severe injuries; a prospective follow-up study. *J Psychosom Res*, *46*(5), 455-64.

Pope, C. A. III, Burnett, R. T., Thun, M. J., Callee, E. E., Krewski, D., Ito. & Thurston, G. D. (2002, March). Lung Cancer, Cardio-Pulmonary Mortality, and Long-term Exposure to Fine Particulate Air Pollution. *JAMA*, *287*:9; 1132-41.

Rice, D. P., MacKenzie, E. J., Jones, A. S., Kaufman, S. R., deLissovoy, G. V., Max, W., McLoughlin, E., Miller, T. R., Robertson, L. S., Salkever, D. S. & Smith, G. S. (1989). *Cost of injury in the United States: A report to Congress*. San Francisco, CA: Institute for Health & Aging, University of California, & Injury Prevention Center, The Johns Hopkins University.

Rubin, D. B., Schafer, J. L. & Subramanian, R. (1998, October). *Multiple Imputation of Missing Blood Alcohol Concentration (BAC) Values in FARS*. (Report No. DOT HS 808 816). Washington, DC: National Highway Traffic Safety Administration.

Selassie, A. W., Zaloshnja, E., Langlois, J. A., Miller, T. R., Jones, P. & Steiner, C. (2008). Incidence of long- term disability following traumatic brain injury hospitalization, United States, 2005. *Journal of Head Trauma Rehabilitation*, *23*(2), 123–131.

Soderstrom, C. A., Birschbach, J. M. & Dischinger, P. C. (1990). Injured drivers and alcohol use: Culpability, convictions, and pre- and post-crash driving history. *Journal of Trauma*, *30*(10), 1208–1213, discussion 1213–1214.

Spicer, R. S. & Miller, T. R. (2010, February). *Uncertainty Analysis of Quality Adjusted Life Years Lost*. Washington, DC: National Highway Traffic Safety Administration.

Spicer, R. S., Miller, T. R., Hendrie, D. & Blincoe, L. J. (2011, October). Quality-adjusted life years lost to road crash injury: Updating the injury impairment index. *Ann Adv Automot Med.; 55*, 365–377. PMCID: PMC3256815

Srinivasan,K. K. & Krishnamurthy, A. Roles of Spatial and Temporal Factors in Variable Message Sign Effectiveness Under Nonrecurrent Congestion. (Report No. 1854). Washington, DC: Transportation Research Record, *Journal of the Transportation Research Board*. Available at http://trb.metapress.com/content/a8j16r51gj7102h7/fulltext.pdf

Terhune, K. W. & Fell, J. C. (1981). The role of alcohol, marijuana, and other drugs in the accidents of injured drivers. Proceedings of the 25th Annual Conference of the American Association for Automotive Medicine, San Francisco, CA, October 1-3, 1981.

Treat, J. R., Tumbas, N. S., McDonald, S. T., Shinar, D., Hume, R. D., Mayer, R. E., Stansifer, R. L. & Castellan, N. J. (1979, May). *Tri-Level Study of the Final Causes of Traffic Accidents, Final Report, Volume 1: Casual Factor Tabulations and Assessments* (Report No. DOT HS 805 085). Washington, DC: National Highway Traffic Safety Administration. Available at http://ntl.bts.gov/lib/25000/25500/25515/DOT-HS-805-085.pdf

Viner, J. G. & Conley, C. (1994). *Consistency of police-reported incapacitating injuries" between States*. Washington, DC: Federal Highway Administration.

Viscusi, W. K. (1993). The Value of Risks to Life and Health. *Journal of Economic Literature*, *31*, 1912-46.

Viscusi, W. K. & Aldy, J. ED. (2003). The Value of a Statistical Life: A Critical Review of Market Estimates Throughout the World. *Journal of Risk and Uncertainty*, *27*.1, 5-76.

Wang, J-S. & Blincoe, L. J. (2001). *BELTUSE Regression Model Update*. Washington, DC: National Highway Traffic Safety Administration.

Wang, J-S. & Blincoe, L. J. (2003). *Belt Use Regression Model – 2003 Update*. (Traffic Safety Facts Research Note, Report No. DOT HS 809 639). Washington, DC: National Highway Traffic Safety Administration. Available at www-nrd.nhtsa.dot.gov/Pubs/809639.pdf

Woolhandler, S., Campbell, T. & Himmelstein, D. U. (2003). Costs of health care administration in the United States and Canada. *New England Journal of Medicine*, *349*(8), 768–775.

Yim, Y. & Ygnace, J-L. (1996). *Link Flow Evaluation Using Loop Detector Data: Traveler Response to Variable-Message Signs*. Washington, DC: Transportation Research Record.

Zaloshnja, E. & Miller, T. R. (2012). How often do catastrophic injury victims become Medicaid recipients?, *Medical Care*, *50*(6), 513–519.

Zaloshnja, E., Miller, T. R., Romano, E. O. & Spicer, R. S. (2004). Crash costs by body part injured, fracture involvement, and threat to life severity, United States, 2000. *Accident Analysis and Prevention*, *36*(3), 415–427.

Zaloshnja, E., Spicer, R. S., Romano, E. O. & Miller, T. R. (2001). Does using AIS85 costs with AIS90 data create serious errors? *Proceedings of the 45th Annual Conference of the Association for the Advancement of Automotive Medicine*, San Antonio, Texas, September 24–26, 2001.

Zaloshnja, E., Miller, T. R., Spicer, R. S. (2000). *Costs of Large Truck and Bus-Involved Crashes*. Washington, DC: Federal Motor Carrier Safety Administration.

End Notes

[1] The Abbreviated Injury Scale (AIS) is an anatomically based, consensus-derived global severity scoring system that classifies each injury by body region according to its relative importance on a 6-point ordinal scale (1=minor and 6=maximal). AIS is the basis for the Injury Severity Score (ISS) calculation of the multiply injured patient. The AIS was developed by the Association for the Advancement of Automotive Medicine (AAAM) See www.aaam1.org/ais/ for further information.

[2] Costs for AIS98 are essentially the same as for AIS90. No data exists to estimate costs accurately for more recent AIS codes except through equivalency tables to older coding. Similarly detailed incidence data for estimating costs in non-CDS strata for MAIS versions other than MAIS85 (notably for heavy truck occupant injury) do not exist.

[3] Note that some MAIS level injuries do not exist for certain body regions. These are noted as NA in this table.

[4] Note that some MAIS level injuries do not exist for certain body regions. These are noted as NA in this table.

[5] Note that some MAIS level injuries do not exist for certain body regions. These are noted as NA in this table.

[6] This consisted of calculating a simple normalizing factor by comparing the results of the main analysis to the sum of the separately calculated reported and unreported analyses. This factor was then applied back to the unit costs. This process maintains the relative differences found in the MDAC analysis, while remaining consistent with the original unit costs and incidence totals, which were derived from a more robust data set.

[7] For crashes with lane closings, crash duration is a proxy for the time lanes are closed. This assumes the crash immediately blocks the traffic lanes or lanes, and that police or emergency vehicle presence remains throughout the timeframe when vehicles are removed.

[8] AADT is the average number of vehicles that travel on a roadway over the course of a day.

[9] This data represents annual average AADT over the 2005-2008 period, the latest data available at the time of this analysis.

[10] $\frac{\sum AADT \cdot SegmentLength}{\sum SegmentLength}$

[11] This data indicated that 65 percent of fatal crashes, 75 percent of injury crashes, and 77 percent of PDO crashes occurred on weekdays. The remainders, 35 percent of fatal crashes, 25 percent of injury crashes, and 23 percent of PDO crashes occurred on weekends.

[12] Personal communication in 2000 from Patricia S. Hu to Ted R. Miller regarding data from 1999 study conducted by Chang-Jen Lan and Patricia S. Hu of urban freeways in Minneapolis-St. Paul.

[13] The jurisdictions included Dade County, FL, Lakewood, CO, Montgomery County, MD, San Antonio, TX, and San Jose, CA.

[14] For example, Lan and Hu found roughly twice the delay hours in crashes involving heavy trucks compared to those that did not involve heavy trucks. Personal communication cited in Zaloshnja et al., 2000.

[15] Figure E simplifies how delay caused by crashes, particularly major crashes, accumulates, since it ignores both the modification of the reduced capacity made when the crash-involved vehicles are moved to the shoulder and the fact that traffic demand is not usually constant during a major crash. However, it is used here only to describe the basic congestion dynamic.

[16] Similar data on crash location was not available in nonfatal injury databases (GES and CDS).

[17] Separate calculations were required for each roadway type to link with AAHT from Table 3-3.

[18] Note: Average Vehicle Capacity Reduction refers to the reduction in the number of vehicles that would have passed the crash site during the duration of the crash, had the crash not occurred.

[19] This basic relationship is commonly found in engineering texts. See for example, Brach and Brach, 2005.

[20] Although there were only 5 data points, we chose a curvilinear model because there is a natural expectation that the decision to divert from a traffic queue would increase in this manner. The longer the queue, the more likely it is that a detour would provide a more efficient route. We also note that although there are a limited number of data points, they are visually consistent with this type of model. Yim and Ygnace only measured diversion on backups of up to 2.5 miles. The curvilinear model only affects queue lengths greater than 2.5 miles under the fatal crash on urban interState expressway scenario. Under these dense traffic circumstances, queues well in excess of 2.5 miles can accumulate. In an urban environment, however, there are typically many opportunities to detour to other routes, so we would expect a relatively high rate of congestion avoidance. It is possible that a curvilinear model could overstate diversion under more extreme conditions. We note that if this model does overestimate the rate of diversion to alternate routes for these extreme conditions, its net impact is to provide a conservative estimate of overall congestion impacts, since those alternate routes would be more efficient than the alternative of waiting for the main roadway to be cleared.

[21] 20 light cycles were observed. This average reflects a variety of observed response behaviors. Most drivers began their acceleration shortly after the vehicle in front of them began to move, but some drivers began to accelerate at virtually the same time as the frontward vehicle, anticipating the forward driver's response and taking advantage of the gap between vehicles as a safety margin for their own simultaneous movement. Offsetting these more efficient behaviors were relatively large time delays caused by drivers who were distracted by cell phones, texting, or other less attentive activities.

[22] Estimate based on engineering judgment. Under this assumption a vehicle would move about 5 feet in the first second and an additional 15 feet in the next second or roughly 20 feet after 2 seconds.

[23] Based on 2010 GES data, commercial trucks are involved in about 5 percent of all police-reported motor vehicle crashes, and make up about 3 percent of all vehicles involved in police-reported motor vehicle crashes. They thus represent a relatively small subset of the universe of crashes addressed in this study.

[24] Personal communication to Ted Miller cited in Zaloshnja, Miller, and Spicer, 2000.

[25] This does not account for assessor and air conditioning loads, which would increase consumption.

[26] http://mctrans.ce.ufl.edu/featured/TSIS/

[27] Derived from the updated data used in production of FHWA publication "Highway Statistics, 2010." See www.fhwa.dot.gov/policyinformation/statistics/2010/mf27.cfm. Accessed on January 8, 2012. This data indicates 78.5 percent of fuel used by motor vehicles is gasoline, and 21.5 percent are alternative fuels, primarily diesel.

[28] Note that separate values are applied to upstream NO_x and PM2.5 emissions. EPA has determined that the different populations exposed to upstream emissions face slightly different health impacts than those exposed through tailpipe emissions. Upstream emissions for NO_x are valued at $4,481 and those for PM2.5 are valued at $211,602. Also note that the CO_2 value used in the CAFE analysis was $22.22 in 2010 dollars per metric ton. This was converted to a short ton value of $20.16 to derive a number compatible with the FMCSA basis.

[29] Note that a small portion of the on-road fleet are diesels, which have different emission characteristics than gasoline engines. Diesel has a slightly higher energy content than gasoline (138,700 BTU/gal versus 125,000) and thus burns about 10 percent more CO_2/gallon. However, they are also more efficient and thus waste somewhat less fuel. These two factors are partially offsetting. This analysis is based on the published EPA values for gasoline.

[30] Excludes State and local taxes, which are a transfer payment from one segment of society to another, and thus are not counted as a societal cost. The $2.46 price results from the FMCSA simulation in which average gasoline prices were $2.43/gallon and average diesel prices were $2.52.

[31] Memorandum to Secretarial Officers and Modal Administrators, "Revised Departmental Guidance on Valuation of Travel Time in Economic Analysis" from Polly Trottenberg, Assistant Secretary for Transportation Policy, Prepared by Peter Belenky, Economist, September 29, 2011.

[32] For all MAIS and fatal injury categories, costs are expressed on a per injured person basis. For PDO crashes, costs are expressed on a per damaged vehicle basis.

[33] We acknowledge that the selection of "half" as a factor to reflect the nature off unreported crashes is somewhat arbitrary. However, lacking specific data, we are hesitant to select values that imply that unreported crashes would have impacts that are more nearly like those of police-reported crashes or closer to zero. We view half as the best way to minimize potential error. Directionally, we only know unreported crashes would cause some level of congestion but that that it is less than reported crashes.

[34] See www.fhwa.dot.gov/policyinformation/statistics/2008/pdf/ps1.pdf All values are for 2008 except percent truck VMT, which was last published for 2005.

[35] All values are for 2008 except percent truck VMT, which was last published for 2005.

[36] VSLY is the value of a statistical life year – a single year of remaining statistical life rather than the full value of all remaining life years as measured by VSL.

[37] After-tax market productivity is inherent to VSLs because it determines the individual's valuation of their potential material consumption. Household productivity is inherent to VSLs because it is a routine activity that is part of life experience. Both aspects are potentially threatened by behaviors that increase risk, and are thus inherently already reflected in the VSL.

[38] Bondy, N., Validation of the National Estimates Produced From NASS GES, NHTSA Research Note, 2014.

[39] Note that even for 2009 data was not available for Montana. Data from 2008 was substituted for Montana. Total crashes for Colorado and Hawaii were estimated based on trends in fatal crashes. This should not significantly impact the accuracy of this analysis.

[40] Elimination of the extreme results for Wake County, for example, would by itself reduce the unreported average to 35 percent.

[41] It is also possible that respondents were not accurately portraying their true medical costs, or that all of their costs were covered by insurance and they misunderstood the question. However, since most policies contain at least some kind of deductible and/or co-pay, this later case seems unlikely.

[42] NHIS covers the civilian, non-institutionalized population living in households, so it would not include military, institutionalized, or homeless population. Neither institutionalized nor homeless would be included in the MDAC survey either, but military living in households could be. This might account for a small portion of the difference between these surveys.

[43] MAIS 0 injuries are allocated a portion of crash costs, primarily congestion and property damage. However, they do not have medical care costs or associated lost market productivity. Allocation of these cases, for which no medical treatment was sought, to MAIS0, is thus done to more accurately account for their lower costs.

[44] For incidence purposes, we used only the 2010 portion of the reweighted hybrid CDS/NASS casualty-level file and the 2010 FARS file

[45] See for example, NHTSA, 2012, September. That publication was based on preliminary data files, whereas this current analysis is based on final 2010 FARS and GES files, and translates KABCO injuries into MAIS equivalents. This results in small but noticeable differences between the previously published data and this current analysis.

[46] This definition is different than previous years and cannot be compared directly to cell phone involvement prior to 2010. See NHTSA, 2012, September for further details.

[47] The discrepancy between the rates for fatal and nonfatal crashes may be a function of police inability to interview survivors in fatal crashes where drivers or occupants are deceased.

[48] There were 81,979 injuries estimated in the 2010 GES. These were adjusted using the same 10.7 percent markup factor discussed in the Incidence chapter to reflect undercounting in GES compared to State total police-reported crashes. It is assumed this same level of undercounting applies to all crash types. This adjustment produces a total of 90,753 nonfatal injuries. However, further adjustment to reflect the more accurate MAIS coding structure indicates over 96,000 nonfatal injuries. See further discussion in this section.

[49] Although VMT is the preferable basis for fatality rates, motorcycle fatality VMT was not recorded reliably until 2007, therefore rate comparisons are based on vehicle registrations.

[50] The heavy truck fatality rate per registered vehicle is much higher than the passenger car and light truck rates because heavy trucks drive many more miles per year.

[51] Motorcycles were included in the MDAVIS survey discussed in the Incidence chapter. However, the survey only contained 6 motorcycle cases. The weighted police reporting rate for those cases (53%) was almost identical to the overall rate for all crashes (54%), but these are too few cases to rely on for a separate motorcycle reporting rate.

[52] See NHTSA, 1988, Blincoe, 1994, and NHTSA, 2011, March.

[53] This analysis includes only police-reported injuries. About 20 percent of MAIS2 injuries and 4 percent of MAIS3 injuries are estimated to be unreported. Belt use rates are unknown for unreported crashes. If belt use rates for

unreported crashes are similar to reported crashes, benefits for each of these two categories would increase proportionally. All MAIS 4, 5, and Fatal injuries are estimated to be reported to police.

[54] Prior years' unit costs were estimated by deflating the 2010 unit costs using the CPI annual average All Items index.

[55] Intersection crashes includes crashes that occur at normal roadway intersections, at driveway or alleyway intersections, and at some highway interchanges.

[56] Roadway departure crashes are defined as crashes in which a vehicle crosses an edge line, a centerline, or otherwise leaves the travelled way. This includes crashes where the first event in the sequence coded for any involved vehicle is run off the road to either the right or left, cross a median, cross a center line, and hit a permanent fixed object, or become airborne, or re-entered the roadway.

[57] This category includes all crashes where a pedestrian or bicyclist was involved. This includes cases where a driver swerves to avoid a pedestrian or bicyclist and crashes his vehicle, causing property damage or injury to the vehicle occupants. It thus includes counts of all fatalities, injuries, or property damage that occur in crashes where a pedestrian or bicyclist was involved, regardless of whether the pedestrian or bicyclist was struck or injured.

[58] This category includes all crashes where a pedestrian was involved. This includes cases where a driver swerves to avoid a pedestrian and crashes his vehicle, causing property damage or injury to the vehicle occupants. It thus includes counts of all fatalities, injuries, or property damage that occur in crashes where a pedestrian was involved, regardless of whether the pedestrian was struck or injured.

[59] This category includes all crashes where a bicyclist was involved. This includes cases where a driver swerves to avoid a bicyclist and crashes his vehicle, causing property damage or injury to the vehicle occupants. It thus includes counts of all fatalities, injuries, or property damage that occur in crashes where a bicyclist was involved, regardless of whether the bicyclist was struck or injured.

[60] Based on 117,538,000 households in the U.S. in 2010 (Source: Statistical Abstract of the United States, 2012)

In: Motor Vehicle Crashes
Editor: Angie Kipling

ISBN: 978-1-61209-225-6

Chapter 2

THE IMPACT OF SAFETY STANDARDS AND BEHAVIORAL TRENDS ON MOTOR VEHICLE FATALITY RATES*

Lawrence Blincoe and Umesh Shankar

ABSTRACT

The National Highway Traffic Safety Administration (NHTSA) is charged with improving the safety of our Nation's roadways. The agency accomplishes this by regulating vehicle safety performance, enforcing existing safety requirements, and promoting programs that influence driver behavior. As a routine part of the agency's planning process, the agency sets goals that reflect its current and future programs. These goals give the agency a focus for setting priorities in its regulatory and behavioral efforts, as well as a basis for measuring the success of these efforts. Typically, the metrics the agency uses to set goals are fatality rates based on exposure to risk. This paper describes the process, assumptions, and methods used by the agency to estimate the impact of its safety regulations and behavioral programs on fatality rates, and measures the impact of these programs on those rates through the year 2020. The study projects an increase in seat belt use rates from 81 percent in 2006 to 94 percent by 2020. This would prevent a growing number of fatalities over this period starting with over 400 additional lives saved in 2007 and increasing to over 3,500 additional lives saved in 2020. Similar savings could come from a continuation of trends in alcohol-involved crashes. Together, these behavioral programs could save over 5,800 additional lives annually by 2020, and reduce the fatality rate by 0.16. Federal requirements for electronic stability control, improved side impact protection, ejection mitigation, and heavy-truck stopping distance will prevent over 3,500 additional fatalities annually by 2020 and reduce the fatality rate by 0.10. If behavioral trends continue, NHTSA's *combined program activity*, including both motor vehicle safety regulations and behavioral programs, will prevent approximately 1,500 additional fatalities by 2008, and nearly 10,000 additional fatalities by 2020, and reduce the fatality rate by 0.27.

---

* This is an edited, reformatted and augmented version of a National Highway Traffic Safety Administration sponsored document, DOT HS 810 777, issued January 2007.

SUMMARY

Analysis Summary

- Fatality rates were estimated for the period from 2006-2020 based on the projected impact of known safety technologies and behavioral trends.
- The impact of both voluntarily installed and federally mandated safety equipment was estimated by computing the rate of penetration into the on-road vehicle fleet.
- Seat belt use was estimated to increase based on its historical pattern – an annual average conversion rate of 8 percent.
- Alcohol-related fatalities were estimated based on their historical trend.
- Motorcycle fatalities were estimated to follow a cyclical pattern defined by generational influence.
- Adjustments were made for overlap among safety benefits.

Results Summary

- Behavioral programs will have a significant impact on fatality rates. If existing trends in seat belt usage and alcohol-involved fatalities are maintained, fatality rates could be reduced by 0.16 by 2020.
- Electronic Stability Control (ESC) will have a strong and growing impact. By 2020, ESC will be in 60 percent of the on-road fleet and it will be responsible for reducing the fatality rate by 0.08.
- Several other anticipated Federal safety standards with later effective dates – side impact upgrade, ejection mitigation, and heavy-truck stopping distance, will impact the fatality rate by 0.02 in 2020.
- If behavioral trends continue, NHTSA's combined program activity, including both motor vehicle safety regulations and behavioral programs, will reduce the fatality rate by 0.27 by 2020.
- The study projects an increase in seat belt use rates from 81 percent in 2006 to 94 percent by 2020. This would prevent a growing number of fatalities over this period starting with over 400 additional lives saved in 2007 and increasing to over 3,500 additional lives saved in 2020. Similar savings could come from a continuation of trends in alcohol-involved crashes. Together, these behavioral programs could save over 5,800 additional lives annually by 2020.
- Federal requirements for ESC, improved side impact protection, ejection mitigation, and heavy-truck stopping distance will prevent over 3,500 additional fatalities annually by 2020.
- If behavioral trends continue, NHTSA's combined program activity, including both motor vehicle safety regulations and behavioral programs, will prevent approximately 1,500 additional fatalities by 2008, and nearly 10,000 additional fatalities by 2020.

INTRODUCTION

The National Highway Traffic Safety Administration (NHTSA) is charged with improving the safety of our Nation's roadways. The agency accomplishes this by regulating vehicle safety performance, enforcing existing safety requirements, and promoting programs that influence driver behavior. As a routine part of the agency's planning process, the agency sets goals that reflect its current and future programs. These goals give the agency a focus for setting priorities in its regulatory and behavioral efforts, as well as a basis for measuring the success of these efforts. Typically, the metrics the agency uses to set goals are fatality rates based on exposure to risk. This paper describes the process, assumptions, and methods used by the agency to estimate the impact of its safety regulations and behavioral programs on fatality rates, and measures the impact of these programs on those rates. Establishing safety goals involves both analytical and policy considerations. The estimates that result from this analysis are intended as guidance for policy makers in establishing safety goals, but the estimates are not, in themselves, recommendations for these goals.

RELEVANT FACTORS

Projections of future fatality rates must take into account the normal turnover of vehicles in the on-road fleet that occurs as older vehicles are replaced by newer vehicles with improved safety characteristics. The difference in safety between newer and older fleets will be influenced by a number of factors including both existing and anticipated safety standards as well as voluntary changes that manufacturers make in vehicle designs. Likewise, trends in driver behavior will influence fatality rates. Demographic factors such as population growth and developmental sprawl will also affect safety by increasing the use of motor vehicles. The effects of the following factors were examined for their influence on future fatality rates:

1) Increased penetration of existing safety standards into the on-road passenger vehicle[1] fleet (ORF).
2) New safety standards certain or likely to be established prior to the outside timeline (2020) but not yet effective.
3) Significant safety innovations voluntarily adopted by manufacturers prior to their establishment as Federal Motor Vehicle Safety Standards (FMVSS).
4) Expected increases in seat belt use rates over the examined timeline.
5) Expected decreases in alcohol-involved fatalities over the examined timeline.
6) Expected increases in vehicle miles traveled (VMT) over the examined timeline.

Externalities such as improvements in roadway design, aging of the population, impact on driving from changes in gas prices, changes in the routine habits of people (e.g., increased walking and bicycling), changes in the mix of vehicle types preferred by consumers, increased telecommuting, demographic shifts, unforeseen economic and technological trends, etc., would also have an effect, but are difficult to predict and were not examined in this study.

Approach

The general approach used to estimate future fatality rates consisted of three steps:

1) Established baseline penetration rates and fatality savings for each safety standard or behavioral factor.
2) Projected the increase in ORF penetration of existing and anticipated standards and the resulting change in safety impacts annually throughout the examined time period.
3) For each year, the impact of all safety factors was then summed and adjusted for both increased levels of VMT and overlap with other safety countermeasures. A resulting fatality rate was then computed based on projected fatalities and VMT.

Penetration Rates

For existing safety standards, current penetration rates within the on-road fleet were derived from Kahane's safety standards benefit study.[2] Kahane supplied 2002 baseline data from a separate run of his database. These data are summarized along with independently derived seat belt data and estimates of remaining potential benefits in Table A-1 in Appendix A. The yearly change in ORF penetration was estimated assuming an annual increase of 1.8 percent in the ORF (based on historical trends), and annual passenger vehicle sales of 17 million (also based on recent history). The annual number of vehicles scrapped was imputed to be consistent with these two assumptions as follows:

$$R_n = ORF_{n-1} + N_n - ORF_n$$

Where: R = Estimated vehicles scrapped (retired)
ORF = On-road fleet
N = New vehicle sales
n = Year

Annual penetration rates were computed by adding sales of current model year vehicles containing new safety equipment to the previous year's total ORF while removing older scrapped vehicles. Older vehicles with safety equipment might be assumed to be removed proportionally to the current rate of equipment penetration. However, vehicle retirements are caused by a combination of damage in crashes and the obsolescence that occurs with use as vehicles break down mechanically or physically to the point where repair costs exceed their value to maintain in running condition. Since this latter factor is a function of age, vehicles that are removed from the fleet actually have, on average, a lower penetration rate of safety standards than found in the current calendar year's on- road fleet.

Whatever presence safety standards have in the ORF is contained within the most recent model years. To the extent that safety equipment is absent from the ORF, its absence is confined to older vehicles that were built prior to its requirement as a safety standard or development as a voluntary feature. Thus, a very old safety technology such as seat belts may already be in nearly 100 percent of the ORF, whereas a more recent safety technology such as antilock brakes may be in less than a third of the ORF. In this example, the third of the ORF that has the recent technology would be vehicles that represent the youngest third of the ORF, which might include, for example, the most recent five years of new vehicle sales. However, because older vehicles are overrepresented in retirements, vehicles of these same ages do not make up a third of the vehicles that are scrapped.

To address this issue, a relationship was developed between the age of the on-road fleet and the age of vehicles that are scrapped. Age specific survival rates developed by NHTSA[3] were combined with age-specific ORF data[4] to develop profiles of cumulative ORF and scrapped fleet retirements by age. The resulting profiles were then regressed to establish a relationship with cumulative retirements as the dependent variable. A polygonal model with an r^2 of 0.992 was chosen as best fit and produced the following relationship:

$$y = 1.0229x^2 + 0.0233x + .0195$$

Where: y = Cumulative retirements associated with x
x = Cumulative on road fleet by age

This process is illustrated in Tables A-2, A-3, and A-4 (see Appendix A), and Figure 1 below.

As applied to this analysis, x is equal to the portion of the ORF in each previous (n-1) calendar year that contains each specific standard. The variable y represents the portion of scrapped vehicles from each current (n) calendar year that contains each specific standard.

For safety equipment required by standards that are not yet effective, as well as for anticipated standards such as Electronic Stability Control, future penetration rates were estimated based on a combination of known voluntary installation rates and proposed or mandated phase-in schedules.

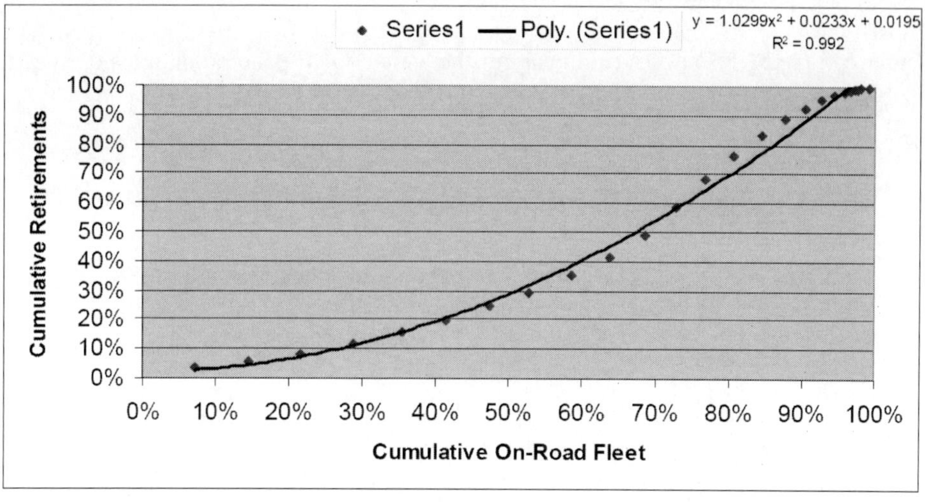

Figure 1. Relationship between Vehicle Retirements and Retirements of Vehicles with Safety Standards.

The formula used to estimate each year's on-road fleet penetration for each specific standard is:

$$P_{n,s}= (P_{n-1,s}*t_{n-1}+N_n z_{n,s}-r_n*y_{n-1,s})/(t_{n-1}+N_n-r_n)$$

Where: P = Penetration rate in on-road vehicle fleet
n = Calendar year (2003-2020)
s = Specific safety standard
t = On-road passenger vehicle fleet size
N = Annual new passenger vehicle sales
z = Portion of new passenger vehicle sales that have a specific safety feature
r = Annual passenger vehicle retirements
y = Portion of retired fleet that contains specific safety standard

Since P is derived for the purpose of calculating calendar year benefits, an additional adjustment is required to reflect the fact that new vehicle sales and annual scrappage totals are spread throughout the year. Both new sales and scrapped vehicles would accumulate during the year and would not reach their annual totals until the end of the year. Therefore, the full safety impact of these shifts in the fleet composition would not be achieved during the current year. To reflect this, a fleet adjustment factor was added. This factor was set to 0.5 for both new sales and current retirements, reflecting an assumption that, on average, half of the potential annual safety impacts would be achieved during the current year.[5]

The penetration formula modified for average safety impacts in year n would thus be:

$$P_{n,s}= (P_{n-1,s}*t_{n-1}+N_n z_{n,s}*L_n-r_n*y_{n-1,s}*L_n)/(t_{n-1}+N_n*L_n-r_n*L_n)$$

Where: L_n = 0.5, but P_{n-1} is recalculated with L_n set at 1.0 to reflect end-of-year penetration rates.

Table A-5 (see Appendix A) summarizes the annual ORF penetration rates derived from this analysis.

PROJECTED INCREMENTAL SAFETY BENEFITS

Using Kahane's 2002 safety benefits and penetration rates as a starting point, full fleet safety benefits at 2002 VMT levels were calculated as follows:

$$F_s = f_{b,s}/P_{b,s}.$$

Where: F_s = Fatalities prevented by specific standard at full implementation
$f_{b,s}$ = Fatalities prevented by specific standard in base year (2002)
$P_{b,s}$ = Penetration rate for specific safety standard in base year (2002)

The incremental fatalities prevented were then calculated as:

$$f_{ns} = F_{s*}(P_{n,s} - P_{b,s})$$

Where: f_{ns} = Incremental fatalities prevented
P_{ns} = Penetration rate of standard s in year n

VEHICLE MILES TRAVELED

The annual increase in VMT was estimated based on the average increase over the most recent 3-year period (2003-2005) or 1.5 percent per year.[6] This factor was applied to both the estimated base fatalities and to the estimated lives saved due to safety standards to reflect the assumption that as travel increases, both potential fatalities and safety benefits increase to reflect the added exposure.

SEAT BELTS

Future seat belt usage rates were estimated by examining the background trend in observed seat belt use rates after eliminating the data spikes caused by the initial institution of primary belt laws. This produces an estimate of the change in belt use expected based on educational and societal trends, but without specific changes in State laws requiring belt use. Currently there are 25 States that have not passed primary belt laws, and data indicate that passage of such laws, which allow for direct enforcement of seat belt wearing, can increase belt use rates significantly. This constraint may thus produce a conservative estimate of belt use increases in specific years when such laws are passed. The metric used to define change in belt use was conversion rates. Conversion rates measure the portion of seat belt nonusers who are converted to belt users each year. Changes in State observation use rates from 1998 through 2005 were examined after eliminating years in which primary laws became effective. No particular trend in conversion rates was apparent so the annual average rate of 8 percent was used for each year. This produced an estimate of about 94-percent usage by 2020 (up from 81% in 2006 – see Table A-5). Seat belt benefits were calculated directly from projected use rates using procedures previously established within NHTSA.[7]

ALCOHOL

The change in alcohol-related fatalities was derived from trend data. The reduction in alcohol-related fatalities was established over the timeline based on these historical trends. This assumes a continuation of the trend in improvements in public awareness and enforcement throughout the time period. Overlap between fatalities prevented by alcohol programs and those prevented by safety standards (including increased belt use rates) was assumed to be proportional to the ratio of additional fatalities prevented due to standards to the total projected fatalities prior to adjusting for the impact of standards. In other words, it was assumed that there is an equal probability that a fatality prevented by reduced alcohol consumption would come from a case that would also benefit from added safety standards or a case that would not benefit from these standards. The relative incidence of these two categories of cases would thus determine the level of overlap.

The overlap formula for alcohol benefits is:

$$OA_n = 1 - \frac{BI_n}{FP_n}$$

$$= 1 - \frac{\sum_{S=1}^{M} f_{n,s}}{FP_n}$$

Where: OA_n = Overlap adjustment factor for alcohol benefits in year n
BIn = Total incremental benefits (fatalities prevented) in year n adjusted for VMT
FPn = Predicted total fatalities after adjustment for increased VMT
This formula was applied separately to each calendar year.

SAFETY STANDARDS BENEFITS OVERLAP ADJUSTMENT

To account for overlap among numerous safety standards, we calculated the combined net effectiveness of all standards against a broad specified target population (in this case, all fatalities) and compared this to the implied net effectiveness that results from a simple summation of savings from all standards (i.e., the total benefits if there was no overlap among any of the examined standards). A separate overlap factor was calculated for each calendar year. The net effectiveness for each standard was calculated by comparing the added annual savings from that standard (over the base year) to the common potential target population, i.e.:

$$e_s = f_{n,s}/BF$$

Where e_s = Net effectiveness of specific standard against base target population
$f_{n,s}$ = added fatalities prevented by specific standard relative to 2002 base year
BF = Base year (2002) total motor vehicle fatalities (43,005)

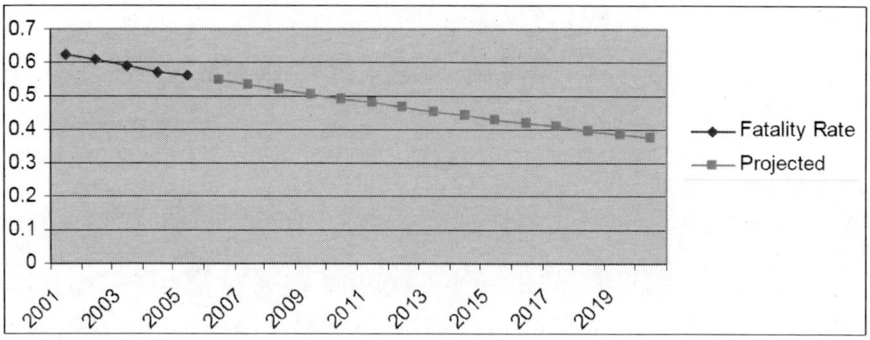

Figure 2. Historical and Projected Alcohol Fatality Rates.

The formula for combined effectiveness is:

$$EC_n = 1 - \prod_{S=1}^{M} (1 - e_s)$$

Where EC_n = Combined effectiveness of all standards net of overlap
e_s = Net effectiveness of standard s

The implied effectiveness if there was no overlap among the standards would be:

$$E_n = \sum_{S=1}^{M} \frac{f_{n,s}}{BF}$$

Where E_n = Implied effectiveness of all standards effective in year n based on simple summation of all benefits

An overlap adjustment factor was then calculated as:

$$O_n = EC_n / E_n.$$

Where O_n = The ratio of effectiveness of combined standards to summed standards.
This factor was then applied to $\sum f_{ns}$ to estimate combined savings net of overlap.

MOTORCYCLES

While significant progress has occurred in preventing fatalities among passenger vehicle occupants, motorcycle fatalities have been rising steadily since the mid-1990s. This trend has been so pronounced that added motorcycle fatalities have essentially offset the impact of higher belt use and improved vehicle safety on passenger vehicle occupants. This has kept the overall motor vehicle fatality total relatively flat with slight increases in the more recent years. Table 1 and Figure 3 illustrate the relative fatality trends in passenger vehicles, nonoccupants, other vehicle occupants, and motorcycles.

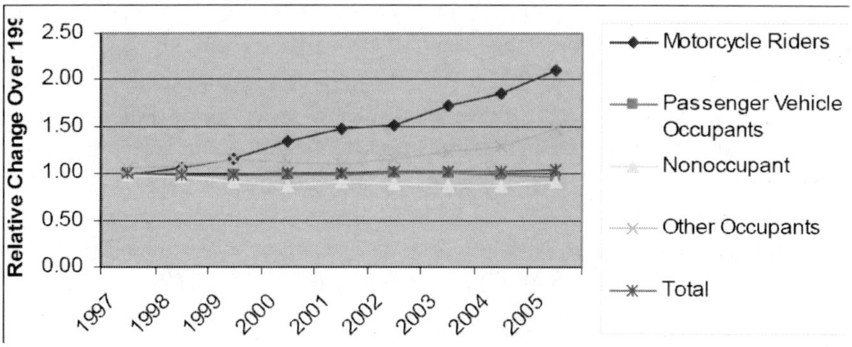

Figure 3. Fatality Trends by Person Type.

Table 1. Motor Vehicle Crash Fatalities by Person Type

	Motorcycle Riders	Passenger Vehicle Occupants	Nonoccupants	Other Occupants	Total
1996	2,161	32,437	6,368	1,099	42,065
1997	2,116	32,448	6,288	1,161	42,013
1998	2,294	31,899	6,119	1,189	41,501
1999	2,483	32,127	5,842	1,265	41,717
2000	2,897	32,225	5,597	1,226	41,945
2001	3,197	32,043	5,756	1,200	42,196
2002	3,270	32,843	5,630	1,262	43,005
2003	3,714	32,271	5,543	1,356	42,884
2004	4,028	31,866	5,532	1,410	42,836
2005	4,553	31,415	5,849	1,626	43,443

When a sharp increase in motor vehicle fatalities occurs for a specific demographic, it should typically involve either an increase in exposure or a change in behavior. The fatality rate trend for motorcycle riders may involve a combination of these factors. To explore the possible causes of this trend, two different exposure measures - VMT and registrations - were examined. Table 2 and Figure 4 illustrate the trend in motorcycle VMT, registrations, and fatalities.

From Figure 4, it is apparent that the increase in motorcycle fatalities involves more than just exposure. The rate of increase in fatalities is significantly higher than either the increase in registrations or the increase in motorcycle VMT. By 2005, fatalities had more than doubled while registrations rose roughly 50 percent and VMT remained virtually unchanged. VMT is usually considered the best measure for exposure since it measures actual miles traveled. However, given that both fatalities and registrations climbed significantly over this period, the lack of change in VMT does not seem credible. Fatality data is collected through FARS and it represents a complete census of all fatal crashes in the United States. Registration data is collected by the States and provided to the Federal Highway Administration which is responsible for the collection and publication of all exposure data (registration, VMT, licensed drivers). The VMT data collected by the Federal Highway Administration are from estimates gathered by individual States. However, State reporting of motorcycle VMT to FHWA is optional. Even in States that report motorcycle VMT, it is often only measured as a standard proportion of total VMT rather than being collected directly through surveys or roadside counters. FHWA estimates VMT for States that do not report based on data from States that do report. The accuracy of these counts is thus quite speculative. If we accept the fatality and registration counts as reasonably accurate, then there are only two possible explanations for the flat trend in VMT. Either:

1) Existing motorcycle riders are buying additional vehicles while maintaining the registration of their old vehicles, but not increasing their driving and new motorcycle riders are buying vehicles they don't use, or
2) The VMT data are flawed.

Table 2. Motorcycle Rider Fatalities, VMT, and Registrations

	Fatalities	VMT (M)	Registrations
1996	2,161	9,920	3,871,599
1997	2,116	10,081	3,826,373
1998	2,294	10,283	3,879,450
1999	2,483	10,584	4,152,433
2000	2,897	10,469	4,346,068
2001	3,197	9,639	4,903,056
2002	3,270	9,552	5,004,156
2003	3,714	9,577	5,370,035
2004	4,028	10,122	5,767,934
2005	4,553	10,770	6,227,146

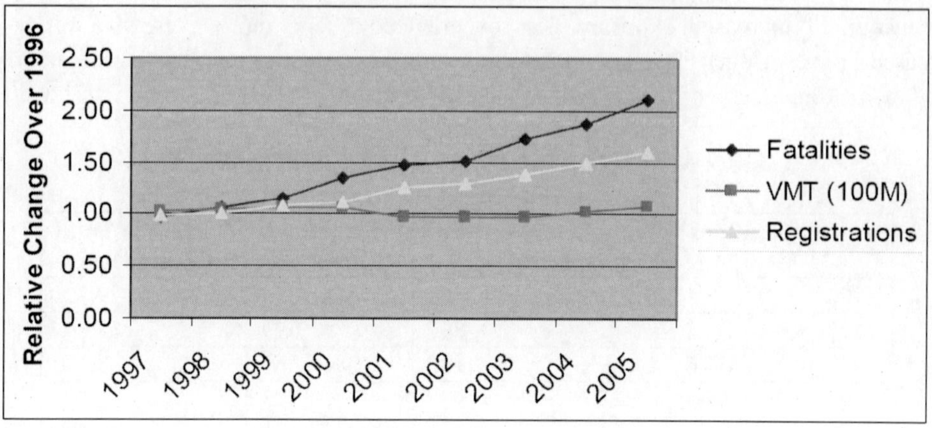

Figure 4. Relative Trends in Motorcycle Rider Fatalities and Exposure.

Although the first two points need not be absolutes (a mixture of *low* marginal VMT by existing riders and *minimal* VMT by new riders is theoretically possible), we believe it is extremely unlikely that a 50-percent increase in registrations would not produce some significant increase in VMT as well. Therefore, for this analysis we have chosen to reject reported VMT as a trend indicator. This leaves registrations as the only viable basis for estimating motorcycle fatality rates.

Estimating the future trend in motorcycle fatalities is problematic. An examination of historical trends back to 1975 indicates that the trend in motorcycle fatalities is highly erratic (see Figure 5). In the late 1970s motorcycles fatalities rose dramatically and remained at a high level for roughly a decade until they began to decline in the late 1980s. In the late 1990s motorcycle fatalities began to climb again. During this roughly 30-year timeframe there is a reasonably strong directional correlation between fatalities and motorcycle registrations. However, as noted above, during the most recent upswing in fatalities, exposure only accounts for a portion of the rise. Another factor may be motorcycle helmet use. NOPUS data indicate that helmet use increased steadily between 1994 and 2000 (climbing from 63% to 71%), but then dropped off sharply in 2001 to 58 percent and continued to decline to the current level of 51 percent. The decline in helmet use corresponded to the most precipitous climb in fatalities during this period.

Another factor that may have contributed to the rise in fatalities is the increase in engine size coupled with an increase in older riders. Table 3 shows that there has been a steady increase in the engine size of motorcycles involved in fatal crashes. Moreover, there has also been a rapid increase in the fatalities among riders ages 40 and older on motorcycles in the 1,001-1,500 cc range. This reflects a growth in the popularity of these large bikes among those in the 40+ age cohort, and particularly within the 50+ age cohort. It appears that much of the upswing in motorcycle sales is driven by middle-aged riders who are either first-time owners or are reacquiring a habit they gave up when they were younger. These heavier and more powerful bikes may be more difficult to control in emergency situations, especially when driven by older riders who are not used to their handling characteristics. Exacerbating this risk is an increased likelihood of riders of larger motorcycles to drive impaired.[8]

Thus, it appears that the rapid increase in motorcycle rider fatalities is due to a combination of increased exposure, less experienced riders, more aggressive bikes, and decreased helmet usage. However, it also appears that exposure may be the larger driving factor in this trend.

Table 3. Motorcycle Fatality Trends Age and Engine Size

	Fatalities Mean Age	Fatal Involved Mean Engine Size (cc)	Fatalities 40 and Over	Percent of Total 40 and Over
1996	33.4	866	641	29.7%
1997	34.8	897	699	33.0%
1998	34.6	902	760	33.1%
1999	36.5	927	973	39.2%
2000	36.8	961	1,178	40.7%
2001	36.3	957	1,261	39.4%
2002	37.9	1,002	1,444	44.2%
2003	38.0	1,014	1,694	45.6%
2004	38.0	1,015	1,854	46.0%
2005	38.8	1,018	2,143	47.1%

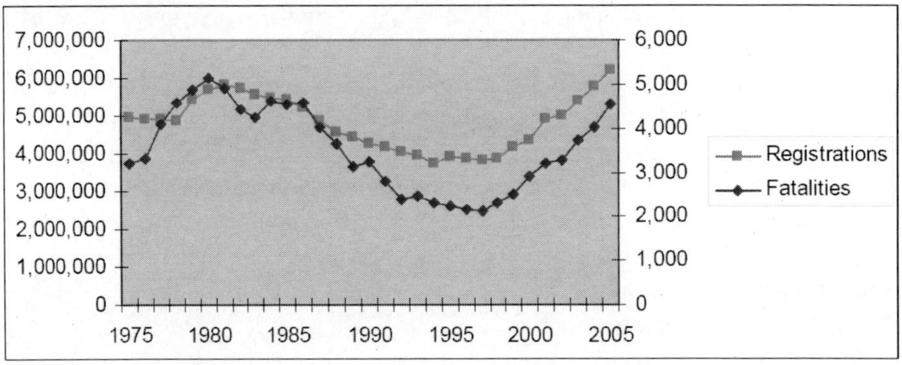

Figure 5. Motorcycle Fatalities and Registrations, 1975-2005.

Table 4. Projected Population of the United States, by Age Group: 2000 to 2050 (In thousands. As of July 1. Resident population.)

Age Group	2000	2010	2020	2030	2040	2050
<5	19,218	21,426	22,932	24,272	26,299	28,080
5-19	61,331	61,810	65,955	70,832	75,326	81,067
Under 20	80,549	83,236	88,887	95,104	101,625	109,147
20-44	104,075	104,444	108,632	114,747	121,659	130,897
45-64	62,440	81,012	83,653	82,280	88,611	93,104
65-84	30,794	34,120	47,363	61,850	64,640	65,844
85+	4,267	6,123	7,269	9,603	15,409	20,861
TOTAL	282,125	308,936	335,805	363,584	391,946	419,854

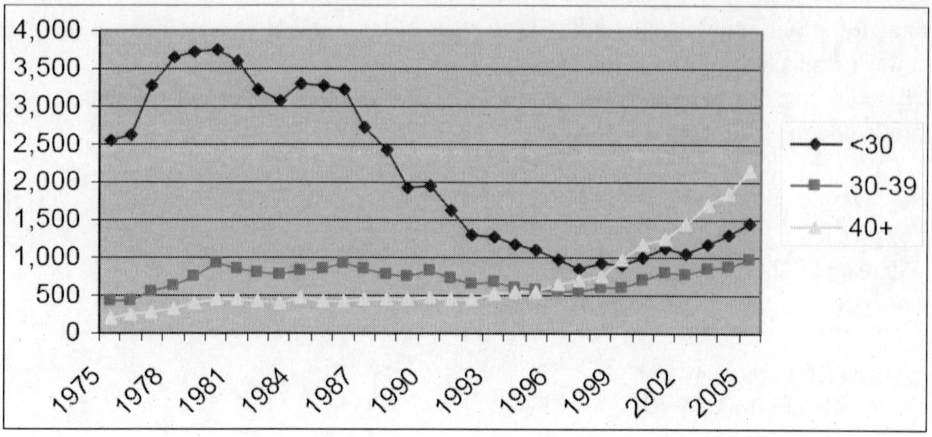

Figure 6. Motorcycle Fatalities by Age Group.

The largest increase in motorcycle rider fatalities have occurred among older riders (age 40+). This is partially a function of the natural expansion due to aging of the post-war baby boom generation. It's not clear whether this mid-life upswing in motorcycle riding is a characteristic of this specific generation or whether it will continue through succeeding generations. Increases in motorcycle fatalities have been trending up in the 30-39 age cohort as well, though to a lesser extent. It thus appears that there will be some level of continued interest in motorcycle riding among middle-aged riders through the next decade. Moreover, Census Bureau projections indicate that the middle-aged (45-64) population that is fueling the current rise in motorcycle sales will continue to grow through 2020, but that most of the growth will occur by 2010 (see Table 4).

To explore the issue of generational interest in motorcycles, the historical trend in motorcycle rider fatalities was plotted by age group. Figure 6 indicates that the previous (1977-1987) surge in motorcycle fatalities was driven by riders in the under-30 age category. Riders in this group would now be in the 40+ category, and many would be in the 50+ category – the exact same age group that is driving the current surge in motorcycle registrations[9] and deaths. These data are thus consistent with the concept that a large part –

possibly the largest part – of the resurgence in interest in motorcycle riding is caused by a rebirth of interest by baby boomers who rode motorcycles in their 20s, abandoned them for passenger vehicles, and then became interested in them again when they became middle-aged. We cannot verify the extent to which the same individuals are involved, but it does appear to be the same generation that is driving both surges.

Another factor to consider is that abrupt changes in fatality trends that are driven by registrations are unlikely. It takes time for consumers to change their driving habits and after investing in an expensive vehicle they are unlikely to abruptly change "en masse" and stop using it.[10] Once a vehicle becomes part of the on-road fleet, it will likely continue to influence the level of exposure that occurs in motorcycles for the remainder of its useful life. With new registrations putting more bikes on the road and increasing exposure, it is likely to take many years, perhaps a decade, before any reversal occurs in this trend. However, given the role of older riders in the current surge, there is likely to be a natural cessation of this surge at some point due to further aging of the middle-aged population that is now acquiring these vehicles. Given this perspective, we have modeled the future trend in motorcycle fatalities as a natural cycle based on the trends seen in the historical record. A sine curve was fit to these historical data producing the following relationship:

$$a*\sin((y-1974)*\pi/(c/2)) + i$$

Where a=1322.081 (S.E. = 72.567)
y = Year
π = 3.14159
c = Cycle length (28 years)
i = Intercept (3480.094) (S.E. = 49.846)

The best fit was provided by a model that assumes a 28-year cycle (r^2 = .9196). Figure 7 shows both the historical data and the resulting trend line.

A similar model was fit to the historical trend in registrations. However, the best fit for registrations was derived from a 26-year cycle (r^2 = .8605). This produced the following relationship:

$$a*\sin((y-1974)*\pi/(c/2)) + i$$

Where a = 946958 (S.E. = 69432)
y = Year
π = 3.14159
c = Cycle length (26 years)
i = Intercept (4726478) (S.E. = 48721)

Figure 8 shows both the historical data and the resulting trend line for registrations.

Although both models do a reasonable job of predicting likely trend behavior, there are several problems. The most obvious is that the history based registration model does not predict a steep enough rise for the recent surge in motorcycle registrations. For 2005, the most recent year of available data, the model predicts a registration total 15 percent below the actual level. The fatality model is more accurate, but it underestimates 2005 fatalities by 5 percent. To adjust for this, both models were normalized to 2005 levels. Thus, the breadth and amplitude of the trend cycle predicted by the models is retained, but they are launched from 2005 levels. A second adjustment was required for fatalities to make them consistent with the global assumption of a 1.5-percent annual increase in VMT.[11] To accomplish this, fatalities were assumed to be the larger of the normalized projection or the previous years' normalized projection increased by 1.5 percent. The results from this process were then combined to produce estimates of motorcycle fatalities/1,000 registrations. The results are shown in Figures 9, 10, and 11.

Note that these estimates were not modified to reflect any specific NHTSA safety programs. There are currently no specific new safety standards targeting motorcycles, and Congress has passed legislation preventing the agency from actively promoting motorcycle helmet use, the most effective known protection measure for motorcycle riders. Moreover, while alcohol-involved crashes have declined among other groups, they have actually increased among motorcyclists, reflecting both the higher rates of impaired driving among middle-aged riders and the overall increase in motorcycle rider fatalities. The projected motorcycle fatality rates found in this analysis thus primarily reflect the normal progression of the cycle initiated in the late 1990s.

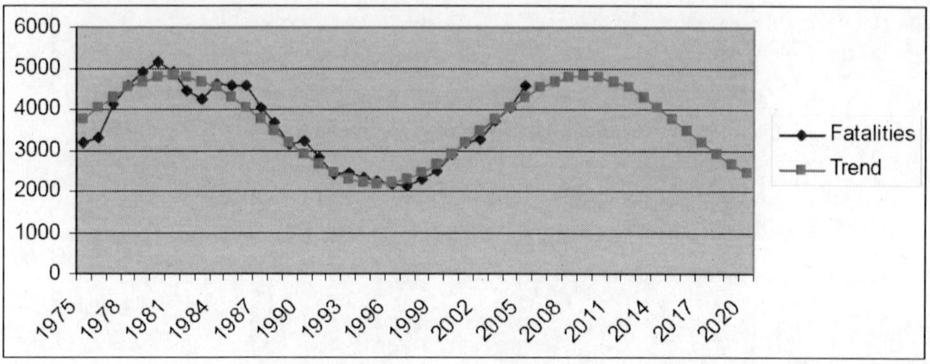

Figure 7. Motorcycle Fatalities Trend Cycle.

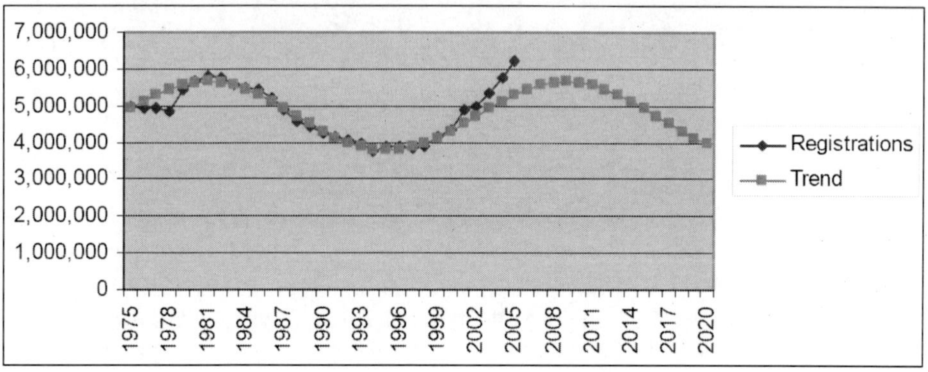

Figure 8. Motorcycle Registrations Trend Cycle.

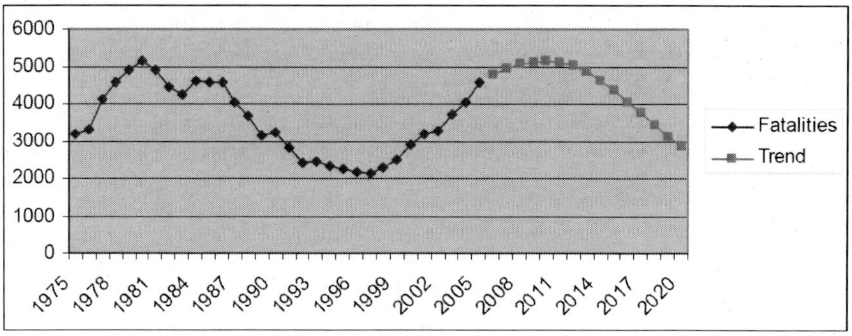

Figure 9. Historical and Projected Motorcycle Fatalities, Normalized.

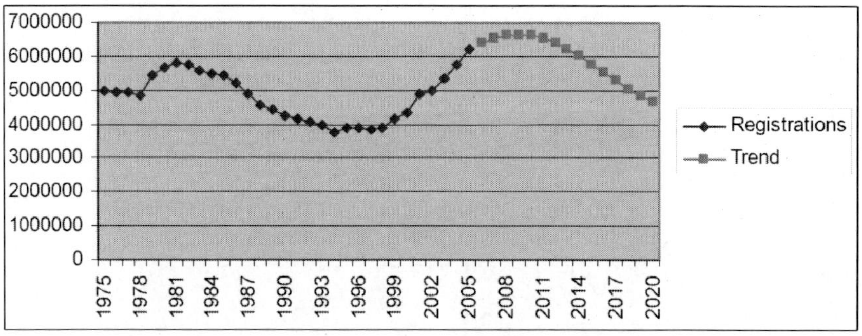

Figure 10. Historical and Projected Motorcycle Registrations, Normalized.

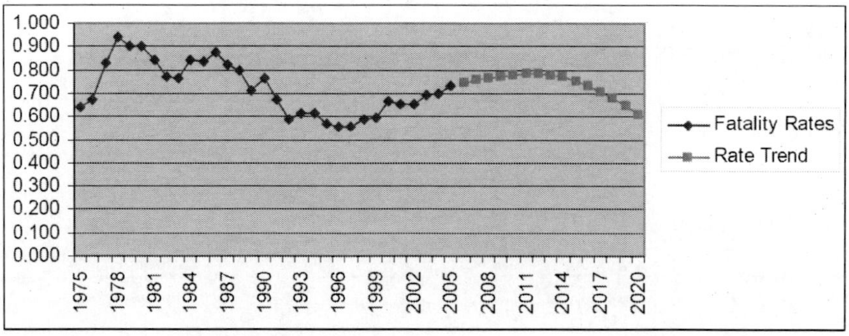

Figure 11. Historical and Projected Motorcycle Fatality Rate.

RESULTS

Projected fatality rates were estimated for five different VMT based metrics: all fatalities, all fatalities excluding motorcycle riders, passenger vehicle[12] occupants, passenger vehicle occupants plus motorcycle riders, and nonoccupants.[13] A separate analysis was also conducted for motorcycles based on registration data rather than VMT (see previous discussion). Table 5 lists the results of these projections for the five VMT based metrics, and Table 6 lists the results for motorcycle riders based on registrations. Figure 12 shows the VMT-based projections together with historical rates. The impact of motorcycles is apparent from these results. When motorcycles are excluded from the overall fatality rate it drops from 1.45 to 1.31 in 2005 and by increasing levels in succeeding years.

Likewise, lumping motorcycles in with passenger vehicles increases the 2005 rate from 1.14 to 1.30. This reflects the upward trend in motorcycle fatalities which have been rising rapidly, while other fatality categories have been declining. This trend is reflected in the rise in motorcycle sales and registrations, which is apparently fueled by their increased popularity among middle-aged riders.

The target years for agency goals will naturally shift over time, but currently the agency is examining goals for the years 2008, 2011, 2016, and 2020. Under the assumptions used in this analysis, by 2008, the overall fatality rate would decline to 1.37, with further declines to 1.27 in 2011, 1.10 in 2016, and 1.01 by 2020. Similar declines occur for the other passenger-vehicle-related metrics. However, the nonoccupant rate declines much more gradually. This reflects the more significant impact that safety standards have on vehicle occupants – a natural outgrowth of safety standards designed primarily to protect vehicle occupants. Alcohol programs are the primary influence on reduction in nonoccupant fatality rates.

Figure 13 shows the historical and projected fatality rates for the four separate component metrics – all fatalities, passenger vehicles, nonoccupants, and motorcycles – that are examined in this study.

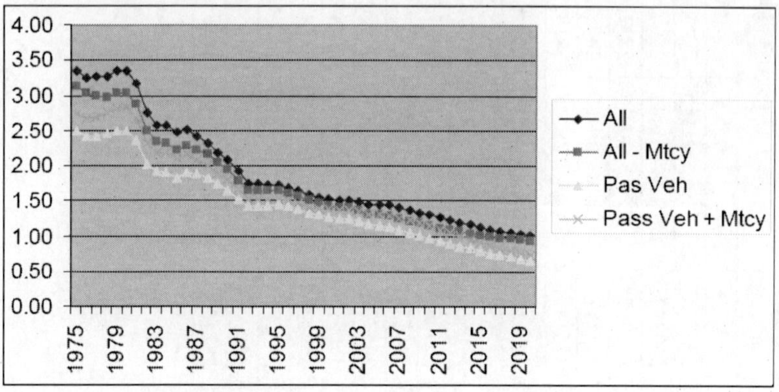

Figure 12. Historical and Projected Fatality Rates.

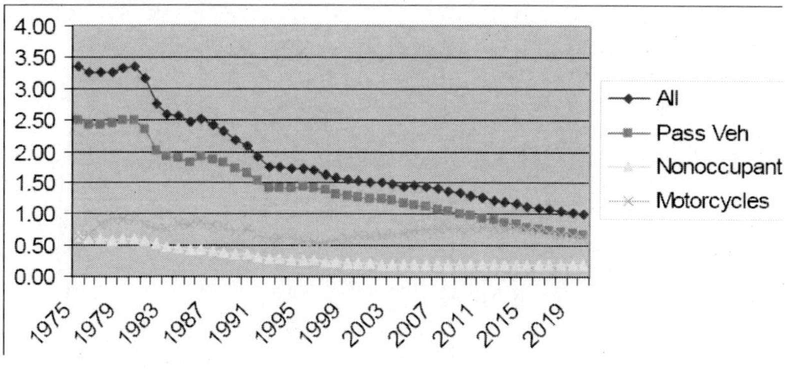

Figure 13. Historic and Projected Fatality Rate Trends Base Scenario.

Table 5. Projected Fatalities, VMT (M), and Fatality Rates, 2005-2020

	All Fatalities			All Fatalities Excluding Motorcycles			Passenger Vehicles Only			Passenger Vehicle and Motorcycles			Nonoccupants		
Year	Fatals	VMT	Rate	Fatals	VMT	Rate	Fatals	VMT	Rate	Fatals	VMT	Rate	Fatals	VMT	Rate
2005	43443	2989807	1.45	38890	2979037	1.31	31415	2749555	1.14	35968	2760325	1.30	5849	2989807	0.20
2006	43751	3034654	1.44	38977	3023723	1.29	31441	2790798	1.13	36215	2801730	1.29	5893	3034654	0.19
2007	43363	3080174	1.41	38422	3069078	1.25	30813	2832660	1.09	35754	2843756	1.26	5953	3080174	0.19
2008	42897	3126377	1.37	37852	3115115	1.22	30169	2875150	1.05	35213	2886412	1.22	6015	3126377	0.19
2009	42410	3173272	1.34	37290	3161841	1.18	29528	2918277	1.01	34648	2929708	1.18	6079	3173272	0.19
2010	41945	3220871	1.30	36789	3209269	1.15	28905	2962052	0.98	34061	2973654	1.15	6144	3220871	0.19
2011	41383	3269184	1.27	36263	3257408	1.11	28182	3006482	0.94	33302	3018259	1.10	6211	3269184	0.19
2012	40827	3318222	1.23	35812	3306269	1.08	27459	3051580	0.90	32474	3063533	1.06	6280	3318222	0.19
2013	40297	3367995	1.20	35451	3355863	1.06	26759	3097353	0.86	31605	3109486	1.02	6351	3367995	0.19
2014	39784	3418515	1.16	35163	3406201	1.03	26071	3143814	0.83	30693	3156128	0.97	6423	3418515	0.19
2015	39306	3469793	1.13	34954	3457294	1.01	25414	3190971	0.80	29766	3203470	0.93	6497	3469793	0.19
2016	38900	3521840	1.10	34848	3509153	0.99	24824	3238835	0.77	28876	3251522	0.89	6573	3521840	0.19
2017	38561	3574668	1.08	34825	3561791	0.98	24297	3287418	0.74	28034	3300295	0.85	6651	3574668	0.19
2018	38285	3628288	1.06	34865	3615218	0.96	23829	3336729	0.71	27250	3349799	0.81	6730	3628288	0.19
2019	38068	3682712	1.03	34947	3669446	0.95	23416	3386780	0.69	26536	3400046	0.78	6812	3682712	0.18
2020	37906	3737953	1.01	35054	3724488	0.94	23053	3437582	0.67	25905	3451047	0.75	6895	3737953	0.18

Table 6. Projected Motorcycle Fatalities, Registrations, and Fatality Rates/1,000 Registrations

	Fatalities	Registrations	Fatality Rate /1,000 Regis.
2005	4,553	6,227,146	0.731
2006	4,774	6,402,760	0.746
2007	4,941	6,534,894	0.756
2008	5,044	6,616,922	0.762
2009	5,120	6,644,730	0.771
2010	5,156	6,616,925	0.779
2011	5,120	6,534,900	0.784
2012	5,015	6,402,769	0.783
2013	4,846	6,227,158	0.778
2014	4,621	6,016,871	0.768
2015	4,352	5,782,454	0.753
2016	4,052	5,535,661	0.732
2017	3,736	5,288,868	0.706
2018	3,420	5,054,450	0.677
2019	3,120	4,844,161	0.644
2020	2,851	4,668,547	0.611

INFLUENCE OF SAFETY COUNTERMEASURES

Projecting future fatality rates involves taking into account the changing influence of existing standards and making assumptions about future trends. By contrast, adopting a goal implies setting a target that will be achieved with additional action. Since these projections are intended as tools in setting agency goals, it is useful to consider the extent to which they reflect existing and future agency activities. The projected rates reflect the natural progression of existing standards into the on-road fleet as well as anticipated safety standards and voluntary changes that manufacturers are making in vehicle designs. They also assume a continuation of past trends in seat belt use and alcohol-involved crashes. The following future agency actions are thus anticipated in these projections:

1) The promulgation of four significant safety standards, three of which the agency has issued a Notice of Proposed Rulemaking (NPRM), but has not yet issued a final rule. These include: FMVSS No.121 Heavy Truck Stopping Distance, FMVSS No.214 Side Impact Pole Test, and FMVSS No.126 Electronic Stability Control. In addition, it assumes the promulgation of a safety standard covering ejection mitigation. Although the agency has conducted considerable research, it has not yet issued an NPRM for this standard.
2) A level of agency effort necessary to sustain the historical rate of change in the seat belt conversion rate. Over the last decade the agency has made seat belt use one of its top priorities and has invested considerable resources in promoting both primary use laws and educational programs to increase voluntary levels of belt use. At current

high use levels (81% nationwide), additional improvements will become increasingly difficult because the remaining nonusers include those most resistant to belt use. Thus, although the pool of nonusers has declined to roughly 19 percent of drivers, sustaining the historical rate of user conversion will require considerable effort.

3) A level of agency effort necessary to sustain the historical downward trend in alcohol-related fatality rates.[14] Again, the self-selection of those most resistant to giving up drinking and driving implies that this will require considerable effort by the agency to maintain the downward trend in alcohol involvement.

Thus, the above projections already anticipate a considerable level of effort by the agency, through both its rulemaking and its behavioral activities. Selection of goals that exceed the timeframes noted in Table 5 would imply an increase in agency activity beyond that noted above. To move the overall fatality rate by a tenth of a point would require preventing approximately 3,000 additional fatalities at current fatality and VMT levels and historically only a handful of agency actions (seat belts and ESC) have had that potential. To move the rate by a hundredth of a point would require the prevention of 300 additional fatalities. It would thus require a highly ambitious program to reach goals that exceed the levels presented in Table 5.

To evaluate the impact of each NHTSA program on fatality rates, future changes in specific programs were eliminated (i.e., the programs' impacts were held constant at current (2006) rates through future years). Under this approach, only demographic factors (as proxied by VMT) and normal safety equipment replacement at current levels would influence future fatality counts. So, for example, installation rates for Electronic Stability Control would be held to their rate in the 2006 new vehicle fleet in future model years as well. The on-road fleet installation percentage would continue to climb due to normal vehicle replacement, but not as swiftly as it would had the new fleet installation rates continued to climb to meet a 100-percent installation rate required by Federal Standard. Likewise, seat belt use rates would be held constant at their 2006 rate, rather than allowing them to increase over time.

Impacts were examined for the following categories:

New Safety Regulations (FMVSS Nos.121, 126, 214, and Ejection Mitigation) FMVSS No.121

FMVSS No.126

FMVSS No.214

Ejection Mitigation

Behavioral Programs (seat belt use and alcohol) Seat Belt Use

Alcohol Programs

All NHTSA Initiatives

Tables B-1 and B-2 (see Appendix B) list the results of this analysis for the All Fatalities rate. For each initiative, future impacts were removed while all other impacts were left in place. Thus, the fatality rate that results under each initiative category is the rate that would result if the future improvement in that particular initiative were *not* achieved, while all other initiatives *were* achieved.

The results in Tables B-1 and B-2 indicate that behavioral programs have significant impacts on the fatality rate in both short- and long-term timeframes. Failure to maintain trends in seat belt use rates would begin to impact the fatality rate immediately in 2007. In 2008 it would raise the rate by .03. By 2011 it would impact the rate enough to increase the fatality rate by .05. By 2016 the rate would rise by .08 and the impact reaches a full tenth point in 2020. Maintaining the status quo on alcohol-involved crashes would also have an aggressive impact on the fatality rate, causing it to increase by .02 by 2008 and .03 by 2011. Through 2020 the impact of not maintaining the rate of decrease in alcohol-involved fatalities increases steadily to .05 in 2016 through 2020. When these two behavioral impacts are combined, the impact is significant, especially in the latter years. Rates rise by .05, .09, and .13 in 2008, 2011, and 2016 respectively. By 2020, failure to show improvement in both seat belt use and drunk driving would increase the fatality rate by .16.

These tables also show significant impacts for the ESC standard. ESC improves the overall fatality rate by .01 by 2010 and progressively increases its impact through 2020 to .08. The other three standards all make gradual contributions, but they do not by themselves have enough impact to move the published fatality rate until the 2017-2020 period. This is primarily due to the projected lead time involved before these standards become effective. Based on NHTSA's NPRMs, ESC installation rates were estimated to be affected by a Federal requirement starting in 2008, whereas side impact standards and ejection mitigation were estimated to be impacted in 2013. Heavy-truck stopping- distance improvements are expected to be required in 2010, but this standard also has a relatively low safety impact compare to the other three standards examined here. When combined, these four standards impact the overall fatality rate in the out years by 0.10 or one tenth – identical to the potential impact from seat belt improvements for that timeframe.

When all future NHTSA activities are removed (i.e., both behavioral and regulatory impacts are held at current levels), the impacts are quite significant. The fatality rate rises rapidly by .05 in 2008, and continues to rise through 2020 to a level that is .27 higher than it would be with NHTSA standards and programs.

Tables C-1 through F-2 (Appendix B) show the results for each of the other metrics examined in this study. These tables show impacts from NHTSA activities similar to those found for the All Fatalities metric.

FATALITY IMPACTS

Although the purpose of this study is to examine trends in fatality rates, trends in total fatalities are a natural product of this investigation. Tables G-1 through K-2 (Appendix B) illustrate the fatality impacts of the safety standards and behavioral programs examined in this study. Improving seat belt use has the largest potential to prevent fatalities. The projected growth in belt use from 81 percent to 94 percent by 2020, would save over 400 lives in 2007 and increase steadily to a savings of over 3,500 lives in 2020. A continued reduction in alcohol-involved crashes also offers significant savings throughout this timeframe. Both programs combined could save over 5,800 lives by 2020. Promulgation of a standard requiring ESC in all light vehicles would began to save a small number of lives in 2008, but would rapidly decrease fatalities as new ESC- equipped vehicles enter the fleet. By 2020, an ESC standard could save over 2,800 lives annually.[15] As with fatality rates, the other three standards have less potential during this timeframe due to their later effective dates. When combined together, these four standards would prevent over 3,500 fatalities by 2020. When all NHTSA activity is combined, nearly 1,500 fatalities are prevented annually by 2008, and these savings accelerate to a level of nearly 10,000 by 2020. These impacts are illustrated in Figures 14 and 15.

SENSITIVITY ANALYSIS

The base scenario used in the above analysis assumes that efforts by the agency and other outside organizations to improve driver behavior will continue to impact safety at the same pace as has been achieved historically. This means that seat belt nonuse will continue to be converted to active use and alcohol-involved fatalities will continue to decline at the same rate that has been observed in the past. While these scenarios are technically achievable, the agency faces an increasingly difficult challenge in maintaining historical rates of progress. As the pool of unbelted and inebriated drivers shrinks, those remaining are typically the least risk-averse and most difficult to influence through either educational programs or legal sanctions.[16]

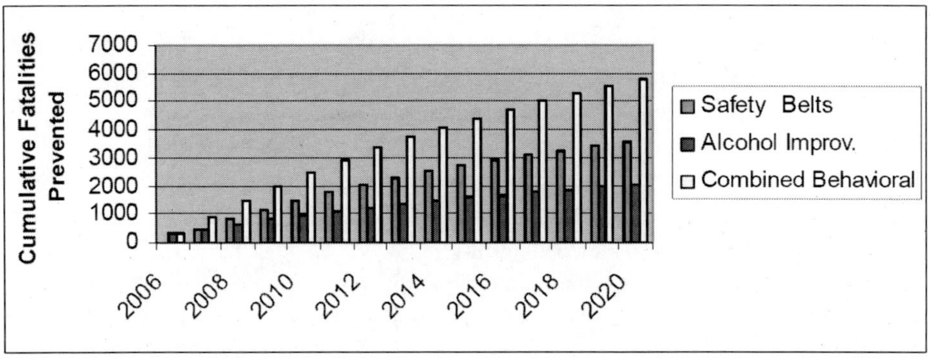

Figure 14. Potential Safety Benefits of NHTSA Programs Over Base Levels.

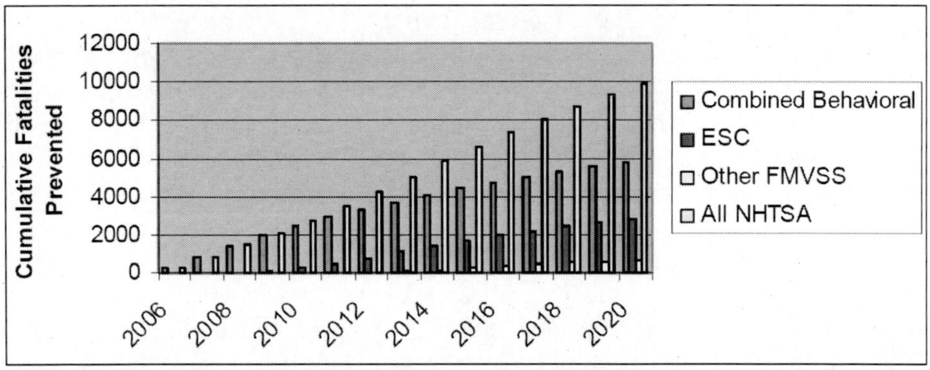

Figure 15. Potential Safety Benefits of NHTSA Programs Over Base Levels.

A further threat to the pace of improvement estimated in the base scenario is uncertainty over the impact of motorcycle fatalities. Data indicate that the primary cause for the recent surge in motorcycle fatalities is increased ridership by middle-aged drivers, especially those in their 50s. As such, there should be a natural decline in motorcycle fatalities as these riders age and reduce their on-road motorcycle exposure or abandon riding altogether. However, data also indicate a small upswing in fatalities among younger riders, and other factors such as oil price instability could cause additional interest in motorcycles and a corresponding upswing in motorcycle fatalities.

The base scenario used in this analysis might thus prove to be optimistic. To explore the impact of these assumptions, alternative scenarios were examined in which more pessimistic assumptions were adopted.

Six different alternative scenarios were examined:

1) Modest Seat Belt Progress scenario in which it is assumed that seat belt use increases at half the rate seen historically (i.e., a conversion rate of 4% instead of 8%).
2) Modest Alcohol Improvement scenario in which it is assumed that there is continued progress at half the historical pace for alcohol-related fatality reduction.
3) Modest Behavioral scenario in which both seat belt use and alcohol involvement in fatalities are assumed to progress at half their historical rate.
4) Continuous Motorcycle Rider Fatality Increase scenario in which it is assumed that motorcycle rider fatalities will increase at the same pace as seen over the past eight years.
5) Trend Average Motorcycle Rider Fatality scenario in which motorcycle rider fatalities are assumed to follow a trend midway between the generational trend used in the base scenario and the continuous increase predicted using the Continuous Motorcycle Rider Fatality Increase scenario.
6) Reasonable Worst Case scenario which combines the assumptions of the Modest Behavioral and Trend Average Motorcycle scenarios. Although a combination using the Continuous Motorcycle Increase scenario would actually be worse, we do not consider such a scenario to be probable.

The agency is proposing to set fatality rate goals for four specific measures – all fatalities, passenger vehicles, nonoccupants, and motorcycle riders. Separate goals will be set for the years 2008, 2011, 2016, and 2020. The results of these alternative scenarios are summarized for these years in Table 7. Detailed results for all years from 2005 through 2020 are contained in Tables L-1 through L-7 in Appendix B. Figure 16 illustrates the gap between the base case and the reasonable worst case for the All Fatalities measure.

The results indicate that there are relatively small potential differences in the early goal years, but that differences become more pronounced in the later years. For the all fatalities measure, there is potentially a .03 shift in the predicted rate for 2008. By 2020, this shift has grown to .08. Similar proportions can be observed in other measures. The most dramatic potential shift occurs for motorcycle riders. Although there is very little difference predicted in earlier years, by 2020 the gap between the base case and the reasonable worst case grows to .17. These results indicate that future fatality rates will be highly influenced by the success of the agency's efforts to positively influence driver behavior, but will also be substantially influenced by market choices that are outside the agency's control.

Table 7. Summary of Alternate Scenario Fatality Rate Estimates

	Base	Modest Seat Belt	Modest Alcohol	Modest Behavioral	Continuous Motorcycle	Average Motorcycle	Reasonable Worst Case
All Fatalities							
2005	1.45	1.45	1.45	1.45	1.45	1.45	1.45
2008	1.37	1.38	1.38	1.40	1.39	1.38	1.40
2011	1.27	1.29	1.28	1.31	1.30	1.28	1.32
2016	1.10	1.14	1.13	1.16	1.16	1.12	1.18
2020	1.01	1.05	1.04	1.08	1.09	1.03	1.09
Passenger Vehicles							
2005	1.14	1.14	1.14	1.14	1.14	1.14	1.14
2008	1.05	1.06	1.06	1.07	1.05	1.05	1.07
2011	0.94	0.96	0.95	0.98	0.94	0.94	0.98
2016	0.77	0.80	0.79	0.83	0.77	0.77	0.83
2020	0.67	0.71	0.70	0.74	0.67	0.67	0.74
Nonoccupants							
2005	0.20	0.20	0.20	0.20	0.20	0.20	0.20
2008	0.19	0.19	0.19	0.19	0.19	0.19	0.19
2011	0.19	0.19	0.19	0.19	0.19	0.19	0.19
2016	0.19	0.19	0.19	0.19	0.19	0.19	0.19
2020	0.18	0.18	0.19	0.19	0.18	0.18	0.19
Motorcycles							
2005	0.73	0.73	0.73	0.73	0.73	0.73	0.73
2008	0.76	0.76	0.76	0.76	0.77	0.76	0.76
2011	0.78	0.78	0.78	0.78	0.80	0.79	0.79
2016	0.73	0.73	0.73	0.73	0.83	0.79	0.79
2020	0.61	0.61	0.61	0.61	0.85	0.78	0.78

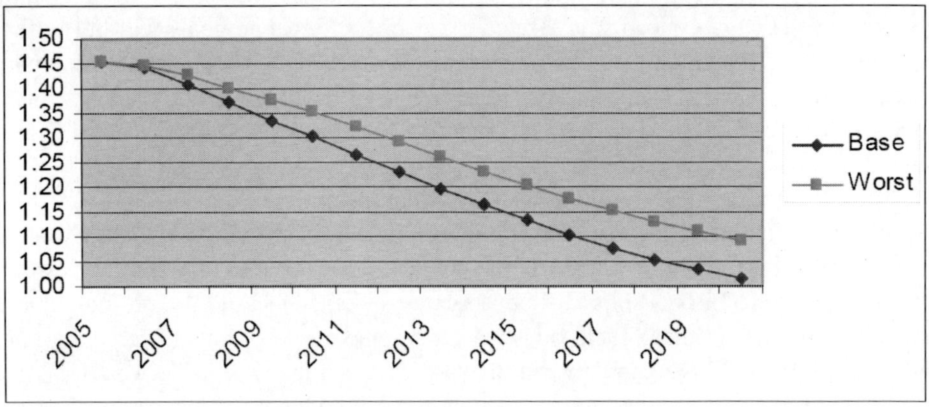

Figure 16. Potential Range of All Fatalities Rate.

DISCUSSION

Projecting fatality rates into the future is an undertaking that is fraught with uncertainty. The results of this analysis are dependent on a series of assumptions and previous evaluations, each of which has its own level of uncertainty. The safety benefits associated with each of the 21 standards examined here were all derived statistically and thus have some level of uncertainty, both individually and in combination. The many assumptions adopted for this analysis, while based on historical data, are dependent on the continuation of both individual and market place behavior. As such, this type of analysis cannot account for shifts in behavior that inevitably occur over time due to unforeseen economic or social influences. The middle-east, which is the primary source of the world's oil reserves, has never been more volatile, and instability in oil prices can lead to changes in consumer behavior that transform the nature of the vehicle fleet and with it, its safety characteristics.

Exposure is a critical metric in estimating future fatalities. This analysis assumes a continuation of current trends in both the level of driving exposure and the type of exposure. By 2020 this assumption by itself leads to an estimated increase in potentially fatal crashes of over 30 percent. If society accelerates or slows this trend, or if it begins to shift its exposure to either more or less risky roadways, future fatality rates may not match the levels predicted here. The relatively smooth declines in fatality rates predicted for the future in Figure 12, while consistent with the trends found in most recent years, are in stark contrast with stair-step declines that were experienced in earlier decades. To some extent this reflects the initial applications of technologies that are now taken for granted such as air bags and seat belts, but it also reflects significant transformations in vehicle design and size, and shifts in economic or social conditions which influence driving behavior. The trends predicted in this analysis are likely to be punctuated by similar unexpected shifts in driver behavior. This would cause short-term deviations away from the predicted trend line. However, over time, the downward trend in fatality rates should continue as the fleet becomes more saturated with vehicles containing better occupant protection and crash-avoidance safety equipment.

We can foresee several possible scenarios that might confound this trend and cause notable deviations from these predictions. The first would be if the current trend in motorcycle fatalities continues to rise at its current pace for an indefinite period. If interest in motorcycles continues to grow indefinitely rather than curtailing over a 26-28 year generational cycle, it could cause a significant increase in fatalities during the later years covered by this analysis.[17] An even greater impact would occur if improvements in seat belt usage were to stall out at current levels. Improvements in seat belt usage account for over 40-percent of fatality reduction in the earlier years to roughly 30 percent in the later years. The agency and the greater safety community face ever increasing challenges in their efforts to improve seat belt use because current non-users are the least risk-adverse drivers. A similar challenge will occur for programs aimed at reducing impaired driving, which account for 10-15 percent of projected safety benefits.

The technological safety improvements that are soon to be required, or that are already being incorporated into the on-road vehicle fleet through both regulation and voluntary installation, assure that significant progress will be made against motor vehicle fatalities and injuries over the next decade. However, the full potential for safety improvement cannot be realized without continued progress in efforts to improve driver behavior.

Appendix A. Data Inputs and Penetration Estimates

Table A-1. Baseline Parameters for Safety Standard Fleet Penetration

FMVSS		Full Fleet Savings	2002 On- Road Fleet %	2002 Savings	Total Remaining Additional Savings
105	Dual Master Cylinders	270	99.80%	269	1
105	Front Disc Brakes	271	99.40%	269	2
201	Instrument Panel Padding	935	99.51%	930	5
203/204	Energy Absorbing Steering Columns	2,682	99.05%	2,657	25
206	Improved Door Locks	1,401	99.80%	1,398	3
208	Driver Frontal Air Bags	3,227	63.57%	2,051	1,176
208	Passenger Frontal Air Bags	821	51.41%	422	399
212	Windshield Adhesive Bonding	353	98.41%	347	6
214	Side Door Beams	689	92.10%	635	54
214	TTI Reduction (Already PC&LTV)	1,143	59.40%	679	464
214	Side Air Bags - Torso (PC&LTV)	896	5.50%	49	846
214	Side Air Bags - Head (PC&LTV)	896	2.60%	23	872
216	Roof Crush Strength (PC Only)	162	99.30%	161	1
126	Electronic Stability Control *	10,000	1.50%	150	9,850
105/121	Medium/Heavy Truck ABS	416	49.22%	205	211
201	Upper Interior Head Impact Protection	959	7.90%	76	883
138	TPMS	120	0.00%	0	120
214	Side Impact Upgrade *	1,791	2.15%	38	1,753
NA	Ejection Mitigation *	1,000	0.00%	0	1,000
121	Heavy Truck Stopping Distance *	180	0.00%	0	180
208	Seat Belts	21,724	75.00%	14,178	7,546

Table A-2. Annual On-Road Fleet Penetration for Existing and Proposed Safety Standards

		2006	2007	2008	2009	2010	2011	2012	2013	2014	2015	2016	2017	2018	2019	2020
105	Dual Master Cylinders	99.85%	99.86%	99.87%	99.88%	99.89%	99.89%	99.90%	99.91%	99.91%	99.92%	99.92%	99.93%	99.93%	99.94%	99.94%
105	Front Disc Brakes	99.55%	99.58%	99.61%	99.64%	99.66%	99.68%	99.70%	99.72%	99.74%	99.76%	99.77%	99.78%	99.80%	99.81%	99.82%
201	Instrument Panel Padding	99.63%	99.65%	99.68%	99.70%	99.72%	99.74%	99.76%	99.77%	99.79%	99.80%	99.81%	99.82%	99.83%	99.84%	99.85%
203/ 204	Energy Absorbing Steering Columns	99.28%	99.33%	99.38%	99.42%	99.46%	99.50%	99.53%	99.56%	99.59%	99.61%	99.64%	99.66%	99.68%	99.70%	99.71%
206	Improved Door Locks	99.85%	99.86%	99.87%	99.88%	99.89%	99.89%	99.90%	99.91%	99.91%	99.92%	99.92%	99.93%	99.93%	99.94%	99.94%
208	Driver Frontal Air Bags	75.18%	77.68%	79.87%	81.79%	83.47%	84.95%	86.24%	87.38%	88.39%	89.27%	90.05%	90.74%	91.35%	91.90%	92.38%
208	Passenger Frontal Air Bags	66.53%	69.87%	72.84%	75.46%	77.77%	79.82%	81.62%	83.21%	84.62%	85.86%	86.97%	87.94%	88.81%	89.59%	90.27%
212	Windshield Adhesive Bonding	98.80%	98.88%	98.96%	99.03%	99.10%	99.16%	99.22%	99.27%	99.31%	99.36%	99.39%	99.43%	99.46%	99.49%	99.52%
214	Side Door Beams	94.04%	94.47%	94.86%	95.22%	95.55%	95.85%	96.12%	96.37%	96.60%	96.81%	97.00%	97.18%	97.34%	97.50%	97.64%
214	TTI Reduction (Already PC&LTV)	72.25%	75.05%	77.51%	79.67%	81.57%	83.24%	84.71%	86.00%	87.14%	88.14%	89.03%	89.82%	90.51%	91.13%	91.69%
214	Side Air Bags -Torso (PC&LTV)	30.04%	36.07%	41.64%	46.75%	51.42%	55.67%	59.53%	63.01%	66.16%	69.00%	71.55%	73.85%	75.91%	77.77%	79.44%
214	Side Air Bags -Head (PC&LTV)	27.51%	33.68%	39.38%	44.64%	49.45%	53.84%	57.83%	61.44%	64.71%	67.66%	70.32%	72.71%	74.87%	76.81%	78.55%
216	Roof Crush Strength (PC Only)	99.47%	99.51%	99.54%	99.57%	99.60%	99.63%	99.65%	99.68%	99.70%	99.72%	99.73%	99.75%	99.76%	99.78%	99.79%
126	Electronic Stability Control	5.43%	7.77%	10.82%	14.34%	18.08%	22.62%	28.06%	33.48%	38.51%	43.18%	47.49%	51.46%	55.10%	58.44%	61.50%
105/ 121	Medium/ Heavy Truck ABS	66.86%	70.18%	73.12%	75.71%	78.00%	80.02%	81.80%	83.38%	84.77%	86.00%	87.09%	88.06%	88.92%	89.68%	90.36%
201	Upper Interior Head Impact Protection	34.37%	40.16%	45.48%	50.34%	54.76%	58.77%	62.39%	65.66%	68.60%	71.24%	73.62%	75.75%	77.66%	79.38%	80.92%
138	TPMS	0.74%	3.97%	9.82%	16.44%	22.70%	28.59%	34.09%	39.20%	43.93%	48.29%	52.30%	55.97%	59.33%	62.39%	65.19%
214	Oblique Pole Test	7.88%	10.11%	12.23%	14.24%	16.13%	17.92%	19.60%	21.66%	25.66%	30.96%	35.91%	40.51%	44.78%	48.73%	52.38%
NA	Ejection Mitigation	0.94%	1.78%	2.91%	4.23%	5.49%	6.70%	7.85%	9.86%	14.27%	20.00%	25.41%	30.49%	35.25%	39.69%	43.82%
121	Heavy Truck Stopping Distance				0.00%	3.41%	9.99%	16.22%	22.12%	27.67%	32.87%	37.73%	42.24%	46.41%	50.27%	53.83%
208	Seat Belt Usage Rate	81.00%	82.52%	83.92%	85.20%	86.39%	87.48%	88.48%	89.40%	90.25%	91.03%	91.75%	92.41%	93.01%	93.57%	94.09%

Table A-3. Cumulative On-Road Fleet and Cumulative Retirements by Vehicle Age Passenger Cars

Vehicle Age	Survival Rate	Retirement Rate	On-Road Fleet Profile	Annual Retirements	Retirements by Age	On- Road Fleet by Age	Cumulative	
							On-Road Fleet by Age	Cumulative Retirements by Age
1	0.9900	0.0100	8,000,000	79,839	1.49%	6.13%	6.13%	1.49%
2	0.9831	0.0069	8,053,834	55,744	1.04%	6.17%	12.31%	2.54%
3	0.9731	0.0100	8,005,049	80,437	1.50%	6.14%	18.44%	4.04%
4	0.9593	0.0138	8,758,284	120,775	2.26%	6.71%	25.16%	6.30%
5	0.9413	0.0180	7,966,759	143,430	2.68%	6.11%	31.27%	8.98%
6	0.9188	0.0225	7,400,233	166,418	3.11%	5.67%	36.94%	12.09%
7	0.8918	0.0270	7,631,081	206,117	3.85%	5.85%	42.79%	15.94%
8	0.8604	0.0313	7,076,943	221,747	4.15%	5.43%	48.22%	20.09%
9	0.8252	0.0352	8,160,683	287,642	5.38%	6.26%	54.47%	25.47%
10	0.7866	0.0386	7,069,872	272,773	5.10%	5.42%	59.89%	30.57%
11	0.7170	0.0696	6,904,614	480,784	8.99%	5.29%	65.19%	39.56%
12	0.6125	0.1045	6,238,595	651,774	12.19%	4.78%	69.97%	51.74%
13	0.5094	0.1031	6,049,746	623,911	11.67%	4.64%	74.61%	63.41%
14	0.4142	0.0951	5,785,594	550,444	10.29%	4.44%	79.04%	73.70%
15	0.3308	0.0834	5,562,610	463,960	8.67%	4.26%	83.31%	82.38%
16	0.2604	0.0704	4,876,582	343,153	6.42%	3.74%	87.05%	88.79%
17	0.2028	0.0577	4,045,657	233,342	4.36%	3.10%	90.15%	93.16%
18	0.1565	0.0463	3,361,057	155,483	2.91%	2.58%	92.72%	96.06%
19	0.1200	0.0365	2,518,942	91,948	1.72%	1.93%	94.65%	97.78%
20	0.0916	0.0285	1,885,341	53,641	1.00%	1.45%	96.10%	98.78%
21	0.0696	0.0220	1,109,748	24,385	0.46%	0.85%	96.95%	99.24%
22	0.0527	0.0169	717,732	12,096	0.23%	0.55%	97.50%	99.47%
23	0.0399	0.0129	591,767	7,609	0.14%	0.45%	97.95%	99.61%
24	0.0301	0.0098	494,398	4,831	0.09%	0.38%	98.33%	99.70%
25+	0.0227	0.0074	2,173,404	16,095	0.30%	1.67%	100.00%	100.00%
Total		0.9773	130,438,525	5,348,376	1.0000	1.0000		

Table A-4. Cumulative On-Road Fleet and Cumulative Retirements by Vehicle Age Light Trucks

Vehicle Age	Survival Rate	Retirement Rate	On-Road Fleet Profile	Annual Retirements	Retirements by Age	On- Road Fleet by Age	Cumulative	
							On-Road Fleet by Age	Cumulative Retirements by Age
1	0.9741	0.0259	7,800,000	201,953	7.19%	9.01%	9.01%	7.19%
2	0.9603	0.0138	7,800,519	107,998	3.85%	9.01%	18.01%	11.04%

Vehicle Age	Survival Rate	Retirement Rate	On-Road Fleet Profile	Annual Retirements	Retirements by Age	On- Road Fleet by Age	Cumulative	
							On-Road Fleet by Age	Cumulative Retirements by Age
3	0.9420	0.0183	7,035,283	128,474	4.58%	8.12%	26.14%	15.61%
4	0.9190	0.0230	7,028,541	161,537	5.75%	8.12%	34.25%	21.37%
5	0.8913	0.0278	6,456,982	179,189	6.38%	7.46%	41.71%	27.75%
6	0.8590	0.0323	5,678,888	183,450	6.53%	6.56%	48.26%	34.28%
7	0.8226	0.0364	5,318,882	193,642	6.90%	6.14%	54.41%	41.18%
8	0.7827	0.0399	4,407,753	175,758	6.26%	5.09%	59.50%	47.44%
9	0.7401	0.0426	4,679,603	199,280	7.10%	5.40%	64.90%	54.54%
10	0.6956	0.0445	4,336,664	192,880	6.87%	5.01%	69.91%	61.41%
11	0.6501	0.0455	3,520,685	160,357	5.71%	4.07%	73.97%	67.12%
12	0.6042	0.0458	2,827,293	129,602	4.62%	3.26%	77.24%	71.73%
13	0.5517	0.0525	2,589,751	136,083	4.85%	2.99%	80.23%	76.58%
14	0.5009	0.0508	2,340,504	118,972	4.24%	2.70%	82.93%	80.82%
15	0.4522	0.0487	2,539,888	123,595	4.40%	2.93%	85.86%	85.22%
16	0.4062	0.0460	2,298,883	105,667	3.76%	2.65%	88.52%	88.98%
17	0.3633	0.0429	1,802,394	77,352	2.75%	2.08%	90.60%	91.74%
18	0.3236	0.0397	1,678,087	66,569	2.37%	1.94%	92.53%	94.11%
19	0.2873	0.0364	1,301,325	47,304	1.68%	1.50%	94.04%	95.79%
20	0.2542	0.0331	1,066,249	35,250	1.26%	1.23%	95.27%	97.05%
21	0.2244	0.0299	641,888	19,173	0.68%	0.74%	96.01%	97.73%
22	0.1975	0.0268	492,438	13,215	0.47%	0.57%	96.58%	98.20%
23	0.1735	0.0240	399,337	9,581	0.34%	0.46%	97.04%	98.54%
24	0.1522	0.0214	333,308	7,119	0.25%	0.38%	97.42%	98.79%
25	0.1332	0.0189	621,374	11,772	0.42%	0.72%	98.14%	99.21%
26	0.1165	0.0168	536,874	8,993	0.32%	0.62%	98.76%	99.53%
27	0.1017	0.0148	415,832	6,142	0.22%	0.48%	99.24%	99.75%
28	0.0887	0.0130	289,805	3,765	0.13%	0.33%	99.58%	99.89%
29	0.0773	0.0114	81,654	931	0.03%	0.09%	99.67%	99.92%
30	0.0673	0.0100	71,447	714	0.03%	0.08%	99.75%	99.95%
31	0.0586	0.0087	61,241	535	0.02%	0.07%	99.82%	99.96%
32	0.0509	0.0076	51,034	390	0.01%	0.06%	99.88%	99.98%
33	0.0443	0.0067	40,827	272	0.01%	0.05%	99.93%	99.99%
34	0.0385	0.0058	30,620	178	0.01%	0.04%	99.96%	99.99%
35	0.0334	0.0051	20,414	103	0.00%	0.02%	99.99%	100.00%
36	0.0290	0.0044	10,207	45	0.00%	0.01%	100.00%	100.00%
		0.9710	86,606,473	2,807,843	100.00%	100.00%		

Table A-5. Cumulative On-Road Fleet and Cumulative Retirements by Vehicle Age Passenger Cars and LTVs

Vehicle Age	Annual Retirements	Retirements by Age	On-Road Fleet Profile	On-Road Fleet by Age	Cumulative	
					On-Road Fleet by Age	Cumulative Retirements by Age
1	281,791	3.45%	15,800,000	7.28%	7.28%	3.45%
2	163,742	2.01%	15,854,353	7.30%	14.58%	5.46%
3	208,912	2.56%	15,040,332	6.93%	21.51%	8.02%
4	282,312	3.46%	15,786,825	7.27%	28.79%	11.49%
5	322,619	3.96%	14,423,741	6.65%	35.43%	15.44%
6	349,868	4.29%	13,079,121	6.03%	41.46%	19.73%
7	399,759	4.90%	12,949,963	5.97%	47.43%	24.63%
8	397,505	4.87%	11,484,696	5.29%	52.72%	29.51%
9	486,921	5.97%	12,840,286	5.92%	58.63%	35.48%
10	465,653	5.71%	11,406,536	5.26%	63.89%	41.18%
11	641,141	7.86%	10,425,299	4.80%	68.69%	49.05%
12	781,376	9.58%	9,065,888	4.18%	72.87%	58.63%
13	759,994	9.32%	8,639,497	3.98%	76.85%	67.94%
14	669,416	8.21%	8,126,098	3.74%	80.59%	76.15%
15	587,555	7.20%	8,102,498	3.73%	84.33%	83.35%
16	448,820	5.50%	7,175,465	3.31%	87.63%	88.86%
17	310,694	3.81%	5,848,051	2.69%	90.33%	92.67%
18	222,052	2.72%	5,039,144	2.32%	92.65%	95.39%
19	139,252	1.71%	3,820,267	1.76%	94.41%	97.10%
20	88,891	1.09%	2,951,590	1.36%	95.77%	98.19%
21	43,558	0.53%	1,751,636	0.81%	96.58%	98.72%
22	25,311	0.31%	1,210,170	0.56%	97.13%	99.03%
23	17,190	0.21%	991,104	0.46%	97.59%	99.24%
24	11,951	0.15%	827,706	0.38%	97.97%	99.39%
25	27,867	0.34%	2,794,778	1.29%	99.26%	99.73%
26	8,993	0.11%	536,874	0.25%	99.51%	99.84%
27	6,142	0.08%	415,832	0.19%	99.70%	99.91%
28	3,765	0.05%	289,805	0.13%	99.83%	99.96%
29	931	0.01%	81,654	0.04%	99.87%	99.97%
30	714	0.01%	71,447	0.03%	99.90%	99.98%
31	535	0.01%	61,241	0.03%	99.93%	99.99%
32	390	0.00%	51,034	0.02%	99.95%	99.99%
33	272	0.00%	40,827	0.02%	99.97%	100.00%
34	178	0.00%	30,620	0.01%	99.99%	100.00%
35	103	0.00%	20,414	0.01%	100.00%	100.00%
36	45	0.00%	10,207	0.00%	100.00%	100.00%
	8,156,219	100.00%	217,044,998	100.00%		

APPENDIX B. RESULTS TABLES

Table B-1. All Fatalities, Fatality Rate Impacts of Behavioral Programs and Proposed Safety Standards

	Base Estimate	Seat Belts	Alcohol Improv.	Combined Behavioral	ESC	Heavy-Truck Brakes	Side Impact	Ejection Mitigation	Proposed Standards	All NHTSA Activities
2005	1.45	1.45	1.45	1.45	1.45	1.45	1.45	1.45	1.45	1.45
2006	1.44	1.44	1.45	1.45	1.44	1.44	1.44	1.44	1.44	1.45
2007	1.41	1.42	1.42	1.44	1.41	1.41	1.41	1.41	1.41	1.44
2008	1.37	1.40	1.39	1.42	1.37	1.37	1.37	1.37	1.37	1.42
2009	1.34	1.37	1.36	1.40	1.34	1.34	1.34	1.34	1.34	1.40
2010	1.30	1.35	1.33	1.38	1.31	1.30	1.30	1.30	1.31	1.39
2011	1.27	1.32	1.30	1.36	1.28	1.27	1.27	1.27	1.28	1.37
2012	1.23	1.29	1.27	1.33	1.25	1.23	1.23	1.23	1.26	1.36
2013	1.20	1.26	1.24	1.31	1.23	1.20	1.20	1.20	1.23	1.35
2014	1.16	1.24	1.21	1.28	1.21	1.17	1.17	1.16	1.21	1.33
2015	1.13	1.21	1.18	1.26	1.18	1.13	1.14	1.14	1.19	1.32
2016	1.10	1.19	1.15	1.24	1.16	1.11	1.11	1.11	1.17	1.31
2017	1.08	1.17	1.13	1.22	1.14	1.08	1.08	1.08	1.15	1.30
2018	1.06	1.14	1.11	1.20	1.12	1.06	1.06	1.06	1.14	1.30
2019	1.03	1.13	1.09	1.18	1.11	1.04	1.04	1.04	1.12	1.29
2020	1.01	1.11	1.07	1.17	1.09	1.02	1.02	1.02	1.11	1.28

Table B-2. All Fatalities Fatality Rate Impacts of Behavioral Programs and Proposed Safety Standards Change from Base Estimate

	Base Estimate	Seat Belts	Alcohol Improv.	Combined Behavioral	ESC	Heavy-Truck Brakes	Side Impact	Ejection Mitigation	Proposed Standards	All NHTSA Activities
2005	0.00	0.00	0.00	0.00	0.00	0.00	0.00	0.00	0.00	0.00
2006	0.00	0.00	-0.01	-0.01	0.00	0.00	0.00	0.00	0.00	-0.01
2007	0.00	-0.01	-0.02	-0.03	0.00	0.00	0.00	0.00	0.00	-0.03
2008	0.00	-0.03	-0.02	-0.05	0.00	0.00	0.00	0.00	0.00	-0.05
2009	0.00	-0.04	-0.03	-0.06	0.00	0.00	0.00	0.00	0.00	-0.07
2010	0.00	-0.05	-0.03	-0.08	-0.01	0.00	0.00	0.00	-0.01	-0.09
2011	0.00	-0.05	-0.03	-0.09	-0.01	0.00	0.00	0.00	-0.02	-0.11
2012	0.00	-0.06	-0.04	-0.10	-0.02	0.00	0.00	0.00	-0.02	-0.13

Table B-2. (Continued)

	Base Estimate	Seat Belts	Alcohol Improv.	Combined Behavioral	ESC	Heavy-Truck Brakes	Side Impact	Ejection Mitigation	Proposed Standards	All NHTSA Activities
2013	0.00	-0.07	-0.04	-0.11	-0.03	0.00	0.00	0.00	-0.04	-0.15
2014	0.00	-0.07	-0.04	-0.12	-0.04	0.00	0.00	0.00	-0.05	-0.17
2015	0.00	-0.08	-0.05	-0.13	-0.05	0.00	0.00	0.00	-0.06	-0.19
2016	0.00	-0.08	-0.05	-0.13	-0.06	0.00	0.00	0.00	-0.07	-0.21
2017	0.00	-0.09	-0.05	-0.14	-0.06	0.00	-0.01	0.00	-0.08	-0.23
2018	0.00	-0.09	-0.05	-0.15	-0.07	0.00	-0.01	-0.01	-0.08	-0.24
2019	0.00	-0.09	-0.05	-0.15	-0.07	0.00	-0.01	-0.01	-0.09	-0.25
2020	0.00	-0.10	-0.05	-0.16	-0.08	0.00	-0.01	-0.01	-0.10	-0.27

Table C-1. All Fatalities Excluding Motorcycles Fatality Rate Impacts of Behavioral Programs and Proposed Safety Standards

	Base Estimate	Seat Belts	Alcohol Improv.	Combined Behavioral	ESC	Heavy-Truck Brakes	Side Impact	Ejection Mitigation	Proposed Standards	All NHTSA Activities
2005	1.31	1.31	1.31	1.31	1.31	1.31	1.31	1.31	1.31	1.31
2006	1.29	1.29	1.30	1.30	1.29	1.29	1.29	1.29	1.29	1.30
2007	1.25	1.27	1.27	1.28	1.25	1.25	1.25	1.25	1.25	1.28
2008	1.21	1.24	1.24	1.26	1.22	1.21	1.21	1.21	1.22	1.26
2009	1.18	1.22	1.21	1.24	1.18	1.18	1.18	1.18	1.18	1.25
2010	1.15	1.19	1.18	1.23	1.16	1.15	1.15	1.15	1.16	1.24
2011	1.12	1.17	1.15	1.21	1.13	1.12	1.12	1.12	1.13	1.22
2012	1.09	1.15	1.13	1.19	1.11	1.09	1.09	1.09	1.11	1.22
2013	1.06	1.13	1.11	1.18	1.09	1.06	1.06	1.06	1.10	1.21
2014	1.04	1.11	1.09	1.16	1.08	1.04	1.04	1.04	1.08	1.21
2015	1.02	1.09	1.07	1.15	1.06	1.02	1.02	1.02	1.07	1.21
2016	1.00	1.08	1.05	1.14	1.05	1.00	1.00	1.00	1.06	1.21
2017	0.98	1.07	1.04	1.13	1.04	0.98	0.99	0.98	1.05	1.21
2018	0.96	1.05	1.03	1.12	1.03	0.97	0.97	0.97	1.05	1.22
2019	0.95	1.04	1.01	1.11	1.02	0.95	0.96	0.96	1.04	1.22
2020	0.94	1.03	1.00	1.10	1.01	0.94	0.95	0.94	1.03	1.22

Table C-2. All Fatalities Excluding Motorcycles Fatality Rate Impacts of Behavioral Programs and Proposed Safety Standards Change from Base Estimate

	Base Estimate	Seat Belts	Alcohol Improv.	Combined Behavioral	ESC	Heavy-Truck Brakes	Side Impact	Ejection Mitigation	Proposed Standards	All NHTSA Activities
2005	0.00	0.00	0.00	0.00	0.00	0.00	0.00	0.00	0.00	0.00
2006	0.00	0.00	-0.01	-0.01	0.00	0.00	0.00	0.00	0.00	-0.01
2007	0.00	-0.01	-0.02	-0.03	0.00	0.00	0.00	0.00	0.00	-0.03
2008	0.00	-0.03	-0.02	-0.05	0.00	0.00	0.00	0.00	0.00	-0.05
2009	0.00	-0.04	-0.03	-0.06	0.00	0.00	0.00	0.00	0.00	-0.07
2010	0.00	-0.05	-0.03	-0.08	-0.01	0.00	0.00	0.00	-0.01	-0.09
2011	0.00	-0.05	-0.04	-0.09	-0.01	0.00	0.00	0.00	-0.02	-0.11
2012	0.00	-0.06	-0.04	-0.10	-0.02	0.00	0.00	0.00	-0.02	-0.13
2013	0.00	-0.07	-0.04	-0.12	-0.03	0.00	0.00	0.00	-0.04	-0.15
2014	0.00	-0.07	-0.05	-0.13	-0.04	0.00	0.00	0.00	-0.05	-0.18
2015	0.00	-0.08	-0.05	-0.13	-0.05	0.00	0.00	0.00	-0.06	-0.20
2016	0.00	-0.08	-0.06	-0.14	-0.06	0.00	0.00	0.00	-0.07	-0.22
2017	0.00	-0.09	-0.06	-0.15	-0.06	0.00	-0.01	0.00	-0.07	-0.23
2018	0.00	-0.09	-0.06	-0.16	-0.07	0.00	-0.01	-0.01	-0.08	-0.25
2019	0.00	-0.09	-0.06	-0.16	-0.07	0.00	-0.01	-0.01	-0.09	-0.27
2020	0.00	-0.09	-0.07	-0.17	-0.08	0.00	-0.01	-0.01	-0.09	-0.28

Table D-1. Passenger Vehicles Only Fatality Rate Impacts of Behavioral Programs and Proposed Safety Standards

	Base Estimate	Seat Belts	Alcohol Improv.	Combined Behavioral	ESC	Heavy-Truck Brakes	Side Impact	Ejection Mitigation	Proposed Standards	All NHTSA Activities
2005	1.14	1.14	1.14	1.14	1.14	1.14	1.14	1.14	1.14	1.14
2006	1.13	1.13	1.14	1.14	1.13	1.13	1.13	1.13	1.13	1.14
2007	1.09	1.10	1.10	1.12	1.09	1.09	1.09	1.09	1.09	1.12
2008	1.05	1.08	1.07	1.10	1.05	1.05	1.05	1.05	1.05	1.10
2009	1.01	1.05	1.03	1.07	1.02	1.01	1.01	1.01	1.02	1.08
2010	0.98	1.03	1.00	1.05	0.98	0.98	0.98	0.98	0.98	1.06
2011	0.94	1.00	0.97	1.03	0.95	0.94	0.94	0.94	0.95	1.05
2012	0.90	0.97	0.93	1.00	0.93	0.90	0.90	0.90	0.93	1.03

Table D-1. (Continued)

	Base Estimate	Seat Belts	Alcohol Improv.	Combined Behavioral	ESC	Heavy-Truck Brakes	Side Impact	Ejection Mitigation	Proposed Standards	All NHTSA Activities
2013	0.86	0.94	0.90	0.98	0.90	0.87	0.86	0.86	0.90	1.02
2014	0.83	0.91	0.87	0.95	0.87	0.83	0.83	0.83	0.88	1.01
2015	0.80	0.88	0.84	0.93	0.85	0.80	0.80	0.80	0.86	0.99
2016	0.77	0.86	0.81	0.90	0.83	0.77	0.77	0.77	0.84	0.98
2017	0.74	0.83	0.78	0.88	0.81	0.74	0.75	0.74	0.82	0.97
2018	0.71	0.81	0.76	0.86	0.79	0.72	0.72	0.72	0.80	0.96
2019	0.69	0.79	0.74	0.84	0.77	0.69	0.70	0.70	0.79	0.96
2020	0.67	0.77	0.72	0.83	0.75	0.67	0.68	0.68	0.77	0.95

Table D-2. Passenger Vehicles Only Fatality Rate Impacts of Behavioral Programs and Proposed Safety Standards Change from Base Estimate

	Base Estimate	Seat Belts	Alcohol Improv.	Combined Behavioral	ESC	Heavy-Truck Brakes	Side Impact	Ejection Mitigation	Proposed Standards	All NHTSA Activities
2005	0.00	0.00	0.00	0.00	0.00	0.00	0.00	0.00	0.00	0.00
2006	0.00	0.00	-0.01	-0.01	0.00	0.00	0.00	0.00	0.00	-0.01
2007	0.00	-0.01	-0.01	-0.03	0.00	0.00	0.00	0.00	0.00	-0.03
2008	0.00	-0.03	-0.02	-0.05	0.00	0.00	0.00	0.00	0.00	-0.05
2009	0.00	-0.04	-0.02	-0.06	0.00	0.00	0.00	0.00	0.00	-0.07
2010	0.00	-0.05	-0.03	-0.08	-0.01	0.00	0.00	0.00	-0.01	-0.09
2011	0.00	-0.06	-0.03	-0.09	-0.02	0.00	0.00	0.00	-0.02	-0.11
2012	0.00	-0.07	-0.03	-0.10	-0.03	0.00	0.00	0.00	-0.03	-0.13
2013	0.00	-0.07	-0.04	-0.11	-0.04	0.00	0.00	0.00	-0.04	-0.15
2014	0.00	-0.08	-0.04	-0.12	-0.05	0.00	0.00	0.00	-0.05	-0.18
2015	0.00	-0.09	-0.04	-0.13	-0.05	0.00	0.00	0.00	-0.06	-0.20
2016	0.00	-0.09	-0.04	-0.14	-0.06	0.00	-0.01	0.00	-0.07	-0.22
2017	0.00	-0.09	-0.04	-0.14	-0.07	0.00	-0.01	0.00	-0.08	-0.23
2018	0.00	-0.10	-0.05	-0.15	-0.07	0.00	-0.01	-0.01	-0.09	-0.25
2019	0.00	-0.10	-0.05	-0.15	-0.08	0.00	-0.01	-0.01	-0.10	-0.26
2020	0.00	-0.10	-0.05	-0.16	-0.08	0.00	-0.01	-0.01	-0.10	-0.28

Table E-1. Passenger Vehicles and Motorcycles Fatality Rate Impacts of Behavioral Programs and Proposed Safety Standards

	Base Estimate	Seat Belts	Alcohol Improv.	Combined Behavioral	ESC	Heavy-Truck Brakes	Side Impact	Ejection Mitigation	Proposed Standards	All NHTSA Activities
2005	1.30	1.30	1.30	1.30	1.30	1.30	1.30	1.30	1.30	1.30
2006	1.29	1.29	1.30	1.30	1.29	1.29	1.29	1.29	1.29	1.30
2007	1.26	1.27	1.27	1.29	1.26	1.26	1.26	1.26	1.26	1.29
2008	1.22	1.25	1.24	1.27	1.22	1.22	1.22	1.22	1.22	1.27
2009	1.18	1.22	1.20	1.24	1.18	1.18	1.18	1.18	1.18	1.25
2010	1.14	1.19	1.17	1.22	1.15	1.14	1.14	1.14	1.15	1.23
2011	1.10	1.15	1.13	1.19	1.11	1.10	1.10	1.10	1.11	1.21
2012	1.05	1.12	1.09	1.15	1.07	1.05	1.05	1.05	1.08	1.18
2013	1.00	1.08	1.04	1.12	1.04	1.00	1.00	1.00	1.04	1.16
2014	0.96	1.04	1.00	1.08	1.00	0.96	0.96	0.96	1.01	1.14
2015	0.91	1.00	0.96	1.05	0.96	0.91	0.92	0.91	0.97	1.11
2016	0.87	0.96	0.92	1.01	0.93	0.87	0.87	0.87	0.94	1.09
2017	0.83	0.92	0.88	0.98	0.90	0.83	0.84	0.83	0.91	1.07
2018	0.79	0.89	0.85	0.95	0.87	0.80	0.80	0.80	0.88	1.05
2019	0.76	0.86	0.82	0.92	0.84	0.76	0.77	0.77	0.86	1.03
2020	0.73	0.83	0.79	0.90	0.81	0.73	0.74	0.74	0.83	1.02

Table E-2. Passenger Vehicles and Motorcycles Fatality Rate Impacts of Behavioral Programs and Proposed Safety Standards Change From Base Estimate

	Base Estimate	Seat Belts	Alcohol Improv.	Combined Behavioral	ESC	Heavy-Truck Brakes	Side Impact	Ejection Mitigation	Proposed Standards	All NHTSA Activities
2005	0.00	0.00	0.00	0.00	0.00	0.00	0.00	0.00	0.00	0.00
2006	0.00	0.00	-0.01	-0.01	0.00	0.00	0.00	0.00	0.00	-0.01
2007	0.00	-0.01	-0.01	-0.03	0.00	0.00	0.00	0.00	0.00	-0.03
2008	0.00	-0.03	-0.02	-0.05	0.00	0.00	0.00	0.00	0.00	-0.05
2009	0.00	-0.04	-0.02	-0.06	0.00	0.00	0.00	0.00	0.00	-0.07
2010	0.00	-0.05	-0.03	-0.08	-0.01	0.00	0.00	0.00	-0.01	-0.09
2011	0.00	-0.06	-0.03	-0.09	-0.02	0.00	0.00	0.00	-0.02	-0.11
2012	0.00	-0.07	-0.04	-0.10	-0.03	0.00	0.00	0.00	-0.03	-0.13
2013	0.00	-0.07	-0.04	-0.12	-0.04	0.00	0.00	0.00	-0.04	-0.16

Table E-2. (Continued)

	Base Estimate	Seat Belts	Alcohol Improv.	Combined Behavioral	ESC	Heavy-Truck Brakes	Side Impact	Ejection Mitigation	Proposed Standards	All NHTSA Activities
2014	0.00	-0.08	-0.04	-0.13	-0.04	0.00	0.00	0.00	-0.05	-0.18
2015	0.00	-0.08	-0.05	-0.13	-0.05	0.00	0.00	0.00	-0.06	-0.20
2016	0.00	-0.09	-0.05	-0.14	-0.06	0.00	-0.01	0.00	-0.07	-0.22
2017	0.00	-0.09	-0.05	-0.15	-0.07	0.00	-0.01	0.00	-0.08	-0.24
2018	0.00	-0.10	-0.06	-0.16	-0.07	0.00	-0.01	-0.01	-0.09	-0.26
2019	0.00	-0.10	-0.06	-0.16	-0.08	0.00	-0.01	-0.01	-0.10	-0.27
2020	0.00	-0.10	-0.06	-0.17	-0.08	0.00	-0.01	-0.01	-0.10	-0.29

Table F-1. Nonoccupants Fatality Rate Impacts of Behavioral Programs and Proposed Safety Standards

	Base Estimate	Seat Belts	Alcohol Improv.	Combined Behavioral	ESC	Heavy-Truck Brakes	Side Impact	Ejection Mitigation	Proposed Standards	All NHTSA Activities
2005	0.20	0.20	0.20	0.20	0.20	0.20	0.20	0.20	0.20	0.20
2006	0.19	0.19	0.20	0.20	0.19	0.19	0.19	0.19	0.19	0.20
2007	0.19	0.19	0.20	0.20	0.19	0.19	0.19	0.19	0.19	0.20
2008	0.19	0.19	0.20	0.20	0.19	0.19	0.19	0.19	0.19	0.20
2009	0.19	0.19	0.20	0.20	0.19	0.19	0.19	0.19	0.19	0.20
2010	0.19	0.19	0.20	0.20	0.19	0.19	0.19	0.19	0.19	0.20
2011	0.19	0.19	0.20	0.20	0.19	0.19	0.19	0.19	0.19	0.20
2012	0.19	0.19	0.20	0.20	0.19	0.19	0.19	0.19	0.19	0.20
2013	0.19	0.19	0.20	0.20	0.19	0.19	0.19	0.19	0.19	0.20
2014	0.19	0.19	0.20	0.20	0.19	0.19	0.19	0.19	0.19	0.20
2015	0.19	0.19	0.20	0.20	0.19	0.19	0.19	0.19	0.19	0.20
2016	0.19	0.19	0.20	0.20	0.19	0.19	0.19	0.19	0.19	0.20
2017	0.19	0.19	0.20	0.20	0.19	0.19	0.19	0.19	0.19	0.20
2018	0.19	0.19	0.20	0.20	0.19	0.19	0.19	0.19	0.19	0.20
2019	0.18	0.18	0.20	0.20	0.18	0.18	0.18	0.18	0.18	0.20
2020	0.18	0.18	0.20	0.20	0.18	0.18	0.18	0.18	0.18	0.20

Table F-2. Nonoccupants Fatality Rate Impacts of Behavioral Programs and Proposed Safety Standards Change From Base Estimate

	Base Estimate	Seat Belts	Alcohol Improv.	Combined Behavioral	ESC	Heavy-Truck Brakes	Side Impact	Ejection Mitigation	Proposed Standards	All NHTSA Activities
2005	0.00	0.00	0.00	0.00	0.00	0.00	0.00	0.00	0.00	0.00
2006	0.00	0.00	0.00	0.00	0.00	0.00	0.00	0.00	0.00	0.00
2007	0.00	0.00	0.00	0.00	0.00	0.00	0.00	0.00	0.00	0.00
2008	0.00	0.00	0.00	0.00	0.00	0.00	0.00	0.00	0.00	0.00
2009	0.00	0.00	0.00	0.00	0.00	0.00	0.00	0.00	0.00	0.00
2010	0.00	0.00	0.00	0.00	0.00	0.00	0.00	0.00	0.00	0.00
2011	0.00	0.00	-0.01	-0.01	0.00	0.00	0.00	0.00	0.00	-0.01
2012	0.00	0.00	-0.01	-0.01	0.00	0.00	0.00	0.00	0.00	-0.01
2013	0.00	0.00	-0.01	-0.01	0.00	0.00	0.00	0.00	0.00	-0.01
2014	0.00	0.00	-0.01	-0.01	0.00	0.00	0.00	0.00	0.00	-0.01
2015	0.00	0.00	-0.01	-0.01	0.00	0.00	0.00	0.00	0.00	-0.01
2016	0.00	0.00	-0.01	-0.01	0.00	0.00	0.00	0.00	0.00	-0.01
2017	0.00	0.00	-0.01	-0.01	0.00	0.00	0.00	0.00	0.00	-0.01
2018	0.00	0.00	-0.01	-0.01	0.00	0.00	0.00	0.00	0.00	-0.01
2019	0.00	0.00	-0.01	-0.01	0.00	0.00	0.00	0.00	0.00	-0.01
2020	0.00	0.00	-0.01	-0.01	0.00	0.00	0.00	0.00	0.00	-0.01

Table G-1. All Fatalities Fatality Impacts of Behavioral Programs and Proposed Safety Standards

	Base Estimate	Seat Belts	Alcohol Improv.	Combined Behavioral	ESC	Heavy-Truck Brakes	Side Impact	Ejection Mitigation	Proposed Standards	All NHTSA Activities
2005	43,443	43,443	43,443	43,443	43,443	43,443	43,443	43,443	43,443	43,443
2006	43,751	43,751	44,045	44,045	43,751	43,751	43,751	43,751	43,751	44,045
2007	43,363	43,778	43,839	44,258	43,363	43,363	43,363	43,363	43,363	44,258
2008	42,897	43,696	43,544	44,355	42,932	42,897	42,897	42,897	42,932	44,392
2009	42,410	43,565	43,218	44,394	42,539	42,410	42,410	42,410	42,539	44,530
2010	41,945	43,428	42,906	44,423	42,203	41,951	41,945	41,945	42,209	44,703
2011	41,383	43,163	42,484	44,311	41,860	41,400	41,383	41,383	41,877	44,841
2012	40,827	42,877	42,058	44,171	41,625	40,853	40,827	40,827	41,652	45,065
2013	40,297	42,594	41,651	44,025	41,426	40,333	40,304	40,305	41,479	45,320

Table G-1. (Continued)

	Base Estimate	Seat Belts	Alcohol Improv.	Combined Behavioral	ESC	Heavy-Truck Brakes	Side Impact	Ejection Mitigation	Proposed Standards	All NHTSA Activities
2014	39,784	42,306	41,253	43,867	41,219	39,829	39,833	39,823	41,356	45,604
2015	39,306	42,032	40,882	43,717	41,023	39,358	39,417	39,386	41,278	45,915
2016	38,900	41,815	40,578	43,619	40,878	38,959	39,067	39,019	41,242	46,252
2017	38,561	41,653	40,338	43,572	40,779	38,627	38,781	38,715	41,247	46,615
2018	38,285	41,542	40,157	43,573	40,725	38,357	38,553	38,473	41,289	47,002
2019	38,068	41,481	40,031	43,620	40,713	38,145	38,380	38,286	41,367	47,412
2020	37,906	41,467	39,957	43,711	40,741	37,988	38,258	38,152	41,480	47,845

Table G-2. All Fatalities Fatality Impacts of Behavioral Programs and Proposed Safety Standards Change from Base Estimate

	Base Estimate	Seat Belts	Alcohol Improv.	Combined Behavioral	ESC	Heavy-Truck Brakes	Side Impact	Ejection Mitigation	Proposed Standards	All NHTSA Activities
2005	0	0	0	0	0	0	0	0	0	0
2006	0	0	-294	-294	0	0	0	0	0	-294
2007	0	-415	-475	-895	0	0	0	0	0	-895
2008	0	-800	-647	-1458	-35	0	0	0	-35	-1495
2009	0	-1155	-808	-1985	-129	0	0	0	-129	-2120
2010	0	-1483	-961	-2478	-258	-6	0	0	-264	-2758
2011	0	-1780	-1101	-2928	-477	-17	0	0	-494	-3458
2012	0	-2051	-1232	-3344	-798	-27	0	0	-825	-4239
2013	0	-2297	-1354	-3729	-1129	-36	-8	-8	-1183	-5023
2014	0	-2521	-1468	-4083	-1435	-44	-48	-38	-1572	-5819
2015	0	-2726	-1576	-4411	-1717	-52	-110	-80	-1971	-6609
2016	0	-2915	-1678	-4719	-1978	-59	-167	-119	-2342	-7352
2017	0	-3092	-1777	-5011	-2218	-66	-220	-154	-2686	-8054
2018	0	-3257	-1872	-5288	-2440	-72	-268	-187	-3004	-8717
2019	0	-3413	-1963	-5553	-2645	-78	-312	-218	-3300	-9345
2020	0	-3561	-2052	-5805	-2835	-83	-353	-247	-3574	-9939

Table H-1. All Fatalities Excluding Motorcycles Fatality Impacts of Behavioral Programs and Proposed Safety Standards

	Base Estimate	Seat Belts	Alcohol Improv.	Combined Behavioral	ESC	Heavy-Truck Brakes	Side Impact	Ejection Mitigation	Proposed Standards	All NHTSA Activities
2005	38890	38890	38890	38890	38890	38890	38890	38890	38890	38890
2006	38979	38979	39270	39270	38979	38979	38979	38979	38979	39270
2007	38419	38834	38898	39317	38419	38419	38419	38419	38419	39317
2008	37838	38637	38499	39311	37873	37838	37838	37838	37873	39347
2009	37299	38453	38138	39315	37428	37299	37299	37299	37428	39450
2010	36849	38328	37861	39378	37106	36854	36849	36849	37112	39658
2011	36365	38140	37543	39370	36840	36381	36365	36365	36857	39900
2012	35946	37989	37284	39396	36741	35973	35946	35946	36768	40291
2013	35605	37891	37098	39472	36728	35640	35612	35613	36782	40767
2014	35321	37827	36965	39579	36748	35365	35369	35359	36883	41316
2015	35099	37805	36890	39725	36804	35151	35208	35178	37056	41923
2016	34961	37851	36897	39938	36922	35019	35127	35078	37283	42571
2017	34886	37947	36968	40202	37082	34951	35104	35039	37546	43245
2018	34855	38076	37082	40499	37268	34927	35120	35041	37826	43928
2019	34850	38219	37222	40811	37460	34926	35158	35065	38107	44603
2020	34852	38361	37369	41123	37646	34933	35200	35095	38374	45257

Table H-2. All Fatalities Excluding Motorcycles Fatality Impacts of Behavioral Programs and Proposed Safety Standards Change from Base Estimate

	Base Estimate	Seat Belts	Alcohol Improv.	Combined Behavioral	ESC	Heavy-Truck Brakes	Side Impact	Ejection Mitigation	Proposed Standards	All NHTSA Activities
2005	0	0	0	0	0	0	0	0	0	0
2006	0	0	-291	-291	0	0	0	0	0	-291
2007	0	-415	-479	-898	0	0	0	0	0	-898
2008	0	-799	-662	-1473	-35	0	0	0	-35	-1509
2009	0	-1153	-839	-2016	-129	0	0	0	-129	-2151
2010	0	-1480	-1012	-2529	-258	-6	0	0	-264	-2810
2011	0	-1775	-1178	-3005	-475	-17	0	0	-492	-3535
2012	0	-2043	-1338	-3450	-795	-27	0	0	-822	-4345

Table H-2. (Continued)

	Base Estimate	Seat Belts	Alcohol Improv.	Combined Behavioral	ESC	Heavy-Truck Brakes	Side Impact	Ejection Mitigation	Proposed Standards	All NHTSA Activities
2013	0	-2286	-1493	-3868	-1124	-36	-8	-8	-1177	-5162
2014	0	-2506	-1644	-4258	-1427	-44	-48	-38	-1562	-5995
2015	0	-2706	-1791	-4626	-1705	-52	-110	-79	-1957	-6824
2016	0	-2890	-1937	-4978	-1961	-59	-166	-118	-2322	-7611
2017	0	-3061	-2082	-5316	-2196	-65	-218	-153	-2659	-8359
2018	0	-3220	-2227	-5643	-2412	-71	-265	-185	-2970	-9072
2019	0	-3369	-2372	-5962	-2611	-77	-308	-215	-3257	-9754
2020	0	-3509	-2518	-6272	-2794	-82	-348	-243	-3522	-10405

Table I-1. Passenger Vehicles Only Fatality Impacts of Behavioral Programs and Proposed Safety Standards

	Base Estimate	Seat Belts	Alcohol Improv.	Combined Behavioral	ESC	Heavy-Truck Brakes	Side Impact	Ejection Mitigation	Proposed Standards	All NHTSA Activities
2005	31415	31415	31415	31415	31415	31415	31415	31415	31415	31415
2006	31441	31441	31685	31685	31441	31441	31441	31441	31441	31685
2007	30813	31226	31208	31625	30813	30813	30813	30813	30813	31625
2008	30169	30964	30706	31511	30203	30169	30169	30169	30203	31547
2009	29528	30677	30199	31367	29657	29528	29528	29528	29657	31501
2010	28905	30382	29704	31209	29163	28911	28905	28905	29168	31487
2011	28182	29955	29096	30909	28657	28198	28182	28182	28673	31435
2012	27459	29504	28483	30578	28255	27486	27459	27459	28282	31466
2013	26759	29050	27884	30239	27885	26794	26766	26767	27938	31524
2014	26071	28588	27292	29885	27504	26116	26120	26110	27640	31608
2015	25414	28136	26724	29536	27129	25466	25524	25494	27383	31716
2016	24824	27736	26219	29236	26800	24884	24991	24943	27164	31848
2017	24297	27388	25774	28983	26514	24363	24517	24452	26982	32001
2018	23829	27086	25385	28774	26269	23901	24097	24017	26833	32176
2019	23416	26830	25047	28608	26062	23493	23728	23634	26717	32370
2020	23053	26617	24758	28482	25891	23136	23406	23300	26630	32583

Table I-2. Passenger Vehicles Only Fatality Impacts of Behavioral Programs and Proposed Safety Standards Change from Base Estimate

	Base Estimate	Seat Belts	Alcohol Improv.	Combined Behavioral	ESC	Heavy-Truck Brakes	Side Impact	Ejection Mitigation	Proposed Standards	All NHTSA Activities
2005	0	0	0	0	0	0	0	0	0	0
2006	0	0	-244	-244	0	0	0	0	0	-244
2007	0	-412	-395	-811	0	0	0	0	0	-811
2008	0	-795	-538	-1343	-35	0	0	0	-35	-1379
2009	0	-1149	-671	-1839	-129	0	0	0	-129	-1973
2010	0	-1477	-798	-2303	-257	-6	0	0	-263	-2581
2011	0	-1774	-915	-2728	-475	-17	0	0	-492	-3253
2012	0	-2044	-1024	-3119	-795	-27	0	0	-823	-4007
2013	0	-2291	-1125	-3481	-1126	-36	-8	-8	-1180	-4766
2014	0	-2516	-1220	-3814	-1432	-44	-48	-38	-1568	-5537
2015	0	-2722	-1310	-4122	-1715	-52	-110	-80	-1969	-6302
2016	0	-2912	-1395	-4412	-1976	-59	-167	-118	-2340	-7024
2017	0	-3090	-1477	-4685	-2217	-66	-220	-154	-2684	-7704
2018	0	-3257	-1555	-4945	-2440	-72	-268	-187	-3004	-8346
2019	0	-3415	-1632	-5192	-2646	-78	-312	-218	-3301	-8954
2020	0	-3564	-1705	-5429	-2837	-83	-353	-247	-3577	-9530

Table J-1. Passenger Vehicles and Motorcycles Fatality Impacts of Behavioral Programs and Proposed Safety Standards

	Base Estimate	Seat Belts	Alcohol Improv.	Combined Behavioral	ESC	Heavy-Truck Brakes	Side Impact	Ejection Mitigation	Proposed Standards	All NHTSA Activities
2005	35968	35968	35968	35968	35968	35968	35968	35968	35968	35968
2006	36216	36216	36459	36459	36216	36216	36216	36216	36216	36459
2007	35750	36162	36149	36566	35750	35750	35750	35750	35750	36566
2008	35198	35991	35751	36556	35233	35198	35198	35198	35233	36592
2009	34578	35723	35279	36446	34707	34578	34578	34578	34707	36581
2010	33902	35372	34748	36253	34158	33908	33902	33902	34164	36531
2011	33053	34816	34038	35850	33525	33070	33053	33053	33542	36376
2012	32139	34169	33257	35353	32929	32165	32139	32139	32956	36240
2013	31188	33461	32437	34792	32305	31224	31196	31197	32359	36077

Table J-1. (Continued)

	Base Estimate	Seat Belts	Alcohol Improv.	Combined Behavioral	ESC	Heavy-Truck Brakes	Side Impact	Ejection Mitigation	Proposed Standards	All NHTSA Activities
2014	30205	32697	31580	34173	31623	30249	30253	30243	31758	35896
2015	29218	31909	30716	33528	30913	29269	29327	29297	31165	35709
2016	28279	31155	29900	32917	30230	28338	28445	28396	30590	35529
2017	27401	30447	29144	32352	29585	27466	27617	27553	30047	35371
2018	26593	29798	28459	31848	28993	26664	26857	26778	29549	35250
2019	25867	29221	27857	31417	28465	25943	26174	26082	29109	35179
2020	25232	28727	27346	31070	28012	25313	25579	25475	28739	35171

Table J-2. Passenger Vehicles and Motorcycles Fatality Impacts of Behavioral Programs and Proposed Safety Standards Change From Base Estimate

	Base Estimate	Seat Belts	Alcohol Improv.	Combined Behavioral	ESC	Heavy-Truck Brakes	Side Impact	Ejection Mitigation	Proposed Standards	All NHTSA Activities
2005	0	0	0	0	0	0	0	0	0	0
2006	0	0	-243	-243	0	0	0	0	0	-243
2007	0	-412	-400	-816	0	0	0	0	0	-816
2008	0	-793	-552	-1357	-35	0	0	0	-35	-1393
2009	0	-1145	-700	-1868	-128	0	0	0	-128	-2002
2010	0	-1470	-846	-2351	-256	-6	0	0	-262	-2629
2011	0	-1763	-985	-2797	-472	-17	0	0	-489	-3323
2012	0	-2030	-1118	-3214	-790	-26	0	0	-817	-4101
2013	0	-2272	-1248	-3604	-1117	-36	-8	-8	-1170	-4889
2014	0	-2492	-1375	-3968	-1419	-44	-48	-38	-1553	-5691
2015	0	-2691	-1498	-4311	-1695	-52	-109	-79	-1947	-6491
2016	0	-2876	-1621	-4638	-1950	-59	-165	-117	-2310	-7250
2017	0	-3046	-1743	-4952	-2185	-65	-217	-152	-2646	-7971
2018	0	-3206	-1866	-5256	-2400	-71	-264	-185	-2956	-8657
2019	0	-3355	-1990	-5551	-2598	-76	-307	-215	-3242	-9313
2020	0	-3495	-2114	-5838	-2780	-81	-347	-243	-3507	-9939

Table K-1. Nonoccupants Fatality Impacts of Behavioral Programs and Proposed Safety Standards

	Base Estimate	Seat Belts	Alcohol Improv.	Combined Behavioral	ESC	Heavy-Truck Brakes	Side Impact	Ejection Mitigation	Proposed Standards	All NHTSA Activities
2005	5849	5849	5849	5849	5849	5849	5849	5849	5849	5849
2006	5893	5893	5937	5937	5893	5893	5893	5893	5893	5937
2007	5953	5953	6026	6026	5953	5953	5953	5953	5953	6026
2008	6015	6015	6116	6116	6015	6015	6015	6015	6015	6116
2009	6079	6079	6208	6208	6079	6079	6079	6079	6079	6208
2010	6144	6144	6301	6301	6144	6144	6144	6144	6144	6301
2011	6211	6211	6396	6396	6211	6211	6211	6211	6211	6396
2012	6280	6280	6491	6491	6280	6280	6280	6280	6280	6491
2013	6351	6351	6589	6589	6351	6351	6351	6351	6351	6589
2014	6423	6423	6688	6688	6423	6423	6423	6423	6423	6688
2015	6497	6497	6788	6788	6497	6497	6497	6497	6497	6788
2016	6573	6573	6890	6890	6573	6573	6573	6573	6573	6890
2017	6651	6651	6993	6993	6651	6651	6651	6651	6651	6993
2018	6730	6730	7098	7098	6730	6730	6730	6730	6730	7098
2019	6812	6812	7205	7205	6812	6812	6812	6812	6812	7205
2020	6895	6895	7313	7313	6895	6895	6895	6895	6895	7313

Table K-2. Nonoccupants Fatality Impacts of Behavioral Programs and Proposed Safety Standards Change from Base Estimate

	Base Estimate	Seat Belts	Alcohol Improv.	Combined Behavioral	ESC	Heavy-Truck Brakes	Side Impact	Ejection Mitigation	Proposed Standards	All NHTSA Activities
2005	0	0	0	0	0	0	0	0	0	0
2006	0	0	-44	-44	0	0	0	0	0	-44
2007	0	0	-73	-73	0	0	0	0	0	-73
2008	0	0	-101	-101	0	0	0	0	0	-101
2009	0	0	-129	-129	0	0	0	0	0	-129
2010	0	0	-157	-157	0	0	0	0	0	-157
2011	0	0	-184	-184	0	0	0	0	0	-184
2012	0	0	-212	-212	0	0	0	0	0	-212
2013	0	0	-238	-238	0	0	0	0	0	-238

Table K-1. (Continued)

	Base Estimate	Seat Belts	Alcohol Improv.	Combined Behavioral	ESC	Heavy-Truck Brakes	Side Impact	Ejection Mitigation	Proposed Standards	All NHTSA Activities
2014	0	0	-265	-265	0	0	0	0	0	-265
2015	0	0	-291	-291	0	0	0	0	0	-291
2016	0	0	-317	-317	0	0	0	0	0	-317
2017	0	0	-343	-343	0	0	0	0	0	-343
2018	0	0	-368	-368	0	0	0	0	0	-368
2019	0	0	-393	-393	0	0	0	0	0	-393
2020	0	0	-418	-418	0	0	0	0	0	-418

Table L-1. Summary of Fatality Rate Estimates for Selected Fatality Rate Measures Base Scenario*

	All Fatalities	Passenger Vehicles	Non-occupants	Motorcycles
2005	1.45	1.14	0.20	0.73
2006	1.44	1.13	0.19	0.75
2007	1.41	1.09	0.19	0.76
2008	1.37	1.05	0.19	0.76
2009	1.34	1.01	0.19	0.77
2010	1.30	0.98	0.19	0.78
2011	1.27	0.94	0.19	0.78
2012	1.23	0.90	0.19	0.78
2013	1.20	0.86	0.19	0.78
2014	1.16	0.83	0.19	0.77
2015	1.13	0.80	0.19	0.75
2016	1.10	0.77	0.19	0.73
2017	1.08	0.74	0.19	0.71
2018	1.06	0.71	0.19	0.68
2019	1.03	0.69	0.18	0.64
2020	1.01	0.67	0.18	0.61

*Assumes continued progress at historical pace for seat belt use and alcohol impairment, and generational cycle for motorcycle fatality surge.

Table L-2. Summary of Fatality Rate Estimates for Selected Fatality Rate Measures Modest Seat Belt Progress Scenario*

	All Fatalities	Passenger Vehicles	Non-occupants	Motorcycles
2005	1.45	1.14	0.20	0.73
2006	1.44	1.13	0.19	0.75
2007	1.41	1.10	0.19	0.76
2008	1.38	1.06	0.19	0.76
2009	1.35	1.03	0.19	0.77
2010	1.32	1.00	0.19	0.78
2011	1.29	0.96	0.19	0.78
2012	1.26	0.93	0.19	0.78
2013	1.23	0.90	0.19	0.78
2014	1.20	0.86	0.19	0.77
2015	1.17	0.83	0.19	0.75
2016	1.14	0.80	0.19	0.73
2017	1.11	0.78	0.19	0.71
2018	1.09	0.75	0.19	0.68
2019	1.07	0.73	0.18	0.64
2020	1.05	0.71	0.18	0.61

* Assumes continued progress at half historical pace for seat belt use.

Table L-3. Summary of Fatality Rate Estimates for Selected Fatality Rate Measures Modest Alcohol Progress Scenario*

	All Fatalities	Passenger Vehicles	Non-occupants	Motorcycles
2005	1.45	1.14	0.20	0.73
2006	1.45	1.13	0.19	0.75
2007	1.42	1.09	0.19	0.76
2008	1.38	1.06	0.19	0.76
2009	1.35	1.02	0.19	0.77
2010	1.32	0.99	0.19	0.78
2011	1.28	0.95	0.19	0.78
2012	1.25	0.92	0.19	0.78
2013	1.22	0.88	0.19	0.78
2014	1.19	0.85	0.19	0.77
2015	1.16	0.82	0.19	0.75
2016	1.13	0.79	0.19	0.73
2017	1.10	0.76	0.19	0.71
2018	1.08	0.74	0.19	0.68
2019	1.06	0.72	0.19	0.64
2020	1.04	0.70	0.19	0.61

*Assumes continued progress at half historical pace for alcohol-related fatality reduction.

Table L-4. Summary of Fatality Rate Estimates for Selected Fatality Rate Measures Modest Behavioral Scenario*

	All Fatalities	Passenger Vehicles	Non-occupants	Motorcycles
2005	1.45	1.14	0.20	0.73
2006	1.45	1.13	0.19	0.75
2007	1.42	1.10	0.19	0.76
2008	1.40	1.07	0.19	0.76
2009	1.37	1.04	0.19	0.77
2010	1.34	1.01	0.19	0.78
2011	1.31	0.98	0.19	0.78
2012	1.28	0.95	0.19	0.78
2013	1.25	0.92	0.19	0.78
2014	1.22	0.88	0.19	0.77
2015	1.19	0.85	0.19	0.75
2016	1.16	0.83	0.19	0.73
2017	1.14	0.80	0.19	0.71
2018	1.12	0.78	0.19	0.68
2019	1.10	0.76	0.19	0.64
2020	1.08	0.74	0.19	0.61

* Assumes continued progress at half historical pace for both seat belt use and alcohol-related fatality reduction.

Table L-5. Summary of Fatality Rate Estimates for Selected Fatality Rate Measures Continuous Motorcycle Fatality Growth Scenario*

	All Fatalities	Passenger Vehicles	Non-occupants	Motorcycles
2005	1.45	1.14	0.20	0.73
2006	1.44	1.13	0.19	0.74
2007	1.41	1.09	0.19	0.76
2008	1.39	1.05	0.19	0.77
2009	1.36	1.01	0.19	0.78
2010	1.33	0.98	0.19	0.79
2011	1.30	0.94	0.19	0.80
2012	1.27	0.90	0.19	0.80
2013	1.24	0.86	0.19	0.81
2014	1.21	0.83	0.19	0.82
2015	1.19	0.80	0.19	0.82
2016	1.16	0.77	0.19	0.83
2017	1.14	0.74	0.19	0.84
2018	1.12	0.71	0.19	0.84
2019	1.10	0.69	0.18	0.85
2020	1.09	0.67	0.18	0.85

*Assumes continued growth of motorcycle fatalities and registrations at previous eight-year average pace.

Table L-6. Summary of Fatality Rate Estimates for Selected Fatality Rate Measures Average of Continuous and Generational Motorcycle Growth Scenario*

	All Fatalities	Passenger Vehicles	Non-occupants	Motorcycles
2005	1.45	1.14	0.20	0.73
2006	1.44	1.13	0.19	0.74
2007	1.41	1.09	0.19	0.76
2008	1.38	1.05	0.19	0.76
2009	1.35	1.01	0.19	0.77
2010	1.31	0.98	0.19	0.78
2011	1.28	0.94	0.19	0.79
2012	1.24	0.90	0.19	0.79
2013	1.21	0.86	0.19	0.80
2014	1.18	0.83	0.19	0.80
2015	1.15	0.80	0.19	0.80
2016	1.12	0.77	0.19	0.79
2017	1.09	0.74	0.19	0.79
2018	1.07	0.71	0.19	0.79
2019	1.05	0.69	0.18	0.78
2020	1.03	0.67	0.18	0.78

*Assumes growth of motorcycle fatalities at pace equal to average of that predicted by generational scenario and continuous scenario.

Table L-7. Summary of Fatality Rate Estimates for Selected Fatality Rate Measures Probable Worst Case Scenario*

	All Fatalities	Passenger Vehicles	Non-occupants	Motorcycles
2005	1.45	1.14	0.20	0.73
2006	1.45	1.13	0.19	0.74
2007	1.43	1.10	0.19	0.76
2008	1.40	1.07	0.19	0.76
2009	1.38	1.04	0.19	0.77
2010	1.35	1.01	0.19	0.78
2011	1.32	0.98	0.19	0.79
2012	1.29	0.95	0.19	0.79
2013	1.26	0.92	0.19	0.80
2014	1.23	0.88	0.19	0.80
2015	1.20	0.85	0.19	0.80
2016	1.18	0.83	0.19	0.79
2017	1.15	0.80	0.19	0.79
2018	1.13	0.78	0.19	0.79
2019	1.11	0.76	0.19	0.78
2020	1.09	0.74	0.19	0.78

*Assumes growth of motorcycle fatalities at pace equal to average of that predicted by generational scenario and continuous scenario, continued progress at half historical pace for seat belt use, and continued progress at half historical pace for alcohol-related fatality reduction.

End Notes

[1] Passenger vehicles include passenger cars, light trucks, SUVs, and vans less than 10,000 lbs. GVWR.

[2] Kahane, C.J,."Lives Saved by Federal Motor Vehicle Safety Standards and Other Vehicle Safety Technologies, 1960-2002 – Passenger Cars and Light Trucks – With a Review of 19 FMVSS and their Effectiveness in Reducing Fatalities, Injuries, and Crashes", National Highway Traffic Safety Administration, DOT HS 809 833, Washington, DC, October 2004.

[3] Lu, S., "Vehicle Survivability and Travel Mileage Schedules, Corrected Document", National Highway Traffic Safety Administration, DOT HS 809 952, Washington DC, September 2006

[4] R.L. Polk Co.

[5] This assumes an even distribution of both sales and retirements throughout the year, with the proportions increasing from zero at the start of the year to 100 percent at the end of the year.

[6] The most recent three-year period was chosen over longer periods because VMT increases have been trending downward. The average for 1975-2005 was 2.8 percent. For 1996-2005 the average increase was 2.1 percent. For 2001-2005 the average was 1.7 percent. For 2003-2005 the average was 1.5 percent.

[7] Blincoe, L.J., "Estimating the Benefits of Increased Safety Belt Use", National Highway Traffic Safety Administration, DOT HS 808 133, Washington, DC, June, 1994 and Wang, J-S, and Blincoe, L.J., "Belt Use Regression Model – 2003 Update", National Highway Traffic Safety Administration, DOT HS 809 639, Washington, DC, May 2003.

[8] 2005 FARS data indicate that 32 percent of riders killed on bikes of 1,001-1,500 cc and 27 percent of those killed on bikes with 1,501 cc or larger were impaired. This contrasts with roughly 22 percent on bikes below 1,001 cc.

[9] Based on Motorcycle Industry Council statistics.

[10] A sudden shift in gasoline prices might affect driver habits fairly quickly. In this case, higher prices would be likely to accelerate interest in motorcycles because they are fuel efficient relative to other passenger vehicles. However, the current upswing in motorcycle popularity began years before the recent rise in gasoline prices and is likely driven by other factors.

[11] Based on historical VMT data.

[12] Passenger vehicles include passenger cars, pickup trucks, vans, and SUVs. It excludes motorcycles, heavy trucks (over 10,000 lbs. GVWR), bicycles, and pedestrians.

[13] Includes pedestrians and bicyclists.

[14] Based on the change in alcohol-related fatality rates from 2000-2004. During this timeframe, alcohol- related fatality rates declined by an average of .01/year.

[15] Note that this represents the difference between the full fleet ESC installation required in NHTSA's proposed safety standard and installation in 42 percent of the new vehicle fleet – the current estimate of pre-standard voluntary installation. Under the agency's proposed standard, 100 percent of the *new* vehicle fleet will have ESC in 2012 and roughly 60 percent of the *on-road* fleet will have it by 2020, resulting in the prevention of 6,000 fatalities by this technology.

[16] After steadily increasing for over a decade, seat belt use declined from 82 percent to 81 percent in 2006.

[17] For example, if we assume a constant increase in motorcycle fatalities equal to the latest eight-year average increase (+ 305 per year), the overall fatality rate would begin to rise in 2008 by .02 and would be .07 higher by 2020. Given the past volatility of motorcycle fatality trends, at least temporary spikes in motorcycle fatalities that are off the predicted trend are likely to occur.

INDEX

A

Abbreviated Injury Scale, 2, 9, 43, 107, 120, 148, 151, 155, 193, 216, 219, 272, 282
access, 135
accounting, 4, 5, 10, 11, 123, 146, 150, 194, 195, 206, 208
acetaldehyde, 82
acid, 82
acute stress, 273
adjustment, 45, 78, 102, 123, 125, 134, 151, 155, 228, 276, 284, 292, 294, 295, 301
adolescents, 279
adults, 204, 226, 232, 278
adverse effects, 82, 83
adverse weather, 115
advocacy, 167
aerosols, 81
age, 44, 81, 100, 101, 139, 151, 158, 202, 205, 221, 222, 223, 224, 225, 226, 228, 229, 230, 231, 232, 233, 290, 291, 298, 299, 309
agencies, 64, 85, 99, 102, 109, 113, 127, 167
agency actions, 305, 306
Alaska, 110, 114
alcohol consumption, 144
alcohol levels, vii, 1
alcohol use, vii, 1, 11, 129, 133, 134, 273, 276, 281
algorithm, 120, 128
ambient air, 81
amplitude, 301
assessment, 134, 136
assets, 277
atmosphere, 81
attitudes, 158
attribution, 9
automobiles, 82
autopsy, 221, 222
average costs, 9, 222, 224, 225, 229
average earnings, 230, 232, 234
avoidance, 110, 283, 311
awareness, 127, 144

B

baby boomers, 154, 300
BAC, vii, 1, 5, 8, 127, 128, 129, 130, 131, 132, 134, 136, 137, 138, 139, 140, 141, 142, 143, 144, 145, 146, 147, 281
base, 86, 182, 292, 293, 294, 308, 309
base year, 86, 292, 294
beams, 155
behaviors, 283, 284
bending, 100
beneficiaries, 227
benefits, 9, 10, 89, 102, 103, 160, 161, 168, 170, 173, 175, 208, 231, 272, 285, 288, 292, 293, 294, 295, 311
benzene, 82
BI, 39
bias, 62, 116, 117, 118, 227, 233, 234
blood, vii, 1, 4, 8, 127, 237, 240, 243, 246, 249, 252, 256, 258, 261, 264, 267, 269
blood vessels, 237, 243, 249, 256, 261, 267
brain, 9, 25, 81, 101, 132, 222, 226
breakdown, 96, 178, 184, 223
BTU, 283
Bureau of Justice Statistics, 276
Bureau of Labor Statistics, 44, 99, 230, 231, 274
burn, 25
business cycle, 231
butadiene, 82
buyer(s), 38
by-products, 82

C

campaigns, 127
carbon, 81, 82, 86, 277
categorization, 176
category a, 137
category d, 25
causation, 143, 144, 193
cell phones, 11, 58, 62, 113, 114, 150, 151, 283
cell size, 225
census, 99, 177, 274, 299
challenges, 311
charities, 3, 7, 206
chemical, 82
Chicago, 279
childhood, 226
children, 226, 232, 234, 279
cities, 58
classification, 184
cleaning, 232
cleanup, 58
climate, 81, 86
climate change, 86
clinical trials, 100
closure, 50, 51, 59, 60, 62, 64, 65, 66, 67, 68, 69, 81
CO2, 81, 83, 84, 85, 86, 87, 88, 89, 91, 94, 283
coal, 81
coding, 24, 25, 27, 120, 132, 232, 282, 284
combined effect, 295
combustion, 81, 82, 83
combustion processes, 81
commercial, 35, 36, 49, 50, 56, 77, 79, 89, 202, 283
communication, 283
community, 311
compensation, 35, 38, 39, 44, 202, 204, 205, 206, 224, 226, 228, 232, 233
complexity, 193
composition, 292
compounds, 82
computer, 109
computing, 9, 51, 203, 288
conditioning, 283
congress, 280, 281, 301
consensus, 282
constant rate, 71
construction, 184
consumer price index, 220
consumers, 89, 98, 102, 289, 300
consumption, 8, 54, 71, 80, 81, 85, 88, 127, 283, 284, 294
control measures, 193
controversial, 203
conversion rate, 288, 293, 305, 309
cooking, 232
cooperation, 177
correlation, 273, 297
cosmetic, 100
cost effectiveness, 100
cost saving, 175
covering, 225, 305
CPI, 220, 285
crops, 82
CT, 229, 274
cycles, 75

D

damages, 35, 39, 86
data collection, 109, 117, 180
data gathering, 97, 136
data set, 24, 132, 148, 152, 179, 193, 227, 234, 282
database, 9, 59, 62, 64, 65, 69, 81, 89, 95, 101, 148, 155, 177, 180, 182, 186, 193, 222, 232, 290
DCI, 224, 226, 228, 229, 232, 233
deaths, 8, 9, 25, 27, 170, 175, 205, 220, 222, 223, 225, 230, 299
deciliter, vii, 1, 4, 8, 127
demographic factors, 306
Department of Agriculture, 99, 176
Department of Health and Human Services, 112, 273
Department of Justice, 276
Department of Transportation, 10, 11, 56, 58, 99, 100, 109, 208, 276
dependent variable, 291
depression, 281
detection, 58, 135
diesel fuel, 85
direct cost(s), 225, 227
direct observation, 77
disability, 6, 9, 27, 43, 100, 204, 206, 230, 231, 232, 233, 234, 272, 281
discharges, 221
discomfort, 9
discriminant analysis, 128
dislocation, 25, 26, 27, 29, 30, 31, 32, 33, 34, 222
disorder, 273, 279
dispersion, 75
displacement, 80, 85
distribution, 26, 55, 66, 85, 92, 95, 108, 125, 137, 139, 179, 180, 186, 188, 203, 206, 334
District of Columbia, 105, 127, 167
disutility, 10, 208
divergence, 52
diversity, 24, 100, 132
doctors, 117
dose-response relationship, 136

DOT, 1, 11, 49, 59, 100, 157, 209, 274, 275, 276, 277, 278, 279, 280, 281, 282, 287, 334
double counting, 102, 220
driver behavior, vii, viii, 11, 49, 105, 124, 287, 289, 308, 310, 311
drug abuse, 234
drug abusers, 234
drugs, 281
DWI, 127

E

earnings, 31, 38, 39, 139, 204, 230, 231, 232, 242, 260
earnings data, 230
economic activity, 105, 106
economic consequences, 6, 10, 273
economic costs, vii, 1, 2, 3, 4, 5, 6, 7, 8, 9, 11, 12, 102, 125, 142, 145, 150, 152, 160, 170, 183, 188, 189, 194, 195, 209, 274
economic downturn, 203
economic losses, 6
economic values, 103
economic welfare, 49
economics, 2, 10, 86, 100, 208
education, 222
educational programs, 193, 305, 308
EIS, 178
emergency, vii, 1, 2, 6, 7, 11, 25, 45, 47, 49, 58, 61, 63, 67, 68, 69, 72, 80, 93, 102, 104, 117, 118, 135, 203, 206, 220, 221, 222, 223, 226, 227, 229, 235, 272, 273, 282, 298
emergency response, 68, 72
emission, 81, 283
emotional distress, 100
employees, 203, 272
employers, 2, 6, 89, 204, 206, 231, 279
employment, 44, 205, 220
EMS, vii, 1, 6, 13, 16, 17, 18, 19, 20, 21, 22, 43, 46, 48, 61, 103, 104, 203, 210, 211, 212, 213, 214, 219
energy, 283
enforcement, 8, 111, 124, 153, 167, 293
engineering, 49, 283
enrollment(s), 225
environment, 157, 176, 180, 283
environmental impact, 3, 7, 80, 81, 272
Environmental Protection Agency, 49, 275
EPA, 81, 82, 85, 86, 87, 88, 99, 283
equipment, 57, 63, 68, 69, 219, 227, 288, 290, 291, 306, 311
estimation process, 49
evidence, 9, 86, 107
exclusion, 123
Executive Order, 277
expenditures, 9, 206, 225
exposure, viii, 36, 52, 54, 55, 56, 80, 86, 111, 136, 137, 176, 187, 189, 190, 287, 289, 293, 296, 297, 298, 300, 309, 311
extraction, 85

F

families, 6, 12, 203, 205, 227
family life, 10
family members, 10
family units, 10
fatality rates, vii, viii, 107, 154, 284, 287, 288, 289, 290, 297, 301, 302, 303, 305, 306, 307, 309, 311, 334
FDA, 99
fear, 36
Federal Highway Administration, 51, 177, 275, 281, 296
FHWA, 51, 78, 96, 109, 177, 178, 180, 185, 283, 296
financial, 9, 98, 134
fiscal year 2009, 204
fluid, 107
food, 82
food chain, 82
force, 125, 139, 205
formaldehyde, 82
formation, 82
formula, 52, 73, 74, 137, 230, 292, 294, 295
fraud, 204
fringe benefits, 204, 230, 231
fuel consumption, vii, 1, 3, 6, 7, 44, 49, 80, 81, 95
funding, 124
funds, 184

G

GAO, 227, 276
geography, 108
Georgia, 56, 58, 110, 114, 167
Germany, 278
goods and services, 8, 102, 230, 272
government revenues, 203
governments, 203
grants, 124
greenhouse, vii, 1, 44, 49, 71, 80, 81, 95, 272
greenhouse gas(es), vii, 1, 44, 49, 71, 80, 81, 95, 272
Gross Domestic Product, vii, 1, 2, 6
grouping, 148, 152, 193, 226, 228

growth, 86, 173, 231, 298, 299, 307, 332, 333
growth rate, 86, 231
guidance, 10, 11, 86, 89, 98, 99, 208, 209, 289
guidelines, 89, 225

H

Hawaii, 110, 114, 284
hazardous waste, 50, 57, 77
hazards, 147
head injuries, 157
health, 3, 7, 38, 49, 81, 82, 83, 86, 87, 95, 100, 117, 202, 203, 204, 205, 206, 225, 226, 229, 233, 272, 273, 278, 280, 282, 283
health and environmental effects, 83
health care, 3, 7, 38, 202, 203, 204, 205, 206, 225, 229, 282
health care costs, 204
health care programs, 203, 205
health effects, 81, 82
health insurance, 205, 226
health problems, 82
health risks, 81
health services, 225
highways, 12, 65, 190, 193, 194
hip fractures, 225
hiring, 272
historical data, 180, 216, 300, 311
history, 155, 281, 290, 300
hospice, 220, 221, 222, 223
hospitalization, 7, 25, 45, 104, 226, 281
host, 202
human, 81, 98, 150, 230, 234
human capital, 98, 230, 234
hybrid, 25, 26, 133, 284
hydrocarbons, 82
hydrogen, 82

I

ideal, 182
identification, 134, 135, 152, 177, 185
illusion, 273
images, 279
impairments, 231
improvements, 8, 86, 155, 193, 289, 293, 306, 307, 311
inattention, 147, 150
incidence, 6, 7, 8, 9, 24, 37, 38, 45, 46, 49, 66, 69, 77, 89, 92, 95, 103, 105, 108, 112, 113, 115, 124, 125, 129, 133, 137, 138, 139, 142, 144, 145, 146, 148, 149, 151, 152, 158, 159, 160, 180, 182, 183, 187, 193, 194, 216, 225, 282, 284, 294
income, 6, 96, 97, 125, 204, 220, 231
income tax, 204
independent living, 227
individuals, 202, 203, 206, 225, 230, 300
industry, 82
inflation, 99
infrastructure, 8, 187
injury claims, 37, 38, 39, 277
injury prevention, 4, 168
institutions, 227
intervention, 273
intoxication, 127
investment, 8, 100
Iowa, 110, 114
IRC, 277
issues, 56, 102, 110, 117, 118, 273

J

jurisdiction(s), 61, 97, 111, 115, 116, 179

L

labor force, 234
labor market, 230, 234
law enforcement, 58, 153
laws, 8, 38, 124, 127, 129, 136, 148, 151, 158, 167, 193, 216, 293, 305
lead, 9, 143, 176, 307, 311
leaks, 85
legislation, 124, 301
liability insurance, 38, 206
life expectancy, 234
lifetime, vii, 1, 2, 5, 8, 27, 204, 220, 224, 226, 227, 228, 229, 230, 231, 232, 233, 234
light, 8, 53, 54, 74, 77, 80, 105, 131, 143, 154, 155, 157, 167, 180, 182, 283, 284, 307, 334
light cycle, 283
light trucks, 80, 154, 155, 167, 334
litigation, 113, 272
local government, 43, 134, 205, 206
location information, 25
longevity, 101
longitudinal study, 278
Louisiana, 110, 114, 167, 222

M

magnitude, 154, 226
MAI, 28

major depression, 273
majority, 36, 80, 81, 120, 129, 138, 184, 226
management, 43, 44, 45, 46, 206
marijuana, 281
marketplace, 230
Maryland, 110, 114, 134, 135, 167, 227
mass, 86, 105
matrix, 25, 46, 180, 222, 233
matter, 80, 82, 85, 86
measurement(s), 10, 49, 77, 98, 208, 233
median, 99, 101, 227, 285
Medicaid, 202, 203, 204, 205, 227, 228, 282
medical assistance, 273
medical care, 4, 6, 7, 8, 44, 47, 98, 102, 105, 113, 117, 134, 170, 175, 192, 203, 204, 206, 220, 227, 284
Medicare, 202, 203, 205, 221, 222, 224, 227, 228, 276
medication, 228, 229
mental health, 220
meta-analysis, 99
metals, 82
methodology, 97, 193
metropolitan areas, 70
Mexico, 110, 114
military, 284
Minneapolis, 49, 79, 283
Missouri, 43, 44, 110, 114, 179
models, 85, 86, 275, 300, 301
modifications, 272
Montana, 110, 114, 284
morbidity, 273
motivation, 68
motor skills, 144
motor vehicle crashes, vii, 1, 2, 3, 4, 5, 6, 8, 9, 10, 11, 24, 25, 49, 65, 90, 95, 98, 100, 102, 105, 108, 117, 127, 132, 142, 147, 150, 151, 152, 158, 189, 193, 194, 202, 203, 205, 206, 208, 272, 274, 279, 283
multiple factors, 150
multiplier, 26, 36, 226, 229

N

narratives, 180
National Highway Traffic Safety Administration, vii, viii, 1, 287, 289
Native Americans, 203
natural disaster(s), 115
neglect, 167
New England, 282
NHS, 178, 179
nitrates, 82
nitrogen, 82
nitrogen dioxide, 82
non-fatal injuries, vii, 2, 3, 64, 121, 219, 227, 228, 230, 231
non-institutionalized, 225, 284
normal distribution, 26
NPR, 307
nursing, 221, 222, 223
nursing home, 221, 222, 223

O

obstruction, 93
Office of Management and Budget, 8, 99
officials, 177
oil, 81, 309, 311
Oklahoma, 110, 114
OMB, 101
operations, 68
opportunities, 89, 176, 283
organ(s), 81, 25, 222, 237, 243, 249, 256, 261, 267
organic chemicals, 82
organic compounds, 82
outpatient, 25
overlap, 58, 111, 115, 135, 150, 288, 290, 294, 295
overlay, 177, 185
overtime, 272
ownership, 225
oxygen, 81
ozone, 82

P

Pacific, 101
pain, 2, 6, 9, 10, 98, 100, 192, 208, 218, 233
parallel, 173, 231
participants, 74, 120, 203
pedal, 80
per capita cost, 125
per capita income, 125
permission, 109
permit, 177, 185
personal communication, 232
pharmaceuticals, 100
photographs, 180
physical therapy, 222, 272
physicians, 25
playing, 150
PM, 82, 86
policy, 10, 35, 36, 38, 39, 42, 100, 205, 289
policy makers, 289
policymakers, 102

pollutants, vii, 1, 44, 71, 81, 85, 86, 91, 95, 272
pollution, 6, 49, 82, 176
population, viii, 2, 9, 81, 96, 97, 105, 107, 108, 117, 124, 125, 157, 167, 168, 176, 177, 178, 187, 205, 225, 234, 284, 289, 294, 299, 300
population density, 187
population growth, 289
population size, 176, 178
portable communication, 62
posttraumatic stress, 273
post-traumatic stress disorder, 275
potential benefits, 161, 290
premature death, 230
preparation, 101
prescription drugs, 225
present value, vii, 2, 204, 220, 231, 232, 234, 272
President, 231, 275
prevention, 6, 306, 334
price instability, 309
primary data, 193, 229
private sector, 6
private sector officials, 6
probability, 50, 64, 65, 66, 77, 81, 93, 95, 120, 131, 144, 147, 227, 230, 231, 232, 233, 293
probability distribution, 65, 66
productivity growth, 231
professionals, 37, 225, 234
programming, 112
project, 44, 49, 134, 220
proliferation, 58, 62, 113, 114
property damage, vii, 1, 2, 3, 6, 7, 35, 36, 37, 38, 39, 42, 43, 45, 47, 93, 95, 98, 102, 104, 105, 108, 112, 113, 115, 123, 131, 203, 205, 206, 233, 284, 285
proposed regulations, 99
prosthetic device, 272
protection, viii, 8, 35, 39, 154, 155, 167, 287, 288, 301, 311
PTSD, 233, 273
public awareness, 127, 148, 151, 158, 193, 216, 293
public education, 8, 167
public health, 6
public interest, 102
public sector, 230
public service, 43
Puerto Rico, 105, 127

Q

quality of life, vii, 1

R

racing, 148
radiation, 81
ramp, 64, 274
rate of change, 305
reaction time, 4, 5, 8, 143
reactions, 74
real income, 113
reality, 41
recall, 117, 118
recession, 8, 106, 107, 154
recognition, 116, 117, 118, 119, 135
recommendations, 9, 289
recovery, 38, 101, 107, 204, 231
recreation, 89
Registry, 228, 280
regression, 74, 136, 231, 233
regression model, 74, 231, 233
regulations, vii, viii, 99, 102, 110, 193, 287, 288, 289
rehabilitation, 2, 9, 220, 224, 227, 272
relevance, 124
remission, 273
repair, 113, 120, 125, 272, 290
representativeness, 36, 39
requirements, vii, viii, 38, 111, 202, 287, 288, 289
researchers, 10, 80, 107
reserves, 311
resources, 6, 98, 125, 177, 305
response, 7, 43, 45, 46, 49, 58, 61, 67, 81, 107, 167, 203, 272, 283
response behaviors, 283
response time, 49, 58, 61
restitution, 272
retirement, 234
retirement age, 234
revenue, 125
risk(s), viii, 10, 36, 37, 38, 98, 99, 136, 137, 144, 167, 208, 234, 272, 284, 287, 289, 298, 308, 311
risk factors, 137, 144
routes, 74, 76, 77, 81, 283
rules, 24, 132
rural areas, 61, 176, 180, 184, 187, 189

S

safety, vii, viii, 7, 8, 10, 45, 100, 102, 103, 106, 109, 124, 134, 143, 155, 167, 168, 170, 177, 184, 193, 204, 208, 283, 287, 288, 289, 290, 291, 292, 293, 294, 295, 301, 303, 305, 306, 307, 308, 311, 334
sample mean, 54
sampling error, 26

sanctions, 308
SAS, 26, 109
savings, viii, 159, 160, 170, 175, 287, 288, 290, 294, 295, 307
scaling, 54, 277
scapula, 237, 243, 250, 256, 262, 267
scope, 64, 77
sea level, 86
seat belt use, viii, 9, 105, 155, 167, 168, 170, 173, 175, 193, 287, 288, 289, 293, 305, 306, 307, 309, 311, 330, 331, 332, 333, 334
sensitivity, 11, 100, 209, 220, 223, 227, 231
services, 2, 6, 7, 43, 45, 47, 49, 80, 102, 104, 203, 206, 222, 225, 230
severity levels, 7, 48, 105, 108, 180
sex, 228, 230, 231, 232, 233
shape, 52
significance level, 26
signs, 43
simulation(s), 49, 50, 77, 78, 283
skull fracture, 232
smog, 82
social influence(s), 311
societal cost, 9, 44, 81, 83, 143, 146, 152, 175, 283
society, viii, 2, 3, 4, 5, 6, 7, 10, 11, 89, 98, 102, 120, 158, 170, 175, 189, 192, 202, 206, 208, 230, 272, 283, 311
software, 26
South Dakota, 110, 114
specialists, 108
specifications, 78
speculation, 97
speed limit, vii, 2, 4, 5, 11, 147, 148
spending, 27, 204, 226
spinal cord, 9, 101, 222, 226, 239, 245, 252, 257, 263, 268, 273
spinal cord injury, 101, 273
spleen, 26
sprain, 226
stability, viii, 8, 106, 287
standard deviation, 99
standard error, 26, 27, 108, 234
state(s), 88, 93
statistics, 35, 51, 52, 76, 283, 284, 334
sternum, 239, 245, 251, 257, 263, 268
stratification, 26, 65, 95, 108, 182
stress, 273, 279
structure, 59, 120, 154, 284
structuring, 6
sulfur, 83
surveillance, 150
survey design, 117
survival, 25, 81, 291
survival rate, 291
survivors, 26, 38, 139, 204, 227, 284
symptoms, 120
synthesis, 50, 280

T

target, 294, 303, 305
target population, 294
taxes, vii, 1, 3, 7, 206, 283
TBI, 224, 226, 227, 233
teams, 45
techniques, 10, 99, 136, 208
technologies, 153, 288, 311
technology, 100, 276, 290, 334
tension, 89
testing, 136
theft, 35
time frame, 54, 116
time periods, 220
total costs, 2, 3, 7, 95, 125, 158, 160, 186, 206, 222, 224
total product, 204
training, 24, 132, 219
transaction costs, 35
transformations, 311
transition period, 180
translation, 111
transmission, 80
transport, 131, 135, 221, 222, 223, 226, 227, 228, 229, 272
transport costs, 222, 227, 228
transportation, 49, 97, 154, 220, 227, 276
trauma, 10, 135, 233, 274
traumatic brain injury, 233, 281
travel delay, vii, 1, 3, 7, 44, 56, 70, 75, 77, 95, 98, 125, 176, 272
treatment, 25, 35, 116, 117, 118, 119, 120, 135, 219, 220, 225, 226, 227, 233, 272, 273, 284
turnover, 289

U

U.S. Bureau of Labor Statistics, 89
uniform, 39
uninsured, 6, 37, 38, 277
unit cost, 2, 7, 8, 24, 25, 26, 27, 44, 45, 48, 92, 95, 102, 104, 105, 120, 139, 145, 146, 157, 160, 219, 236, 255, 282, 285
United States, vii, 1, 2, 4, 6, 97, 99, 108, 131, 205, 206, 219, 273, 274, 276, 277, 278, 279, 280, 281, 282, 285, 296, 299

universe, 50, 58, 65, 77, 118, 119, 283
updating, 44, 153
urban, 12, 49, 51, 52, 55, 56, 57, 58, 60, 61, 62, 67, 69, 70, 78, 79, 96, 97, 125, 176, 177, 178, 179, 180, 182, 184, 186, 187, 188, 189, 190, 193, 225, 283
urban areas, 177, 178, 184, 189
Urban Institute, 202, 278, 279
US Department of Health and Human Services, 280

V

validation, 97
valuation, 5, 10, 100, 102, 208, 275, 284
vapor, 85
variables, 25, 49, 152, 179
vehicle safety performance, vii, viii, 287, 289
velocity, 73
vessels, 240, 246, 252, 258, 264, 269
victims, 3, 6, 7, 9, 11, 24, 25, 43, 98, 102, 107, 139, 160, 206, 219, 227, 230, 231, 234, 273, 282
violence, 226
vision, 225
vocational rehabilitation, 43, 44, 206, 227
volatile organic compounds, 82
volatility, 334

W

wage rate, 49, 80, 89, 98, 232
wages, 9, 10, 100, 102, 204, 208, 231, 272
walking, 289
war, 299
Washington, 43, 46, 110, 114, 176, 179, 273, 274, 275, 276, 277, 278, 279, 280, 281, 282, 334
waste, 283
wealth, 97
wear, viii, 2, 4, 9, 11, 12, 143, 157, 158, 160, 161, 163, 167, 168, 175
welfare, 43, 206
Wisconsin, 110, 114, 179
workers, 52, 204, 205, 224, 226, 228, 232, 233
working population, 234
workload, 222
workplace, vii, 1, 2, 6, 44, 98, 102, 125, 203, 206, 272
worldwide, 99

Y

yield, 89, 116, 223